Next Generation AI Language Models in Research

In this comprehensive and cutting-edge volume, Qureshi and Jeon bring together experts from around the world to explore the potential of artificial intelligence models in research and discuss the potential benefits and the concerns and challenges that the rapid development of this field has raised.

The international chapter contributor group provides a wealth of technical information on different aspects of AI, including key aspects of AI, deep learning and machine learning models for AI, natural language processing and computer vision, reinforcement learning, ethics and responsibilities, security, practical implementation, and future directions. The contents are balanced in terms of theory, methodologies, and technical aspects, and contributors provide case studies to clearly illustrate the concepts and technical discussions throughout. Readers will gain valuable insights into how AI can revolutionize their work in fields including data analytics and pattern identification, healthcare research, social science research, and more, and improve their technical skills, problem-solving abilities, and evidence-based decision-making. Additionally, they will be cognizant of the limitations and challenges, the ethical implications, and security concerns related to language models, which will enable them to make more informed choices regarding their implementation.

This book is an invaluable resource for undergraduate and graduate students who want to understand AI models, recent trends in the area, and technical and ethical aspects of AI. Companies involved in AI development or implementing AI in various fields will also benefit from the book's discussions on both the technical and ethical aspects of this rapidly growing field.

Next Generation AI Language Models in Research

Promising Perspectives and Valid Concerns

Edited by Kashif Naseer Qureshi
and Gwanggil Jeon

CRC Press
Taylor & Francis Group
Boca Raton London New York

CRC Press is an imprint of the
Taylor & Francis Group, an **informa** business

Designed cover image: Ai tech, businessman show virtual graphic Global Internet connect Chatgpt Chat with AI, Artificial Intelligence. using command prompt for generates something, Futuristic technology transformation (stockphoto)

First edition published 2025
by CRC Press
2385 NW Executive Center Drive, Suite 320, Boca Raton FL 33431

and by CRC Press
4 Park Square, Milton Park, Abingdon, Oxon, OX14 4RN

CRC Press is an imprint of Taylor & Francis Group, LLC

ISBN: 978-1-032-66793-5 (hbk)
ISBN: 978-1-032-66788-1 (pbk)
ISBN: 978-1-032-66791-1 (ebk)

DOI: 10.1201/ 9781032667911

Typeset in Minion
by Deanta Global Publishing Services, Chennai, India

Contents

Preface, vii

Acknowledgments, viii

About the Editors, x

Contributors, xii

Introduction, xv

CHAPTER 1 ▪ Introduction and History of Language Models in Research 1

KASHIF NASEER QURESHI AND GWANGGIL JEON

CHAPTER 2 ▪ The Advantages of Artificial Intelligence in Research 25

USMAN AHMAD AND KASHIF NASEER QURESHI

CHAPTER 3 ▪ AI Models and Data Analytics: Transforming Research Methods 45

AMNA KHATOON, ASAD ULLAH, AND KASHIF NASEER QURESHI

CHAPTER 4 ▪ Advanced AI-Based Healthcare Systems Applications and Services 86

AYESHA ASLAM, KASHIF NASEER QURESHI, AND ADIL HUSSAIN

CHAPTER 5 ▪ AI Models in Environmental Social Science Research: Leveraging Technology for Sustainable Solutions 114

CUONG PHAM-QUOC, TRAN VU PHAM, NGUYEN CAO TRI, VU HIEN PHAN, AND KASHIF NASEER QURESHI

Chapter 6 ■ Artificial Intelligence in Natural Science Research: Innovations and Limitations 129

Adil Hussain, Peng-Cheng Xu, Wang Shixin, and Kashif Naseer Qureshi

Chapter 7 ■ AI Models and Ethical Considerations in Research 170

Amreen Shafique, Muhammad Shahid Saeed, Danish Javeed, and Kashif Naseer Qureshi

Chapter 8 ■ AI Models and Applications of AI Models 191

Seemab Kareem, Sidra Zubair, and Kashif Naseer Qureshi

Chapter 9 ■ The Role of Human Expertise in AI-Aided Research 218

Mamoona Amin, Muhammad Mueed Hussain, and Kashif Naseer Qureshi

Chapter 10 ■ Limitations and Challenges of Language Models in Research 242

Amna Iqbal, Majid Hussain, Hina Zafar, and Muhammad Anwar

Chapter 11 ■ Ethical Implications of Language Models 273

Hanaa Omar Nafea, Kashif Naseer Qureshi, and Noman Arshad

Chapter 12 ■ Security and Privacy Concerns in AI Models 293

Muhammad Muneer, Faisal Rehman, Muhammad Hamza Sajjad, Muhammad Anwar, and Kashif Naseer Qureshi

INDEX, 327

Preface

THE RAPID EVOLUTION OF artificial intelligence (AI) has ushered in a new era of innovation and possibility. At the forefront of this technological revolution stand next-generation AI language models, representing a convergence of advanced algorithms, vast amounts of data, and computational power. These models, such as OpenAI's GPT series and its successors, have captured the imagination of researchers, practitioners, and the general public alike with their ability to understand, generate, and manipulate human language in ways previously unimaginable. *Next Generation AI Language Models in Research: Promising Perspectives and Valid Concerns* seeks to explore the frontiers of this transformative technology, offering a nuanced examination of both its potential and its pitfalls. In these pages, we embark on a journey that traverses the exciting landscape of AI language models, delving into their applications, implications, and the ethical considerations that accompany their deployment. As editors and contributors to this book, we have endeavored to provide a holistic perspective on the landscape of next-generation AI language models. Drawing upon the expertise of researchers, practitioners, and thought leaders from diverse backgrounds, we aim to offer a comprehensive exploration that is both informative and thought-provoking. Our hope is that this book serves as a catalyst for informed dialogue, inspiring readers to contemplate the profound implications of AI language models and to contribute to shaping their future trajectory. We extend our gratitude to the contributors who have generously shared their insights and expertise, as well as to the readers who embark on this intellectual journey with us. It is our collective responsibility to navigate the promises and pitfalls of next-generation AI language models with wisdom, empathy, and a steadfast commitment to the greater good.

Kashif Naseer Qureshi
Ireland

Gwanggil Jeon
Korea

Acknowledgments

THIS WORK RECEIVED SUPPORT from the Higher Education Authority (HEA) under the Human Capital Initiative-Pillar 3 project, Cyberskills.

CYBER SKILLS

Cyber Skills is a HEA Human Capital Initiative funded under pillar 3 that brings together Ireland's leading experts in cybersecurity education. Munster Technological University, University of Limerick, and Technological University Dublin are collaborating to provide pathways and micro-credentials to address skill shortages in the area of Cybersecurity. Our programs are designed to provide industry-based learners with the knowledge and skills designed to enhance their careers with cybersecurity expertise.

Cyber Skills is the only resource in Ireland where you will find courses and modules that have been specifically designed and created by industry and academic experts. Working closely with our industry partners, including Dell, Mastercard, ADI, J&J, etc. we have designed courses informed by the needs of the workplace to enhance the skills of IT Cybersecurity professionals in a wide variety of industries.

Cyber Skills benefits from the first of its kind world-class cloud-based Cyber Range and mobile Cyber Range unit. These Cyber Ranges provide a secure, sandboxed area which simulates real-world feel scenarios and environments where students can test their new skills in an environment that can replicate their work-based systems. Labs and assignments will be used to reinforce the content from the lectures and a full range of scenarios will provide the opportunity to test the vast array of techniques required to keep ahead in this challenging and ever-changing environment.

Since its foundation, Cyber Skills has recruited over 14 academic staff who come from multi-disciplinary backgrounds, with a passion for cybersecurity. Combining years of experience with expert knowledge our lecturers enable students to achieve their academic and career progression goals.

All programs delivered by Cyber Skills are aligned to the Internationally recognized NIST-NICE Cybersecurity Workforce Framework. This framework allows Cyber Skills to provide clear guidance to learners on the work role tasks, knowledge, and skills being covered in their selected course so that they may select a course that suits the work role they are in or planning to apply for.

About the Editors

Prof. Kashif Naseer Qureshi is an associate professor in Cybersecurity in the Department of Electronic and Computer Engineering at the University of Limerick, Ireland. He received a PhD. degree from the University of Technology Malaysia (UTM) in 2016. He is the Co-principal investigator in Cyber Reconnaissance and Combat project in the higher education commission. His research interest focuses on the Internet of Everything (IoE), Internet of connected Vehicles (IoV), Electronic Vehicles charging management planning and recommendation (EV), and Internet of Things (IoT) use cases implementation in wireless and wired networks and cybersecurity. His research interest includes many topics under the general areas of cyber security, wireless networks, security perspectives for IoE and IoT networks, privacy-preserving data aggregation, computer forensics, and cloud security. His name listed in the top 2% of scientists for three consecutive years from Stanford University, USA. He has published various high-impact factor papers in international journals and conference proceedings and served on several conferences IPCs and journal editorial boards. He has five books related to cybersecurity and AI fields. He has completed various projects related to routing and cyber-security domains in the UK, China, Ireland, Malaysia, and Pakistan.

Prof. Gwanggil Jeon received BS, MS, and PhD (summa cum laude) degrees from the Department of Electronics and Computer Engineering, Hanyang University, Seoul, Korea, in 2003, 2005, and 2008, respectively. From September 2009 to August 2011, he was with the School of Information Technology and Engineering, University of Ottawa, Ottawa, ON, Canada, as a post-doctoral fellow. From September 2011 to February 2012, he was with the Graduate School of Science and Technology, Niigata University, Niigata, Japan, as an assistant professor. From December 2014 to February 2015 and June 2015 to July 2015, he was a visiting scholar at

Centre de Mathématiques et Leurs Applications (CMLA), École Normale Supérieure Paris-Saclay (ENS-Cachan), France. From 2019 to 2020, he was a prestigious visiting professor at Dipartimento di Informatica, Università degli Studi di Milano Statale, Italy. He is currently a full professor at Incheon National University, Incheon, Korea. He was a visiting professor at Sichuan University, China, Universitat Pompeu Fabra, Barcelona, Spain, Xinjiang University, China, King Mongkut's Institute of Technology Ladkrabang, Bangkok, Thailand, and the University of Burgundy, Dijon, France. Dr. Jeon is an IEEE senior member, and associate editor of *Sustainable Cities and Society*, *IEEE Access*, *Real-Time Image Processing*, *Journal of System Architecture*, and *MDPI Remote Sensing*. Dr. Jeon was a recipient of the IEEE Chester Sall Award in 2007, the ETRI Journal Paper Award in 2008, and the industry-Academic Merit Award by the Ministry of SMEs and Startups of Korea Minister in 2020.

Contributors

Usman Ahmad is a lecturer in Bath Spa University Academic Centre RAK, UAE. His research interests are language models and AI tools.

Mamoona Amin is a lecturer in the Department of Computer Science, Institute of Management Sciences, Lahore, Pakistan.

Muhammad Anwar is an assistant professor in the Department of Information Sciences, Division of Science and Technology, University of Education, Lahore, Pakistan.

Ayesha Aslam is a PhD scholar in the School of Information Engineering, Chang'an University, Xi'an, Shaanxi, China.

Adil Hussain is a PhD scholar in the School of Electronic and Control Engineering, Chang'an University, Xi'an, Shaanxi, China.

Majid Hussain is a lecturer in the Department of Computer Science, University of Faisalabad, Faisalabad, Pakistan.

Muhammad Mueed Hussain is a lecturer in the Institute of Management Sciences, Lahore, Pakistan.

Amna Iqbal is a lecturer in the Department of Computer Science, University of Faisalabad, Faisalabad, Pakistan.

Danish Javeed is a PhD scholar in Software College, Northeastern University, Shenyang, China.

Gwanggil Jeon is a professor at Incheon National University, Incheon, Korea. His research interests are cybersecurity, AI, and machine and deep learning systems.

Seemab Kareem is a lecturer in the Faculty of Computing, Riphah International University, Gulberg green campus Islamabad.

Amna Khatoon is a PhD scholar in the Department of Information Engineering, Chang'an University Xi'an, China.

Muhammad Muneer is a lecturer in the Department of Statistics and Data Science, University of Mianwali, Pakistan.

Hanaa Omar Nafea is an assistant professor in the Department of Computer Science, College of Computer Science and Engineering, Taibah University, Al-Medina Al-Munawwarah, Saudi Arabia.

Tran Vu Pham is a scholar at Ho Chi Minh City University of Technology (HCMUT), Ho Chi Minh City, Vietnam.

Vu Hien Phan is a scholar at the International University (HCMIU), Thu Duc, Ho Chi Minh City, Vietnam.

Cuong Pham-Quoc is an associate professor at Ho Chi Minh City University of Technology (HCMUT), Ho Chi Minh City, Vietnam.

Kashif Naseer Qureshi is an associate professor in the Department of Electronic and Computer Engineering at the University of Limerick, Ireland. He is an active researcher in the field of cybersecurity and AI.

Faisal Rehman is part of the Department of Statistics and Data Science, University of Mianwali, Pakistan, and Department of Robotics and Artificial Intelligence, National University of Science and Technology (NUST), Islamabad, Pakistan.

Amreen Shafique Saeed is a PhD scholar in the School of Software, Dalian University of Technology, Dalian, China.

Muhammad Shahid Saeed is a PhD scholar in the School of Software, Dalian University of Technology, Dalian, China.

Muhammad Hamza Sajjad is a lecturer in the Department of Statistics and Data Science, University of Mianwali, Pakistan.

Wang Shixin is a PhD scholar in the School of Electronic and Control Engineering, Chang'an University, Xi'an, Shaanxi, China.

Nguyen Cao Tri is a scholar at the Vietnam National University – Ho Chi Mih City (VNUHCM), Thu Duc, Ho Chi Minh City, Vietnam.

Asad Ullah is PhD scholar in the Department of Information Engineering, Chang'an University Xi'an, China.

Peng-Cheng Xu is a PhD scholar in the School of Electronic and Control Engineering, Chang'an University, Xi'an, Shaanxi 710064, China.

Hina Zafar is a lecturer in the Department of Computer Science, University of Faisalabad, Faisalabad, Pakistan.

Sidra Zubair is a lecturer in the Faculty of Computing, Riphah International University, Gulberg Green Campus, Islamabad.

Introduction

In today's hyperconnected world, where technology permeates every aspect of our lives, Artificial Intelligence (AI) has emerged as a transformative force. In the landscape of artificial intelligence, the emergence of next-generation language models represents a significant milestone. These models, often leveraging deep learning architectures, have showcased remarkable capabilities in understanding, generating, and manipulating human language. Their applications span across various domains, from aiding in natural language understanding and generation tasks to facilitating human-computer interactions and automating repetitive processes. *Next Generation AI Language Models in Research: Promising Perspectives and Valid Concerns* explores the forefront of this technological advancement, offering a comprehensive exploration of both the potential benefits and the associated challenges. As we embark on this journey of innovation, it becomes imperative to critically examine the promises these models hold and the concerns they raise.

The editors of this book, with their deep expertise and extensive experience in the field of AI, provide valuable insights into the unique risks associated with AI. At the heart of this book lies an acknowledgment of the transformative potential of next-generation AI language models. They have revolutionized the way we interact with technology, enabling more intuitive and efficient communication between humans and machines. Through their ability to comprehend context, infer meaning, and generate coherent text, these models have opened doors to new possibilities in fields as diverse as healthcare, education, finance, and beyond.

Yet, amidst the excitement surrounding these advancements, it is crucial to recognize and address the valid concerns that accompany them. Ethical considerations, privacy implications, biases in data and algorithms, and the potential for misuse are among the complex issues that demand careful scrutiny. As stewards of AI progress, we are tasked with navigating these challenges responsibly, ensuring that the benefits of these technologies are equitably distributed while mitigating their risks.

This book serves as a beacon for researchers, practitioners, policymakers, and enthusiasts alike, offering insights, reflections, and guidance on the multifaceted landscape of next-generation AI language models. Through a balanced examination of their promises and concerns, it seeks to foster a deeper understanding of the opportunities and responsibilities that accompany their adoption.

I believe that *Next Generation AI Language Models in Research: Promising Perspectives and Valid Concerns* will be useful to readers who are starting to approach this complex technical topic since it puts together many different perspectives, application examples, and specific solutions. At the same time, it will be a useful reference for the more experienced researcher who aims at going deeper into a specific vertical application of AI, or who looks for possible open questions and/or future research topics to be explored.

Dr. Kashif Naseer Qureshi,
Department of Electronic and Computer Engineering, University of Limerick, Limerick, Ireland

CHAPTER 1

Introduction and History of Language Models in Research

Kashif Naseer Qureshi and Gwanggil Jeon

1.1 INTRODUCTION

Language Models (LM) have been adopted in various fields to generate the human language. These computational models have capabilities to predict the word sequence and generate text [1]. The Language Models achieved significant improvements and trained on massive amounts of text data with high accuracy. These models are based on transformer architectures to tackle the long-range dependencies from languages [2]. The large language models have made significant advancements and trained on massive amounts of data to generate human-like responses such as Generative Pre-trained Transformer (GPT-3). These models use transformer architecture and improve the language models' ability to handle the natural language texts [3]. The transformer architecture is based on self-attention method and makes better relationship between sentences, words, and their positions. This architecture also used the pre-trained model by using large datasets and performing a wide range of language tasks. One of the well-known examples is Bidirectional Encoder Representation (BERT) which is pre-trained by using Natural Language Processing (NLP) tasks

DOI: 10.1201/9781032667911-1

like classification and entity recognition. These models are also capable of learning complex patterns, semantic relationship, and linguistic nuances. The latest ChatGPT uses OpenAI and is trained on internet-sourced data such as books, articles, and websites. The latest version, GPT-4, has more advanced features like visual information and generates responses with more rich information and content. There is a large range of applications where these language models are implemented such as healthcare, education, finance, engineering, and social sciences.

In the healthcare sector, these language models are used for radiological decision-making, patient care, and clinical genetics. The advanced AI-based tools have potential benefits and improvements for decision-making. These tools are used for investigating the different aspects of data such as integration of biotechnology, decision support, diagnosis, analysis, and evaluation [3]. There are many examples of known AI tools designed for healthcare like Babylon Health and Buoy, and Ada Health. These models are also adopted by the education sector where these models are trained on educational material and help the teachers and students to generate required output such as curriculum development, course development, lecture preparation, assignments, and quiz development. These tools also provide personalized experience to the users to check their strengths and weaknesses. The finance sector also adopted AI-based tools for better decision-making, prediction, and data analysis. Well-known AI-based tools in the finance sector are BloombergGPT and FinGPT.

Language models are showing tremendous performance in modeling the code and integration with AI-based programming assistance [4]. Codex is one of the largest models deployed in real-world productions such as in GitHub Copilot. This is an IDE developer assistance platform for automatically generating the code by using the user's content. The large language models by using autonomous agents are expected to perform diverse tasks and provide valuable insights across various domains.

1.2 BACKGROUND OF LANGUAGE MODELS

The language models are pivotal tools for research to process and generate the human language. These models are rooted in fields of Machine Learning (ML) and NLP and achieved significant progress over the years. These models offer various solutions in the fields of AI, data science, and linguistics to address the complex challenges. The concept of language models was started in the mid-20th century when researchers explored the new machine translation and computation linguistics [5]. At that time, the language models utilized rule-based methods with limited knowledge. Then these models changed into statistical language models to predict the phrases and words from context and overall improved the machine translation and speech recognition systems. Neural network is another step forward in language models to change the traditional statistical approaches such as Recurrent Neural Networks (RNNs). These models achieved significant improvement to capture sequential dependencies in language data.

Language models have evolved along with the development of natural language processing and machine learning techniques. Initially, language models were simple rule-based systems that aimed to analyze and generate text based on grammatical rules stipulated by the designers. However, they failed to comprehend the natural language due to its complexity and ambiguity. With the development of the NLP, multiple tools and frameworks, both statistical and non-statistical, have been created. Statistical models such as n-gram models have introduced the concept of probability in language generation. In doing so, they assess the likelihood of a word to appear given the context of a sequence of other words. While such models have outperformed rule-based instruments, they were still limited by short-range dependencies and lack of semantic understanding. Moreover, they required large corpora to be effective.

1.3 SIGNIFICANCE OF LANGUAGE MODELS IN RESEARCH

Language models have gained popularity with a positive impact on research and development in different fields. The new language models discover and accelerate information and knowledge. The knowledge is extracted from large databases such as text sources, relevant research papers, and desirable findings. The usage of language models in research accelerates the research process and decreases the amount of time and effort. Language models are also useful to generate hypotheses and research insights to analyze the data. These models are useful in identifying the patterns for better results which are difficult to find by humans. Another significant contribution of language models in research is developing computational tools for researchers such as for translation, text summarization, and equation solving. These tools are helpful to understand the data more effectively. Language models are useful to understand the human language to understand human behavior to develop new models and theories in different fields like linguistics, sociology, and psychology. In the field of healthcare, the language models are used to identify the drug targets to develop new therapies such as identifying the patterns from medical literature for better prediction. These processes improve healthcare services and reduce serious disease risks. The findings and results are further used for new diagnostic tests for the disease. The language models are also useful in academic and education fields where these models are used to personalize learning experiences for students such as for reading recommendations based on students' reading levels and their interests. In the area of astronomy, language models are used to analyze massive datasets like telescopic images and satellite data. The analyzed data is further used to identify new objects from planets and galaxies. Language models have been adopted for continued research development to make new discoveries.

1.4 EARLY LANGUAGE MODELING APPROACHES

The early language models were introduced in the 1950s and 1960s as statistical processes. Statistical language models like conditional random fields and maximum entropy models are used to capture the dependencies between words. At that time the language models were simple and used for generating natural-sounding text. These were the foundation or start of the language models. The early modeling methods were N-gram language model, Hidden Markov Models (HMMs), and Probabilistic Context-free Grammars (PCFGs). The N-gram language models are used to find the probability of a word occurring in a sentence and generate the text by predicting the next word sequence. These models are also useful to fix the content size and address the long-range dependencies. HMMs are used to capture the long-range dependencies in language. These models are also used in various applications such as machine translation, speech recognition, and parsing. These models are also useful for sequences of observations. PCFG models are useful for capturing the syntactic structure and associated probabilities. Rule-based system is another example of early language models which are rule-based systems, especially for predefined grammatical rules syntactic structures. These models are limited to handle the natural languages and their complexities. Latent Semantic Analysis (LSA) is another example that represents the documents and words in a high-dimensional semantic space. These models are helpful in capturing the semantic relationship between words.

Other language models have been introduced with more sophisticated approaches such as Recurrent Neural Networks (RNNs), Long Short-Term Memory (LSTM), and Generative Pre-Trained Transformer (GPT). The RNN models are neural networks that process speech and text or sequential data. These models are used in a variety of language models and applications like text summarization and machine translation. The LSTM networks are a type of RNN and are useful for better capturing of long-range dependencies in language. Transformers are also used for specific parts of the input sequence to generate the output. This is also one of the types of neural networks used in different language modeling applications. The GPT models are another attractive example of transformer language model to train the massive text and code datasets. These models are able to generate the human-level text, write different types of creative content, and translate the languages.

1.5 MILESTONES IN LANGUAGE MODEL DEVELOPMENT

The development of language models started in the 1950s and achieved some remarkable milestones in generating and understanding human languages.

- **In 1950s:** This is the started era of languages models when the N-gram models were introduced by using the idea to check the probability of words occur in sentence is influenced by the words that have come before it. These models were simple in processes but presented the base of language models.
- **In 1960s:** After ten years, another achievement was observed with the name of HMMs with a more sophisticated method for language models. These models presented the hidden states that represent the underlying structure of a sentence.
- **In 1970s:** With the passage of time, language models were developed, and in this era another achieved model was PCFGs with a more formal model. These models represented the structure of languages with more accuracy in sentence patterns and grammatical rules.
- **In 1980s:** In this era, new connectionist models were introduced. These models were designed with the inspiration of human brains especially for complex relationships of phrases and words.
- **In 1990s:** RNNs was introduced as a powerful method in language models by using the long-range dependencies in language.
- **In 2000s:** The LSTM was introduced as one of the types of RNN. This method solved the vanishing gradient issue and provided a significant enhancement in language modeling accuracy.

- **In 2010s:** The DL was introduced as a revolutionized language model like GPT and BERT. These models capture the difficult relationships between phrases and words and provide better accuracy to generate the human quality text.
- **In 2020s:** The language models have been continuing to evolve. The new techniques and methods are developed, especially as large language models like Jurassic-1 Jumbo and Megatron-Turning NLG. These models are trained on massive datasets like translating language, text formatting, and creative contents.

Language modeling has evolved continually with more impressive and advanced features to communicate, interact, and learn. Figure 1.1 shows the language models' evolution process.

1.6 TYPES OF LANGUAGE MODELS

Language models are categorized into two main categories: statistical and neural network-based models.

1.6.1 Statistical Models

These models use the idea where the probability of words occurring in a sentence is influenced by the words that have come before it. These models are used to generate the text and predict the word sequence by using the previous N words. These models are simple and easy to train. However, these models do not perform well to capture the long-range dependencies in language. The common statistical models are N-gram, Bigram, and Trigram models. The N-gram models are simple to count the word frequencies in large text datasets. The extracted information is further utilized to check the phrases and word probability. In addition, the bigram models are used to check the probability of words by checking the previous word and are able to capture the long-range dependencies in language. The trigram models are used to consider the probability of words by checking the two previous words and are able to capture the long-range dependencies in languages.

1.6.2 Neural Network Models

On the other hand, the neural network models are more efficient compared to statistical models and are able to learn more complex phrases and word relationship. These models are trained on massive datasets to

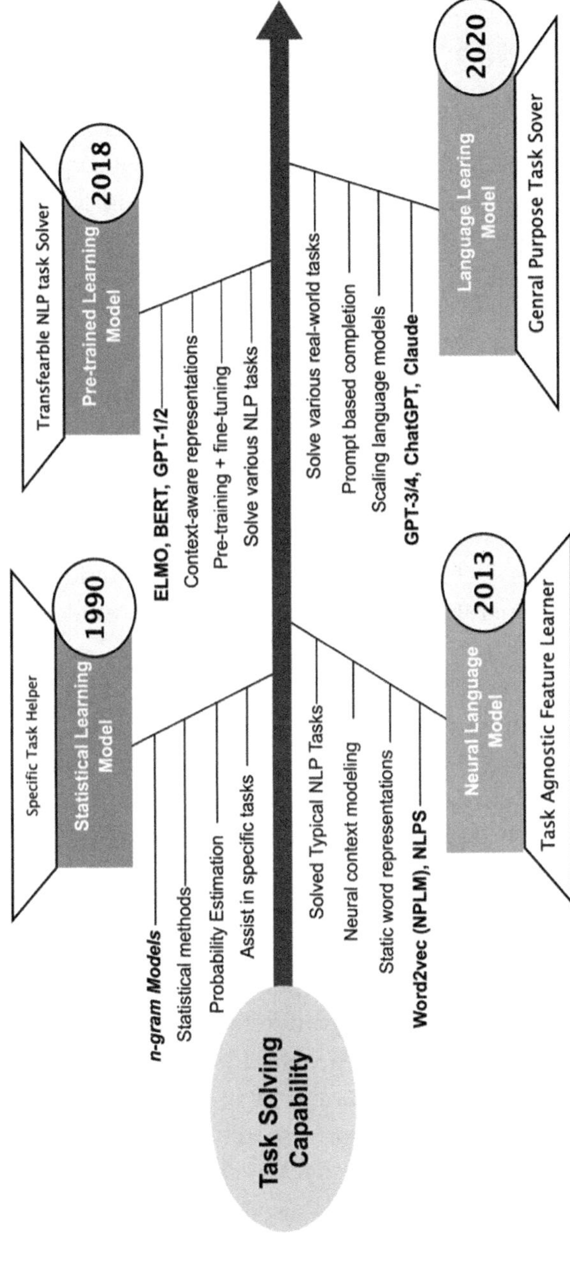

FIGURE 1.1 Language models' evolution process.

learn the statistical patterns to understand human languages. The common types of neural networks are RNNs, LSTM, and transformers. The RNN models are types of neural networks developed for speech, text, and sequential data. These models are able to learn long-range dependencies in languages. The LSTM is another type of neural network designed for vanishing gradient issues to process the long-range dependencies in language compared to traditional RNN models. The transformer model is a type of neural network designed for specific parts of the input sequence to generate the output. These models are able to learn the complex relationships between phrases and words compared to RNNs.

1.6.3 Generative Pre-Trained Transformer (GPT) Models

These models are used to train massive datasets containing text and code. The GPT models are transformer-based models used for content and creative writing. These are powerful models and tools used for a variety of NLP tasks. These models train on datasets and then trained datasets are used to generate new text by using the knowledge of statistical patterns and generate the next word or sentence. These models are used in a wide range of applications such as natural language understanding (NLU) and natural language generation (NLG). In NLU, these models are used to understand the user query [6]. Whereas the NLG model generates the text for content writing or code. These models are also useful for text summarization from long documents. The GPT models are useful for questions and answers such as book author name, definition, and abbreviation details. These models are also used to generate different creative text formats like email, musical pieces, code, scripts, and letters.

1.6.4 Large Language Models (LLMs)

Large language models (LLMs) are used to train the large datasets of text compared to GPT. These models perform different tasks like generating creative text formats, writing different types of content, and being used in research and development fields. LLMs are also used in Artificial Intelligence (AI) to train massive amounts of data. These models process the data in the real world like Google search to generate the text. These models are still under development to interact with devices and machines. These models are trained on large datasets like websites, books, articles, and codes. After training the data, models generate the text and languages and write different types of creative content [7]. These language models are used in many applications such as text summarization, machine translation, and creative content writing Compared to traditional language models, LLMs perform a wide range of tasks with better accuracy due to handling of complex learning features compared to traditional language models.

1.7 APPLICATIONS OF LARGE LANGUAGE MODELS

There are a wide range of applications where the Large Language Models (LLMs) are used. Most well-known applications have adopted Natural Language Processing (NLP) such as word-level, sequence-level, sentence-level, relation extraction, and text generation. The applications for word- and sentence-level tasks are word clustering, sentence matching, and sentiment classification [8]. To address these work- and sentence-level issues, the model understands the semantic information from sentences. The applications for sequence tagging also perform fundamental tasks. Information extraction applications are used to extract useful structural information from unstructured text data such as event and relation extraction. Text generation applications are also useful for machine translation and automated summarization. These applications are flexible to effectively handle the special needs in real-time applications. LLM applications have achieved significant improvements to achieve competitive performance, especially for high-quality labeled data. Other types of application are used for Information Retrieval (IR) to discover information from documents. These applications retrieve the relevant information from large-scale corpus to acquire the most relevant information. The existing IR models can capture the text semantics to improve the existing dense models.

Recommender systems are used to analyze user queries to retrieve the relevant information from documents. The recommendation models are based on recency-focused and in-context learning to improve the systems' performance and alleviate the model biases. The recommendation simulators are also used by using the user's real behavior. These simulators are equipped with encompasses relevant identify information. The multimodal large language model is also used to process information like text, audio, and video. Knowledge Graphs (KGs) enhanced LLM applications are used to handle complex tasks to improve the tasks. There are two applications in this category: single-agent and multi-agent applications. Single-agent applications are based on single-agent mode to solve the tasks from user requests such as AutoGPT [9], long short-term memory management, and search engines. The AutoGPT understands the user requests from its memory and performs like decomposing, reasoning, and detailing plan execution to assist the tools. Some other applications like GPT-Engineer, WebGPT, and XAgent process resolve the user requests. Whereas multi-agent applications use the agent's collaboration to unleash collective intelligence.

There are a number of GPT plugins and applications that have been developed with advanced functionalities and features. These applications are used to create and customize the contents as per users' requirements. GPT plugins are used to make new contents by accessing the external data sources, enhancing user experiences, and automating tasks. Plugins use AI generative tools and provide extended features. Users input their queries and tools extract and generate the data. There are many browsers that are compatible with these AI plugins, such as Bing which is compatible with ChatGPT by using API. The companies are working to design and develop more compatible tools integrated with language models and AI methods to generate more attractive and useful outputs. The well-known companies to develop the plugins are FiscalNote, Speak, Wolffram, Zapier, Klarna, OpenTable, and Shopify. The designed plugins are flexible to develop new applications and answer user queries. There are some other tools also available for coding like Code interpreter which is one of the built-in Python code interpreters used to perform logical calculations and write codes. This tool is useful for problem-solving and develop the problem solution. Another example is the third knowledge-based plugin useful for developers. This tool uses ChatGPT for data access and gathers relevant and useful information. The useful contents should be emails, notes, and files. Third-party plugins like ShowMe are also available which generates the diagrams. The ScholorAI is another example to access academic journals.

LLMs have a wide range of applications in various domains. They excel at generating coherent and contextually relevant text, making them ideal for creative writing, storytelling, dialogue generation, and poetry. Additionally, LLMs can accurately translate text from one language to another by leveraging their understanding of context and semantics. They can also analyze text for sentiment, which is useful in market research, social media monitoring, and customer feedback analysis. Furthermore, LLMs can summarize large bodies of text, making it easier to comprehend and retrieve information quickly. They are also capable of answering questions posed in natural language, ranging from factual inquiries to complex reasoning tasks. Moreover, LLMs power chatbots, enabling human-like dialogue for customer support, virtual assistance, and entertainment purposes. Additionally, LLMs can analyze user preferences to recommend personalized content such as articles, videos, music, and products. They can even generate code snippets or entire programs based on natural language descriptions of desired functionality. Moreover, LLMs can sift through legal documents, contracts, and case law to extract relevant information for legal research and decision-making. Furthermore, LLMs can assist in medical diagnosis by analyzing symptoms, recommending treatment options, and enhancing patient data with medical literature insights. They can also support language learners with personalized exercises and feedback to improve vocabulary acquisition and grammar proficiency. LLMs are also useful in automated content creation for marketing, advertising, journalism, and creative endeavors. Moreover, LLMs can analyze unstructured data, such as text documents and social media posts, to extract insights, trends, and patterns for decision-making purposes. Lastly, LLMs can power virtual simulations and training environments by generating realistic scenarios, dialogues, and interactions.

1.8 GPT-SERIES MODELS

ChatGPT has gained popularity due to its excellent capabilities in communication with humans. ChatGPT is designed by using a powerful GPT (Generative Pre-Training) model by using optimized conversations. These models used world knowledge and served as task solvers. These models accurately predict the next word by using decoder-only transformer language models and scaling the model size. Initially, OpenAI developed GPT-1 and GPT-2 models and considered them as foundation models. GPT-1 was introduced in 2018 by using the neural network architecture. Then, the GPT-2 was introduced by using large datasets with better multi-task-solving capabilities. The technical evolution of the GPT series started in 2018 and continuously has evolved with its versions 2, 3, and Codex. These versions are equipped with decoder-only architecture generative pre-training, unsupervised multi-task learner scaling, in-context learning exploration with scale limit, and code pre-training capabilities. In 2022, the new version of GPT 3.5 was introduced which is capable of code model, instruction follow-up, human alignment, and comprehensive abilities such as code-davinci-002, text-davinci-002, text-davinci-003, and GPT 3.5 turbo. GPT-4 was recently introduced with a more advanced context window, strong reasoning abilities, and multimodal ability (Figure 1.2).

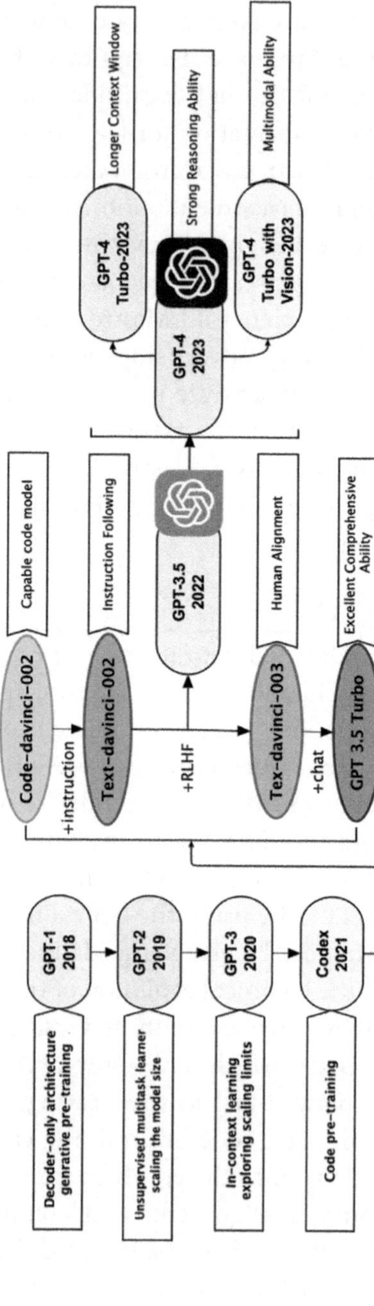

FIGURE 1.2 Technical evolution of GPT.

Table 1.1 presents a detailed comparison of various LLMs, focusing on key technical details such as model size, architecture, pre-training data sources, fine-tuning capabilities, and open-source availability. The parameters column specifies the number of model parameters, while the architecture column specifies the underlying structure, typically Transformer-based. The pre-training data column lists the datasets used for initial training, and the fine-tuning data column indicates the sources used for task-specific adaptation. The "Open Source" column indicates whether the models are publicly accessible for use and modification. This comparison is valuable for gaining insights into the LLM landscape and assists in making informed decisions for different natural language processing tasks. Table 1.1 shows the comparisons of various LLMs.

1.9 OPPORTUNITIES FOR LEARNING

There are various opportunities that have been observed after the adoption of language models in different fields. These opportunities are identified in the education sector due to its wide range of applications for teaching, learning, and training. Adoption of language models improved the traditional systems processes where students and teachers have new experiences for learning and teaching. These opportunities have a positive impact on all levels of education, from primary to higher studies. The large language models assist with writing skills, writing styles, and critical thinking processes as well as improving grammatical and syntactic errors. These models also assist in the development of reading comprehension skills like making summaries and making complex text into easy reading stuff. The middle and higher education sectors also used these models for learning, conducting scientific experiments, and generating practice problems and quizzes for a better grip on subjects and understanding reasoning behind the solutions. These language models explored new doors to empower learners with disabilities where they use speech-to-text or text-to-speech applications to complete their tasks with ease and comfort. These models are very useful for professional studies or training like for programming, report writings, decision-making, and project management. The currently used language-based applications in education sectors are GPT, BERT, XLNet, T5, RoBERTa, and GPT-3. These models are based on transformer architecture and pre-trained datasets [10].

TABLE 1.1 Comparison of Various LLMs

Model Name	Architecture	Parameters	Pre-Trained Data	Open Source	Fine-Tuning Data
GPT-3 (OpenAI)	Transformer (T5)	175 billion	Common Crawl, BooksCorpus, Wikipedia	×	Custom datasets
BERT (Google)	Transformer (BERT)	340 million–1.3 billion	BooksCorpus, English Wikipedia	√	Task-specific corpora
RoBERTa (Facebook)	Transformer (BERT)	125 million–355 million	Common Crawl, BooksCorpus, Wikipedia	√	Task-specific corpora
T5 (Google)	Transformer (T5)	11 billion - 11 trillion	C4, WebText, BooksCorpus	√	Task-specific corpora
GPT-2 (OpenAI)	Transformer (GPT)	1.5 billion	WebText, BooksCorpus, Wikipedia	√	Custom datasets
XLNet (Google)	Transformer (BERT)	340 million–570 million	BooksCorpus, English Wikipedia	√	Task-specific corpora
DistilBERT (Hugging Face)	Transformer (BERT)	66 million	English Wikipedia	√	Task-specific corpora
ELECTRA (Google)	Transformer (BERT)	110 million–340 million	BooksCorpus, English Wikipedia	√	Task-specific corpora
GPT-3.5 (OpenAI)	Transformer (T5)	175 billion	Common Crawl, BooksCorpus, Wikipedia	×	Custom datasets
BART (Facebook)	Transformer (BART)	406 million	Common Crawl, BooksCorpus, Wikipedia	√	Task-specific corpora

1.10 WORKING PROCESS OF LANGUAGES MODELS

The working process of language models is based on statistical models and DL methods where these models process extensive amount of data. These models use AI tools to generate human-like output. The language models learn the intricate patterns and relationships which exist in data. These models generate and resemble the style and characteristics. The first working step is pre-training phase where models' language models are exposed to various data sources including websites, articles, and books. By using the unsupervised learning methods, the models learn to predict the next word in a sentence with the help of the context of preceding words. This strategy helps the model to design and understand grammar, syntax, and semantic relationships. The pre-training phase is further categorized into two other categories including general and specialized [11]. Due to the large amount of data in the system, it is difficult to pre-process the data and remove noise, unnecessary material, and redundant and poisonous data. After this stage, the next step is quality filtering by checking the low-quality material or unwanted data. For this process, filtering, keyword filtering, and statistic filtering techniques are used. When the quality of the data is finalized, then the next step is deduplication where duplicate data in the corpus decreases the diversity of language models. The next phase is privacy reduction where privacy concerns are taken into account when data is used for pre-training. The data is checked in terms of any user-generated content like personal and sensitive information to address privacy concerns. Any type of personally identifiable information is also removed at this stage. At last, tokenization is performed where segments are created from raw data and enable sequence methods of individual tokens and further fed into language models.

In pre-training phase, the models undergo a fine-tuning process where the model is trained on a specific domain or task. This phase is also helpful for accuracy because of its labeled and guided strategy to generate more accurate and appropriate responses for the target task. Fine-tuning strategy makes these language models feasible for various domains such as questions and answering, language translation, and text generation. The language models have the ability to capture the statistical patterns existing in the training data. These models gain a comprehensive understanding and are able to generate the desired response. When the user gives any input or query from the application, then the model generates the response by using probabilistic techniques. Figure 1.3 shows the data preprocessing steps to train the language models.

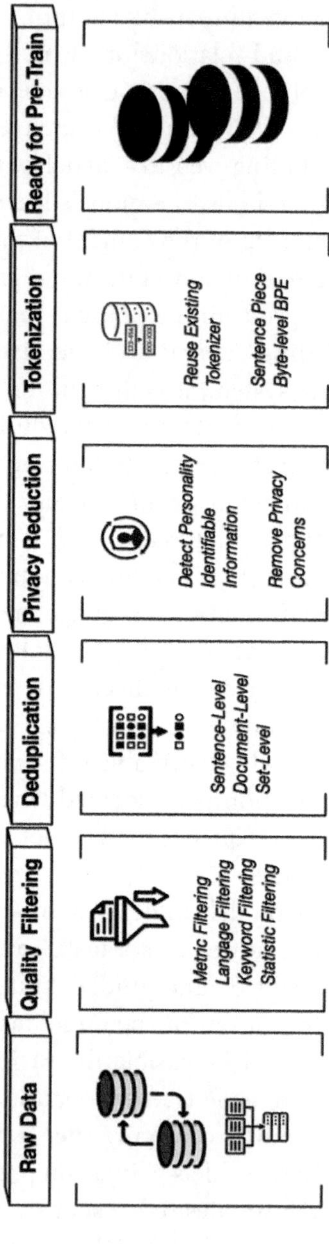

FIGURE 1.3 Steps involved in data processing to train the language model.

1.11 CHALLENGES AND LIMITATIONS

Language models are powerful tools and have a number of positive features in the research and development sector. As technologies have some limitations as well, the language models have their own limitations like data biases, lack of context, lack of common-sense reasoning, explainability and interpretability, generation of misinformation, and ethical consideration. The data biases is another challenge due to massive amount of text data from real world. These data biases lead to discriminatory or offensive content for researchers. Researchers use different techniques to handle the data biases such as using evaluation metrics or developing fair and diverse datasets. The lack of context and common-sense reasoning is another limitation of language models for researchers to understand the language and apply the common-sense meaning or intent. This limitation is handled by using the improvement strategies for contextual understanding and integrating the common-sense reasoning into models. Interpretability and explainability are also one of the limitations to understand the models because sometimes it is difficult to understand the model predictions. Language models have some limitations in terms of generating misinformation and creating deepfakes. The new models should be able to detect and prevent the generation of malicious content. Ethical consideration is another challenge and need to develop actively engaged discussions about ethical concerns and consideration [12]. Some other limitations have been observed in large language models such as lack of interpretability, limited domain-specific knowledge, limited memory, limited scalability, lack of causality, and limited attention span.

Language models demonstrated the potential for bias when training the data for development. There is a need to design more sophisticated language models for better discern between assumptions and factual information and deal with novel training methods and datasets. Another limitation of language models is explainability because of the huge number of parameter count. As language models grow in size, with some reaching billions of parameters like GPT-3 which has around 175 billion parameters, they become increasingly powerful. This complexity makes these models for humans to understand the decision-making process. This complexity also leads to lack of transparency to gain insight into all inputs and outputs. In addition, the data was also collected from diverse sources which make decision-making challenging. This issue is more serious in some sensitive domains like healthcare and finance fields. The stakeholders are still reluctant to trust these language models, especially for these fields' sensitive applications. Reasoning errors are also noticed in these language models where they make mistakes in logical reasoning due to inherent limitations or ambiguities to understand the difficult logical operations. Spelling errors are also noticed in these models like in GPT-4 due to its statistical nature. Security is also one of the limitations and challenges for these language models due to various adversarial attacks such as jail break, data poisoning, and prompt injection attacks. The malicious actors inject misleading information and try to conduct jail break attack to hack the sensitive data to manipulate the model output. Data privacy and security in language models raise concerns due to users' privacy violations. Lack of security and privacy policies make this challenge more serious to mitigate security and privacy concerns. Transparency needed during data collection, storage and protect this data from unauthorized access, unethical usage, and breaches.

Large language models need high computation processing which leads to high energy consumption. This issue makes these models challenging in terms of sustainability and scaling. Copyright issues are also key challenges for large language models during training the models in terms of designing any specific documents. Bias and unfairness is another key issue of language models whenever a trained model is biased toward certain groups or people where the results are discriminative in nature. The learners are heavily depending on language models without using their own critical thinking and skills because models are generating and simplifying the information. The new language models are accurate and effective but are not replacing the creativity, skills, and thinking processes of human beings. Lack of understanding and expertise is another challenge to understand these language models and integrate with new technologies. Some technical challenges also exist to train the large language models such as cost, maintenance, and integration with new technologies. The large language models' training is a minifacial burden on institutions. The cloud-based infrastructure and services are expensive and need more expertise to handle. Some other issues have been noticed where the language models may generate information without considering the training data which leads to factually incorrect information and misleading output.

Some society and human challenges have been observed in large language models because their outputs are similar to humans. Some environmental challenges also exist during training the data such as usage of water in data centers to regulate the server's temperature. Usually, fresh water is utilized to avoid any sort of bacterial growth. The deployment of these language models also needs significant energy resources. As per one report, ChatGPT development needs thousands and millions of kWh during training. The heavy machineries and their processes consume more energy due to high processing and data transfer requirements. To address these environmental factors, there is a need to develop more energy-efficient algorithms and methods to reduce the amount of energy to train these AI systems. Singularity is another concern where the AI system will be more intelligent than humans. Several Turing tests have been conducted to check the capabilities of AI systems, and till now these systems have not surpassed human-level intelligence. However, in the near future when these AI systems reach that level, then these systems will replace humans and lead to uncertainty and survival issues. OpenAI system is a competition matte for companies and for-profit generation. The massive growth of AI has started in the form of tools, systems, and robots. The competition among organizations leads to another economic challenge for the world due to excessive control on investors.

1.12 ENCOUNTER THE CHALLENGES AND RISKS

A number of challenges exist which need important steps to address and mitigate the challenges. This section discusses the important steps to address the existing challenges and issues of large language models.

1. **Copyright Issues:** The documents' originality should be transparent and according to the policies of data usage. Permission should be obtained for further data training and use in the models. . The copyrights terms and compliance should be followed to generate the model. The policies will be clear for users' awareness and further information.

2. **Fairness and Bias:** The diverse data will be used to train the model which is not biased with any particular group. Evaluation and monitoring mechanisms should be strong to monitor the overall model performance and address any biases. Pre- and post-processing and other methods should be adopted for bias-correction. Transparency methods need to be adopted to deal with bias and fairness issues. The training sessions will be adopted to tackle the potential biases and any sort of model failure.

3. **Model Adoption:** Learners heavily rely on language models without evaluating the limitations and unexpected errors of language models. There is a need to re-evaluate models again when generating the hypotheses and other different objectives. Always need to cross-check the model-generated data and compare that with other resources like books and articles.
4. **Relevancy and Accuracy:** To check the language model output, there is need for critical thinking and other problem-solving activities to ensure the relevancy and accuracy of the generated output. The methods are needed for evaluation of the generated output, especially in information-sensitive fields like education, healthcare, finance, and mass communication.
5. **Lack of Expertise:** Demand-based professional training is required for users, educators, and experts to understand the language models in terms of originality, ethics, and professionalism. The feedback and regular analysis require to train the professionals and ensure the language models' effectiveness in any field.
6. **Comparison:** In some cases, the comparison is difficult between model-generated and human-generated output. To address this issue there is a need for analysis techniques to measure machine- and human-generated text.
7. **Maintenance and Cost:** Financial support is required to develop large language models to perform field-specific tasks. The cost should be shared to design and develop more effective cloud-based systems for powerful computational resources. Research and development sector needs more attention on the maintenance processes of language models, especially to reduce the size and cost and to improve the performance.
8. **Privacy and Security:** Policies needed for the development and implementation of large language models which clearly outline the storage, processing, and collection policies. Transparency is required for the usage of these languages' models. The data should be protected from unauthorized access, unethical usage, and breaches. Regular audits are required for data privacy and security.
9. **Sustainability:** The language models should be sustainable in terms of resources such as energy-efficient hardware, feasible architecture,

and frameworks. The data collection, analysis, annotation, and processing are involved in language models which need regulatory compliance and ethical consideration.

10. **Lack of Adaptability:** Some fields need customization to adapt the language models as per their usage and requirement. The language model should be hybrid in nature to align with any specific domain and field to personalize the material. There is a need to design and develop more advanced models for better adaptation in any field.

1.13 CURRENT RESEARCH

In recent years, a number of large language models have been developed such as BERT, GPT, RoBERTa, XLNet, T5, and GPT-3 and 4. These models have been designed by using pre-trained methods applied to massive datasets of text. The main objectives of these models are to generate human-like text, assist in the translation and summarization services, and perform the tasks. Another latest addition is BLOOM developed by BigScience-community as an open-source project. This version covered around 46 natural languages and 13 programming languages and offered enormous new opportunities and applications for research and industrial fields [13]. As per existing research in this area, so many existing research has been conducted to identify the potential of these language models. Authors in [14] proposed AI generative models to generate the flashcards and quizzes for better learning experiences. In another research, the authors [15] suggested the GPT-3 for this purpose where the users generate the quizzes, multiple choice questions for students and teachers for assignments, and exam preparations. In another study [16], authors introduced the GPT-3 as pedagogical agent to enhance the students' skills. In another study [17], the authors suggested the GPT-3 as code generator to improve computer programming skills.

Some other studies discussed the AI generative models' impact in the finance sector where these applications are utilized for prediction, risk assessments, and financial reporting. Authors in [18] discussed the BloombergGPT which is trained on a large financial corpus to perform tasks like classification, question answering, and entity recognition. The AI-generative models are suggested for the engineering field, especially for software engineering, for code generation and software testing and debugging. Text-audio generation model is proposed in [19], to customize the new voices by using speech data and utilizing the acoustic and conditional layer in decoders. Text-to-music generative model is proposed in [20], by using the deep cross-model correlation learning architecture for audio and lyrics. In this model, the intermodal canonical correlation analysis is utilized for similarity calculation between audio and lyrics. Another effort in [21] where social media data was taken to generate visual information by using cross-model fusion and attentive pooling method. Text graph generation is another area where the tools are used to generate relationships among semantic and internal states as graph structure.

Text code generation is a significant feature of language models that aim to automatically generate valid programming code by using natural language descriptions and providing coding assistance. CodeBert is proposed in [21], as a bimodal transformer-based pre-trained text-code model aiming to capture the semantic connections between PL and NL. CuBERT [22] was designed by using similar architecture working without the sentence separation between natural languages and body of sentences. Another effort with the name of CodeT5 presented in [23] is based on pre-trained encoder. LLMs are one of the most innovative tools introduced in the last few years. BERT, GPT, RoBERTa, XLNet, T5, GPT-3, and numerous others are works based on a few pre-training schemes done on massive text data. The main aim is to make human-like text that creates transformation, controlled text, and plenty of other tasks. Another exciting figure in this generation-based area is BLOOM, provided by a community called BigScience. BLOOM refers to 46 natural languages and 13 programming languages and opens immense possibilities and applications in both science and industry fields. Work on these models has been actively carried out for potential publication.

1.14 CONCLUSION

To conclude, this chapter has provided a starting explanation and historical overview of language models as part of the research. Language models are, without a doubt, one of the most fundamental aspects of natural language processing and have made significant advancements over time. The chapter began by providing an explanation of language models and their importance as well as mentioning some of the various applications they have, including but not limited to machine translations, speech recognition, and generation of text among many others. It also explained modern trends and possible directions language model research will take, focusing on the transformer models and the rapid advances in unsupervised and self-supervised learning. This overview constitutes an invaluable guide through the labyrinth of modern language models' research and their significance in natural language processing for both scholars and practitioners.

REFERENCES

1. B. Min *et al.*, "Recent advances in natural language processing via large pre-trained language models: A survey," *ACM Computing Surveys*, vol. 56, no. 2, pp. 1–40, 2021.
2. E. Kasneci *et al.*, "ChatGPT for good? On opportunities and challenges of large language models for education," *Learning and individual differences*, vol. 103, p. 102274, 2023.
3. K. N. Qureshi and T. Newe, "Artificial internet of things: A new paradigm of connected networks," in *Artificial Intelligence of Things (AIoT)*. CRC Press, 2024, pp. 3–20.
4. F. F. Xu, U. Alon, G. Neubig, and V. J. Hellendoorn, "A systematic evaluation of large language models of code," in *Proceedings of the 6th ACM SIGPLAN International Symposium on Machine Programming*, 2022, pp. 1–10.
5. Y. Liu *et al.*, "Summary of chatgpt-related research and perspective towards the future of large language models," *Meta-Radiology*, vol. 1, p. 100017, 2023.
6. A. Aslam, K. N. Qureshi, and T. Newe, "Future privacy and trust challenges for AIoT networks," in *Artificial Intelligence of Things (AIoT)*. CRC Press, 2024, pp. 198–216.
7. Y. Chang *et al.*, "A survey on evaluation of large language models," *arXiv preprint arXiv:2307.03109*, 2023.
8. S. Minaee, N. Kalchbrenner, E. Cambria, N. Nikzad, M. Chenaghlu, and J. Gao, "Deep learning--based text classification: A comprehensive review," *ACM computing surveys (CSUR)*, vol. 54, no. 3, pp. 1–40, 2021.
9. "Significant-Gravitas/Auto-GPT." https://github.com/.
10. A. Radford, K. Narasimhan, T. Salimans, and I. Sutskever, "Improving language understanding by generative pre-training," 2018.
11. I. T. Javed and K. N. Qureshi, "Role of blockchain models for AIoT communication systems," in *Artificial Intelligence of Things (AIoT)*. CRC Press, 2024, pp. 122–139.
12. M. U. Hadi *et al.*, "A survey on large language models: Applications, challenges, limitations, and practical usage," *Authorea Preprints*, 2023.
13. B. Workshop *et al.*, "Bloom: A 176b-parameter open-access multilingual language model," *arXiv preprint arXiv:2211.05100*, 2022.
14. S. Moore, H. A. Nguyen, N. Bier, T. Domadia, and J. Stamper, "Assessing the quality of student-generated short answer questions using GPT-3," in *European Conference on Technology Enhanced Learning*. Springer, 2022, pp. 243–257.
15. R. Dijkstra, Z. Genç, S. Kayal, and J. Kamps, "Reading comprehension quiz generation using generative pre-trained transformers," in *iTextbooks@ AIED*, 2022, pp. 4–17.

16. R. Abdelghani *et al.*, "Gpt-3-driven pedagogical agents to train children's curious question-asking skills," *International Journal of Artificial Intelligence in Education,* vol. 34, pp. 1–36, 2023.
17. S. MacNeil, A. Tran, D. Mogil, S. Bernstein, E. Ross, and Z. Huang, "Generating diverse code explanations using the gpt-3 large language model," *Proceedings of the 2022 ACM Conference on International Computing Education Research,* vol. 2, pp. 37–39, 2022.
18. S. Wu *et al.*, "Bloomberggpt: A large language model for finance," *arXiv preprint arXiv:2303.17564,* 2023.
19. M. Chen *et al.*, "Adaspeech: Adaptive text to speech for custom voice," *arXiv preprint arXiv:2103.00993,* 2021.
20. D. Paul, M. P. Shifas, Y. Pantazis, and Y. Stylianou, "Enhancing speech intelligibility in text-to-speech synthesis using speaking style conversion," *arXiv preprint arXiv:2008.05809,* 2020.
21. H. Liang *et al.*, "JTAV: Jointly learning social media content representation by fusing textual, acoustic, and visual features," *arXiv preprint arXiv:1806.01483,* 2018.
22. A. Kanade, P. Maniatis, G. Balakrishnan, and K. Shi, "Learning and evaluating contextual embedding of source code," in *International Conference on Machine Learning.* PMLR, 2020, pp. 5110–5121.
23. Y. Wang, W. Wang, S. Joty, and S. C. Hoi, "Codet5: Identifier-aware unified pre-trained encoder-decoder models for code understanding and generation," *arXiv preprint arXiv:2109.00859,* 2021.

CHAPTER 2

The Advantages of Artificial Intelligence in Research

Usman Ahmad and Kashif Naseer Qureshi

2.1 INTRODUCTION

Research has always served as the foundation for progress and innovation in new fields, whether it is in the realms of science, medicine, and finance or beyond the pursuit of knowledge through inquiry for better understanding and solving problems. New areas of research have been expanding and emerging with the help of new integrated Artificial Intelligence (AI) technologies [1, 2]. AI is reshaping the systems and processes to gather, analyze, and process information. AI acts as a catalyst for discovery and facilitates collaboration among researchers for better decision-making. The new tools can solve complex and intensive problems and open new opportunities for researchers and scientists to adopt modern ways to obtain desired outcomes [3]. AI-based systems are reshaping and transforming the business to maximize productivity and profits [4]. In medical science research, AI opens new data analysis methods as per the experiment needs. From genomics to precision medicine, AI is pushing boundaries by achieving new findings and better results. The AI facilitates researchers for better collaboration by enabling knowledge sharing

DOI: 10.1201/9781032667911-2

within their communities such as conducting literature reviews, predicting results, and data analysis.

With many advantages, AI also introduced ethical considerations and challenges. This chapter explores the advantages, limitations, and possible biases of AI algorithms and ethical obligations. This chapter also summarizes the AI processes involved in research. Ever-evolving nature of AI is paving the way for modern research in different fields. This can revolutionize the research process and can be positive for society.

2.2 ENHANCING DATA ANALYSIS USING AI TOOLS AND TECHNIQUES

Tools and technologies can help to collect, organize, and process large datasets. These tools have been used in the past as well, but with AI integration in these tools, the speed and accuracy of the research are quite good. The nature of the tool depends on the type of data and research goals. There is progressive adaptation of AI tools to process large and complex datasets.

2.2.1 Data Processing with AI

Data is critical for academic or applied research as it has key information about different processes and workflows. The emergence of AI has provided academics with a formidable ally in their pursuit of knowledge [4]. One of the primary benefits of AI in the field of research is its exceptional capacity to analyze extensive volumes of data efficiently and accurately. The capability of AI tools and technologies is quite high in terms of processing different data types. It can also deal with structured and unstructured data formats [5]. In terms of data types, new tools are highly accurate and fast to process textual, visual, and sensor data values. Using manual data processing techniques, researchers were limited to only a specific type of data, but with advancements in AI, they can extend their research to various types of datasets. The extraction of useful data points in complex datasets is now doable only with the availability of AI tools. This is not only making the job easier for the researchers but also increasing the efficiency and speed of research and development.

2.2.2 Exploring Patterns and Trends

The human brain is quite impressive in terms of pattern recognition, but in this era of big data, there is a limitation when confronted with large and complex datasets. The data is being produced at a very fast pace so

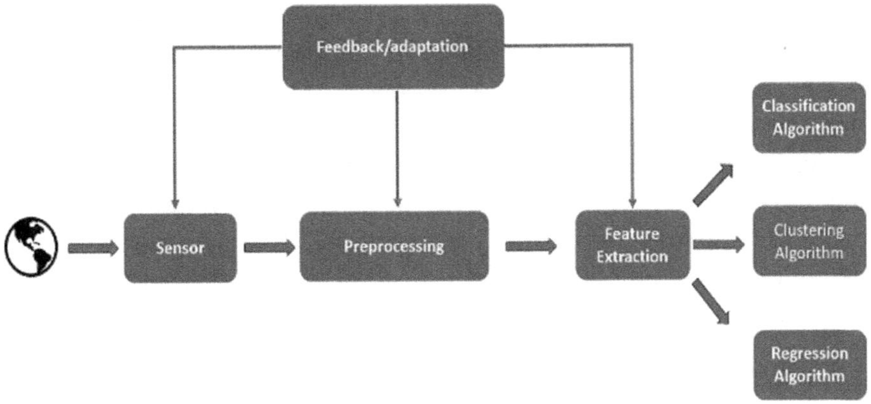

FIGURE 2.1 Pattern recognition using AI.

traditional methods are insufficient to find patterns in the data. Tools with AI features are now very vital to handle these kinds of datasets and find relevant patterns in the data [6]. Deep learning techniques are very helpful to find trends in complex datasets. Multidimensional datasets are very hard to process, and there is also difficulty predicting future trends in these datasets. To solve this problem, AI tools are helpful for predicting future trends in these datasets. In machine learning, machines are trained on complex datasets to extract information and then use it for prediction. This method is handy to extract key variables and features that might change specific trends in future. In medicine and diagnostics, this is helpful to find the exact information for specific diseases.

Agricultural research is also benefiting from the latest AI tools to determine the exact parameters of temperature and water for the optimal growth of plants and crops. With the integration of AI in agriculture, crop growth is increasing exponentially. The genomics industry also uses AI to identify interesting patterns and trends in genetic data. This helps scientists to map specific data points to different characteristics in humans and animals. Figure 2.1.

Latest AI tools with machine learning and deep learning techniques can process key features from complex datasets [7]. Applications of these techniques are widely ranged across different fields of studies. New researchers are using these tools and technologies to ease the process of finding key features in the data. The capability encompasses a wide range of study fields, including but not limited to economics, social sciences, healthcare, marketing, social sciences, biology, and environmental studies. Academics have the potential to enhance their comprehension of events, enhance the precision of their forecasts, and develop evidence-based approaches to tackle intricate problems. The ability to find patterns that are often hidden in complex datasets makes AI very useful for researchers. In real life, there are examples of complex datasets, which need to be processed to find useful information and trends. AI powered with advanced data processing tools can solve issues and strengthen the research and development process. By identifying and analyzing patterns and trends researchers can obtain insights that guide their investigations toward the most promising and fruitful paths.

2.2.3 Real-World Examples of AI in Data Analysis

To demonstrate the significant impact of AI in the field of data analysis, it is necessary to examine a few concrete instances from real-world scenarios. The field of healthcare has undergone a significant transformation with the advent of AI, particularly in the realm of medical image analysis [8]. This technological advancement has facilitated the timely identification of diseases such as cancer by using sophisticated imaging techniques. Additionally, it can evaluate patient records to identify potential dangers and provide recommendations for individualized treatment programs. The finance industry uses AI-powered algorithms to evaluate real-time market data, enabling rapid trading choices, optimization of investment portfolios, and identification of fraudulent transactions [9].

In the field of Environmental Science, climate researchers utilize AI techniques to analyze extensive information obtained from satellites and sensors. This application of AI aids in the monitoring of environmental changes, prediction of natural disasters, and development of sustainable policies. The field of social sciences benefits greatly from the utilization of AI as it enables the analysis of extensive social media data. This application of AI allows for the identification of public attitudes, monitoring of trends, and comprehension of human behavior [10]. Consequently, AI proves to be an excellent tool for sociological and psychological study. These examples illustrate the extensive implications that AI's capacity to augment data analysis holds for diverse study disciplines.

2.3 ACCELERATING DISCOVERY

With the adoption of AI now research is going in multiple directions for different areas [11]. This is helping scientists to work on different areas of research that were impossible before AI. With the latest tools and technologies, new concepts and problem statements are found which is making new ways for researchers to work on. The speed of the research and new findings is now fast due to the usage of AI data processing tools and technologies. It was quite a difficult and time-taking process to extract useful information from a large dataset manually. This is now replaced by AI tools which are making this process much faster than traditional processing methods. This section will explore how AI is making the discovery process fast and efficient.

2.3.1 Accelerating the Research Process

Traditional research methods were very slow as researchers had to wait for years to conclude a study due to the lengthy process of data collection and processing. They had to run experiments for all different hypotheses for findings. However, AI tools and technologies can simulate this experimentation process. With the latest AI tools researchers can simulate the models for fast data processing. This is not only helping fast conclusion but also making this process more reliable as AI tools are intelligent to find exact information in the complex dataset [12]. With a less time-consuming process, AI is helping researchers to find timely discoveries in life-saving scenarios such as drug development and diagnostics. Although traditional methods made the foundation of the research process, with the integration of AI tools, the results and findings are possible in a shorter time. Researchers can also apply hit and trial methods to different hypotheses in less time to discover different possibilities in research problems. This technique is very helpful in making new discoveries for the researchers so they can look up multiple aspects of a specific research problem.

2.3.2 AI Role in Medical Sciences

AI has changed the traditional research methods in the medical field. Genomic data is now being processed with AI-driven algorithms to extract key information. This is helping scientists to diagnose diseases and discover new drugs. Pharmaceutical companies are using AI tools to do tests of patients for diagnostics. The research quality is improved by the integration of AI methods in medical research. Diagnostic methods are now much faster and more efficient with the usage of AI. Large and complex datasets can be processed by these tools which makes the discovery process fast. This is also helping to reduce the workload on healthcare workers. Automation technology is helping healthcare workers to deploy this technology for automated processes and monitor the conditions of patients. Healthcare data can be processed at a very large scale using large language models. With technology being used, healthcare is now accessible in remote areas where people can access healthcare services. This is also reducing the cost of healthcare services. The process of drug discovery is also fast with the usage of AI. For experimental analysis, AI can help in simulating different experiments to collect data and then process the data.

2.3.3 Data Automation for Research Efficiency

To fasten the research and discovery process, automation can be very helpful. Automation was not possible until the development of AI tools and technologies. One of the major advantages of AI in research is its ability to automate complex tasks and procedures [13]. Data collection and sample handling can be automated in laboratories and research centers that can make this process fast and efficient. AI-based tools can go beyond automation to analyze and predict trends in the data. Literature review is very important to formulate questions for research statements. With AI-powered tools, literature review can be automated. This will save time as AI tools can analyze volumes of research papers in less time compared to humans. This is also used to extract key information from these papers to find research gaps. Reproducible research is now very much possible with the help of AI tools. Table 2.1 shows the fields where AI improved the automation process.

Automation is also making it easy for the researchers to focus on extended research areas by saving their time. AI-based tools are utilizing intelligent algorithms to complete manual tasks. Just like any other industry, the research field is also benefiting from the AI-based machine and systems.

2.4 IMPROVING DECISION-MAKING

Making decisions in the world of research often involves navigating scenarios, handling resources, and assessing risks. For researchers and companies, AI has developed into a decision-making tool. It helps researchers make wise choices, determine risks, and allocate resources as well [14]. With the rapid growth of AI technologies, human decision-making abilities have grown significantly. These days, AI systems are exceeding human capabilities in several fields, and it is crucial to know the effects of AI on decision-making [17]. To help researchers, AI-powered systems that support systems for choices use knowledge based on data. Rather than depending just on intuition, these tools can evaluate data, current

TABLE 2.1 Data Automation in Different Fields

Fields	Description
Enhanced Data Analysis [6]	Processes vast amounts of data efficiently, uncovering patterns and trends that might be difficult or impossible for humans to detect
Accelerated Discovery [12]	Speeds up the research process, particularly in drug discovery and scientific experiments, by rapidly analyzing vast datasets and predicting outcomes
Improved Decision-Making [14]	Assists in informed decision-making by analyzing complex data, providing insights, and predicting outcomes, aiding in risk assessment and resource allocation
Personalized Research and Medicine [15]	Tailors research and medical treatments to individual profiles, optimizing efficacy and minimizing potential side effects
Collaboration and Knowledge Sharing [16]	Facilitates research collaboration and interdisciplinary knowledge sharing by providing tools for literature review, and data sharing
Automated Repetitive Tasks [17]	Automates repetitive and mundane tasks, allowing researchers to focus on high-level thinking and more creative aspects of their work
Predictive Analytics and Forecasting [18]	Employs predictive modeling and forecasting based on historical data, aiding researchers in anticipating trends, future outcomes
Pattern Recognition and Image Analysis [19]	Excels at pattern recognition and image analysis, assisting in medical imaging interpretation, biological imaging

trends, and relevant factors to give evidence-based recommendations. To reduce risks and boost the likelihood of outcomes, this analytical strategy plays a vital role. For instance, in the field of drug discovery, AI can predict drug effectiveness by analyzing its framework and historical records. That's how researchers can focus their energies on those who are eligible for additional research.

2.4.1 Assisting Informed Decision-Making

In this field, we will face the results of our efforts. These choices are crucial as they effectively select different tools. AI provides decision support via data-driven insights, predictive analysis, and scenario simulations. These AI-powered systems can examine all relevant factors to offer solid suggestions instead of depending only on intuition. It not only enhances the decision-making quality but also minimizes the risk of making errors. AI can help researchers by providing them with guidelines to boost the possibilities of effectiveness. It is crucial in research scenarios to evaluate and split resources due to financial time and human resources limitations. No doubt, in these areas by assessing risks and improving the use of resources, AI is playing a significant role.

2.4.2 Risk Assessment

To assess risks, AI examines data as well as highlights errors. In clinical trials, for example, AI can detect the probability of these events or assess the individuals who are in an elevated danger of harm. It enables researchers to adopt appropriate measurements. By observing trends in large amounts of historical data, AI can enhance the evaluation of these risks. Those patterns provide solutions to risks that are usually hard for analysts to analyze. Furthermore, by allowing for assistance and successful methods for risk mitigation, this approach enables it to be easier to detect new risks. Moreover, a better understanding of these possible impacts regarding those risks is also provided by AI-powered simulations that create these situations by utilizing the data. AI has proven useful in industries like finance for the detection of activities by recognizing these patterns of transaction and errors that drastically enhance the efforts of risk assessment. To identify metrics affecting risk levels and allow the use of targeted methods for risk management, AI has driven the sensitivity analysis of models.

2.4.3 Resource Allocation

Researchers can use AI to maximize the usage of resources such as lab equipment, personnel, and funding. The resources are allocated where they are most required by using AI. By reducing any waste, it accelerates the outcome of research as well. We can forecast what resources are needed by considering the pattern of usage and the next project's needs by using AI algorithms. This type of approach ensures that all needed resources are always there, avoiding delays. Additionally, AI allows flexible utilization of resources, which allows us to modify project goals and current time requirements. It accelerates productivity and enhances resource utilization. One more field in which AI excels is budget optimization. AI supports us in promoting the worth of resources by assessing spending patterns and suggesting settlements. While assigning individuals to appropriate tasks, it also adds to personal allocation by taking into consideration expertise, workload, and project assignments. It assures employee happiness and progress in productivity as well.

2.5 CASE STUDIES IN AI-DRIVEN DECISION-MAKING

Some real-life examples showcase the impact of AI on decision-making in research.

2.5.1 Drug Discovery

The medical field is utilizing AI for drug discovery based on automated chemical reactions. Drug candidates are classified for different use cases to study the effects of specific drugs. AI-based methods are applied for testing molecules and predicting their effectiveness. The good thing about this process is the ability of automation that can pave new ways for drug discoveries. With the usage of machine learning, we can predict the relation between the physical and chemical properties of molecules at quantum mechanics level with less time as compared to traditional methods [19]. Due to the high cost of these experiments, AI is helping researchers in the automation of these experiments to reduce overall cost. Figure 2.2 shows the process of AI-based data-driven decision-making.

2.5.2. Astronomy

Astronomy data is being processed with the latest AI-based models to identify celestial objects and anomalies. Machine learning algorithms like Random Forest (RF), Gradient Boosting Machine (GBM), K-nearest Neighbors (KNN), and Depth-First Search (DFS) are applied for discovery of new planets, quasars, and transient objects [20]. For example, the Large Synoptic Survey Telescope (LSST) works on AI algorithms to analyze the

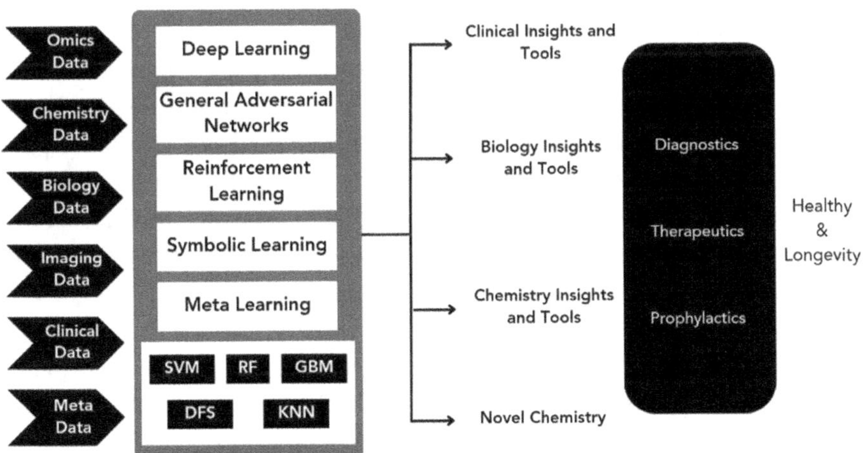

FIGURE 2.2 The process of AI-based data-driven decision making.

data and enable astronomers to make discoveries more effectively. Digital image processing techniques with the integration of machine learning can detect patterns in astronomical image data.

2.5.3 Financial Research

Market trends and investment portfolios can be predicted with the usage of AI-driven techniques. Real-time processing of financial data can help predict future trends in the stock exchange. Financial technology (FinTech) is very famous nowadays as it bridges the gap between traditional funds management systems with advanced intelligent systems. Powerful and intelligent tools are helping humans to find irregularities in financial transactions [21]. It also helps investors to make timely decisions and gain profits in their investments. AI is helping banks to detect fraudulent activities by integrating machine learning methods. Future funds investment can be done based on the results of AI-driven models. Financial challenges are now shared mostly with the latest AI tools to reduce the burden on humans.

2.6 PERSONALIZED RESEARCH AND MEDICINE

With the advancements in AI, the age of personalized medicine and research has arrived. The fields of genomics and precision medicine are evolving, focusing on patients with personalized medicine and diagnostics. Distinctive characteristics of individual patients can be studied using AI tools. This leads to new research possibilities in medicine [22]. AI is being used in the transformation by processing complex datasets such as genomic information, patient records, and clinical data [15]. With this analysis, new research approaches and treatment plans are developed to meet the needs of individual patients. One of the major aspects of AI in research and medicine is the possibility of personalized medicine portfolio. Unlike traditional approaches where one method was applied to all different patients, AI is helping doctors to consider an individual's unique genetic composition. AI tools can also analyze the medical history of the patient to extract important features. This data will help to provide a precise and effective treatment plan that will optimize healthcare results. Based on individual profiles, research can be done using previous data. AI-based tools can suggest new research problems, methodologies, and collaboration possibilities that align with the researcher's expertise. This will help the community to share ideas and diagnose fatal diseases. Wireless health monitoring devices are being used now to collect data on a disease in the patient. This data is very crucial for the timely remedy of the disease. AI is now reshaping the drug discovery process by the integration of intelligent systems to process the data at much faster speed as compared to traditional methods. Literature review of previous techniques is also easy with large language models to extract useful information.

2.6.1 Genomics

AI is revolutionizing the field of genomics by decoding an individual's genetic makeup. This is helping researchers to find predispositions toward diseases, develop therapies, and predict outcomes of treatments. Genes are being studied to find effective treatments for diseases using AI tools. Human capacities are limited to a specific level; therefore, there is a need to develop automated machine learning systems for data analysis. Advanced AI models can be trained on large medical datasets, and automation can be applied for personalized medical research [18].

Genomics and precision medicine are prime examples of AI in the medical field. By processing complex datasets, AI can detect genetic patterns linked to diseases of specific treatments. The information is very critical to design personalized treatment plans for individuals based on their genetic properties. For the treatment of cancer, AI can examine the profile of patients to provide recommendations for therapies that are more likely to produce positive results. Individual health profiles can be built based on genetic and lifestyle factors. These profiles can be useful to study different types of patients at different stages to get more insights about the effectiveness of the medicine.

AI-powered genomics empowers researchers to:

- **Detect Disease Vulnerability:** AI improves the examination of information to identify mutations leading to illnesses. By processing datasets and historical records, AI can find patterns that help in detecting vulnerabilities at an early stage [23]. This early detection allows for interventions and the adoption of measures. In the field of medicine, risk-free healthcare measures for individuals who have a higher chance of developing specific conditions.
- **Uncover New Drugs:** AI has changed the field of drug discovery. It plays a vital role in finding drug targets and predicting how specific medicine reacts with an individual's unique genetic properties [24]. This change speeds up the process of drug-making and makes it

more accurate and efficient. By processing large amounts of data and detecting patterns, AI helps in personalized medications that are not only more effective but also have fewer side effects.

- **Tailored Cancer Treatments**: AI has been vital in customizing cancer treatments by using its capacity to process patient data. By applying AI-based methods of a patient's molecular characteristics, oncologists can make better decisions regarding the most suitable treatment plans. This personalized way of medication improves the effectiveness of therapy while minimizing any risks. It shows a breakthrough in fighting cancer with the potential to enhance results and overall quality of life.
- **Precision Medicine:** In healthcare, precision medicine uses AI to personalize treatment plans. By identifying lifestyle, genetic profile, and medication history AI can help design personalized treatment plans for individual patients [25]. This strategy leads to personalized medicines as per the demand and conditions of the patient. AI improves diagnostic accuracy by considering a patient's genetic characteristics, medical records, and other factors. It also helps to make treatment plans, reducing negative side effects and improving patient recovery. It also helps in predicting disease progression, enabling active intervention and personalized healthcare strategies.

2.7 ETHICAL CONSIDERATIONS OF AI-DRIVEN RESEARCH

While personalized research and medicine is beneficial, it is critical to prioritize its aspects. Key issues are privacy protection, consent, data security, and having access to personalized care. It's very important to build strong ethical frameworks and regulations to protect sensitive data. AI systems should protect the user's data. Patients should be informed about the usage of their shared data. They should also be aware of the possible outcomes of the treatment plans. AI algorithms should be monitored to ensure fairness and usefulness of the systems. This will help to solve any bias issues in the AI models. Trust should be built by discussing the decisions driven by AI. The issue of data ownership is still debatable as research and medicine powered by AI is progressing.

As AI continues its progress in personalized research and healthcare, keeping a balance between the advantages and ethical concerns becomes critical [26]. This will ensure that patients receive care while safeguarding their data and privacy. There is a need to make guidelines about the procedures of AI algorithms used in medicine and health to help users understand the implications of sharing their data.

2.8 COLLABORATION AND KNOWLEDGE SHARING

Collaboration and knowledge sharing are very important aspects of the research community. AI is becoming vital in affecting how researchers collaborate, exchange ideas, and promote research liaisons [16]. AI is completely changing research collaboration and knowledge sharing with the support of citations and real-world case studies. Online platforms powered by AI have become tools for researchers. Platforms like Microsoft Teams and Slack use AI algorithms for tasks such as chatbots, document management, and project tracking. For example, Microsoft Teams uses AI to offer meeting transcriptions, allowing researchers to easily review and search through meeting discussions [27]. These tools efficiently overcome limitations enabling research teams from around the world to share knowledge easily.

The latest recommender systems use AI-based algorithms to find researchers with similar interests and goals. One of the examples of these recommender systems is ResearchGate platform, that is a big online platform for researchers. On this platform, researchers can make teams and groups based on similar research interests. This is helping researchers to share ideas and work with other research community members easily [28]. Most of these recommendations are guided by AI to play a vital role in the exchange of knowledge between researchers. Researchers can now connect with each without any barrier. Researchers can easily work on joint projects for the greater benefit of the community. AI-based algorithms can analyze researcher's publication, expertise, and partnerships to recommend collaborators for a project. Communication and collaboration are greatly improved by the integration of AI among teams conducting research [29].

These online applications seamlessly integrate features such as real-time messaging, shared document editing, and project management, enabling researchers to easily work together regardless of their locations. This online collaboration helps minimize delays and speeds up the progress of projects. AI algorithms can analyze areas of specialization within a field. By analyzing research papers and contributions AI helps in finding experts in domains. This encourages cooperation by guiding researchers towards the experts, for consultations or collaborations. Literature review is very important in research, and it can become exhausting and time-consuming due to the availability of large number of published articles. To help reduce this burden AI can be used to simplify this process by analyzing and summarizing research articles to extract comprehensive understanding of existing knowledge about a specific topic. AI-based algorithms are used to scan the volumes of papers. AI can also analyze citations and connection between research papers to make it easy for the researchers to identify works and their impact on field.

AI can serve as a connection between researchers across fields. It can help in identifying shared interests, approaches, or obstacles thus facilitating introductions and collaborations. This encourages researchers to combine their knowledge and address problems from different angles, which often leads to the development of solutions. For example, AI can bring together a biologist studying gene expression and a computer scientist proficient in ML enabling them to collaborate on analyzing data. Many worldwide issues, like the impact of climate change, public health emergencies, and urban sustainability, necessitate the use of AI-based approaches. AI can play a role in formulating and executing these approaches by incorporating knowledge from various disciplines. For instance, when creating infrastructure, AI can examine the information provided by city planners, environmental experts, economists, and engineers to develop optimal plans that account for social welfare, environmental concerns, and economic viability.

Moreover, AI fosters data sharing and collaboration by offering solutions for managing data. Through AI algorithms data can be labeled, making it easily accessible and searchable for researchers. This promotes the sharing of data, encourages its reuse, and facilitates collaboration on research projects ultimately nurturing an open and collaborative research community.

2.8.1 Overcoming Language Barriers in Real-Time Translation

Distinctions in language might often render collaboration difficult. These AI-powered translation tools, like Google Translate and DeepL, have become the source of progression to provide translations across multiple languages [30]. These resources enable researchers from several fields to collaborate effectively, exchange innovative ideas, and play an active part in global research programs. By removing language barriers and promoting interaction between researchers from different fields, the skills of AI's language translation facilitate a multidisciplinary approach in a productive atmosphere.

2.8.2 AI Tools for Literature Review and Data Sharing

AI certainly influenced the literature review. NLP algorithms are used to evaluate research articles and extract key ideas and relationships by AI tools like Iris.ai and Semantic Scholar, specifically aiding researchers in literature discovery [29]. AI is also affecting data management and sharing. AI algorithms are utilized by tools like Figshare and Zenodo to classify and recognize studies, making it easier for researchers to share, find, and utilize data. Furthermore, AI strengthens data security standards by ensuring research data safety through sharing. For analyzing data from many resources and offering cross-border insights, AI is extremely effective. As an example, in studies, the AI can analyze geographic and ecological data to recognize that ecosystems have been affected by climate change. These inclusive ideas can be extremely useful for solving everyday problems. A technique called data fusion by AI is used to combine information from different sources and fields to give an outlook on a given issue. To generate customized therapies, data fusion in the field of healthcare might include genetic data, medical data, and environmental elements. Innovative research is at the forefront due to these broad strategies. Through automating procedures and producing research content, AI-driven platforms such as OpenAI's GPT3 allow expert knowledge exchange and disciplinary cooperation.

2.9 CHALLENGES AND ETHICAL CONSIDERATIONS

In the research field, AI has future potential. The recognition of the difficulties and ethical concerns associated with its integration is more fundamental. Effective datasets are required for training and analyzing AI algorithms. On the other hand, getting comprehensive data can be quite difficult where data availability seems poor [31]. Bias in training data or algorithm design may result in exclusive as well as inaccurate results. Different patient groups might get the same or different treatment depending on their needs when AI models are used in the medical field. So, there is a barrier to their acceptance in research, just because of the poor accessibility of AI models. It could be challenging to see how AI arrives at outcomes, specifically in fields where interpretability and transparency are extremely important.

When we talk about AI, transparency and explainability are mandatory. Transparency demonstrates that an AI system's workings and decision-making processes are quite vivid and understandable, like having an eye on its operations. It allows us to clarify how specific outcomes are achieved. It's accessible to a broader audience, including those without expertise in the field. It also goes beyond transparency by revealing the decision-making process. Explainable AI systems allow users to understand the outcomes by executing techniques such as decision trees or linear regression and explaining their decisions logically [29]. To fill in the gap between technical AI outputs and human comprehension, AI models can be explained, portrayed using heat maps and translated into human language.

2.9.1 Addressing Biases and Ethical Concerns in AI

To create more techniques to detect and fix biases in AI systems, researchers are working so hard. There are two methods, fairness ML and bias audits, for ensuring that the outcomes generated by AI models are equitable and impartial [32]. The field of academia plays a role in the progress of AI research. Organizations such as the Association for Computing Machinery (ACM) are providing guidelines, rules, and regulations for the implementation of institutions and researchers. For research purposes, concerns regarding ethics have been incorporated into the design, development, and execution of AI systems. Researchers have been becoming more aware of the implications of data collection and use. Informed consent, data anonymization, and privacy protection have recently become vital elements of ensuring that research data is collected and used ethically. AI has become so crucial to maintaining quality and minimizing bias. Models and misleading results can be triggered by concerns such as inaccurate data, imbalances, and bias in sampling. Biases can also be generated by all these data, selection process, algorithmic design, confirmation bias, or implicit biases.

Furthermore, these biases exhibit the tendency to sustain discrimination and uphold societal inequalities. Getting over these obstacles needs a special emphasis on diverse and representative data collection methods, transparency, ongoing evaluation practices in AI systems, and proactive techniques to identify and address biases. We may attain results that are helpful to all members of society by pursuing AI systems that promote fairness and accuracy. A critical concern of ethics is that all of humanity, regardless of economic status, is benefited by AI. Before applying AI technologies, it is crucial to deal with the impact of AI on society and jobs through responsible innovation and thorough testing [33]. With AI, we can collectively tackle challenges like pandemics and climate change by working collaboratively and exchanging information. Ultimately, the long-term effects and future implications of AI are critical considerations. All ethical or environmental concerns that might occur have to be planned and dealt with.

2.10 CONCLUSION

The whole transformation of research in different domains by AI systems has been discussed in this chapter. It starts by highlighting the significance of research and, after that, introduces AI as a tool that creates research environments. This unit also sheds light on the improvement of data analysis via AI and how it handles large amounts of data in detecting patterns and implementing them in real-life scenarios. It also tells us about the progress of research discoveries via AI in different fields, such as drug development, scientific experiments, and predictive modeling. Moreover, this unit emphasized the contribution of AI in decision-making via offering ideas and predictive analytics, as well as examples of its effects on risk assessment, resource allocation, and practical case studies. This unit also discussed research and medicine, focusing on AI's ability to modify approaches based on profiles, particularly in genomics and precision medicine. Ethical concerns regarding AI-driven research are cautiously tackled, with a focus on implementation. Furthermore, this unit focused on how AI creates interaction and exchange of information for researchers, boosting effectiveness and cross-disciplinary interaction within the research community. Ultimately, this unit highlighted challenges such as data precision, prejudices, and the significance of practices in education. As a whole, the chapter provided an illustration of how AI can impact analysis by increasing efficiency, quality, and morality in different fields.

REFERENCES

1. Y. Xu, X. Liu, X. Cao, C. Huang, E. Liu, S. Qian, X. Liu, Y. Wu, F. Dong, and C.-W. Qiu, "Artificial intelligence: A powerful paradigm for scientific research," *The Innovation,* vol. 2, no. 4, p. 100179, 2021.
2. S. Naseem, A. Alhudhaif, M. Anwar, K. N. Qureshi, and G. Jeon, "Artificial general intelligence-based rational behavior detection using cognitive correlates for tracking online harms," *Personal and Ubiquitous Computing,* vol. 27, no. 1, pp. 119–137, 2023.
3. Y. K. Dwivedi, L. Hughes, E. Ismagilova, G. Aarts, C. Coombs, T. Crick, Y. Duan, R. Dwivedi, J. Edwards, and A. Eirug, "Artificial Intelligence (AI): Multidisciplinary perspectives on emerging challenges, opportunities, and agenda for research, practice and policy," *International Journal of Information Management,* vol. 57, p. 101994, 2021.

4. P. R. Daugherty and H. J. Wilson, *Human+ machine: Reimagining work in the age of AI.* Harvard Business Press, 2018.
5. W. Liang, G. A. Tadesse, D. Ho, L. Fei-Fei, M. Zaharia, C. Zhang, and J. Zou, "Advances, challenges and opportunities in creating data for trustworthy AI," *Nature Machine Intelligence,* vol. 4, no. 8, pp. 669–677, 2022.
6. A. Haleem, M. Javaid, M. A. Qadri, R. P. Singh, and R. Suman, "Artificial intelligence (AI) applications for marketing: A literature-based study," *International Journal of Intelligent Networks,* vol. 3, pp. 119–132, 2022.
7. S. Iqbal, H. Maryam, K. N. Qureshi, I. T. Javed, and N. Crespi, "Automised flow rule formation by using machine learning in software defined networks based edge computing," *Egyptian Informatics Journal,* vol. 23, no. 1, pp. 149–157, 2022.
8. M. Rehman, I. T. Javed, K. N. Qureshi, T. Margaria, and G. Jeon, "A cyber secure medical management system by using blockchain," *IEEE Transactions on Computational Social Systems,* vol. 10, pp. 1–14, 2022.
9. L. Cao, "Ai in finance: challenges, techniques, and opportunities," *ACM Computing Surveys (CSUR),* vol. 55, no. 3, pp. 1–38, 2022.
10. A. Kuzior and A. Kwilinski, "Cognitive technologies and artificial intelligence in social perception," *Management Systems in Production Engineering,* vol. 2, pp. 109–115, 2022.
11. E. O. Pyzer-Knapp, J. W. Pitera, P. W. Staar, S. Takeda, T. Laino, D. P. Sanders, J. Sexton, J. R. Smith, and A. Curioni, "Accelerating materials discovery using artificial intelligence, high performance computing and robotics," *npj Computational Materials,* vol. 8, no. 1, p. 84, 2022.
12. D. Q. Wang, L. Y. Feng, J. G. Ye, J. G. Zou, and Y. F. Zheng, "Accelerating the integration of ChatGPT and other large-scale AI models into biomedical research and healthcare," *MedComm–Future Medicine,* vol. 2, no. 2, p. e43, 2023.
13. D. S. Battina, "Application research of artificial intelligence in electrical automation control," *International Journal of Creative Research Thoughts (IJCRT), ISSN,* vol. 3, pp. 2320–2882, 2015.
14. Y. Alufaisan, L. R. Marusich, J. Z. Bakdash, Y. Zhou, and M. Kantarcioglu, "Does explainable artificial intelligence improve human decision-making?," *Proceedings of the AAAI Conference on Artificial Intelligence,* vol. 35, no. 8, pp. 6618–6626, 2021.
15. B. Lin and S. Wu, "Digital transformation in personalized medicine with artificial intelligence and the internet of medical things," *Omics: A Journal of Integrative biology,* vol. 26, no. 2, pp. 77–81, 2022.
16. M. H. Jarrahi, D. Askay, A. Eshraghi, and P. Smith, "Artificial intelligence and knowledge management: A partnership between human and AI," *Business Horizons,* vol. 66, no. 1, pp. 87–99, 2023.
17. M. Shin, J. Kim, B. van Opheusden, and T. L. Griffiths, "Superhuman artificial intelligence can improve human decision-making by increasing novelty," *Proceedings of the National Academy of Sciences,* vol. 120, no. 12, p. e2214840120, 2023.

18. C. Caudai, A. Galizia, F. Geraci, L. Le Pera, V. Morea, E. Salerno, A. Via, and T. Colombo, "AI applications in functional genomics," *Computational and Structural Biotechnology Journal,* vol. 19, pp. 5762–5790, 2021.
19. H. S. Chan, H. Shan, T. Dahoun, H. Vogel, and S. Yuan, "Advancing drug discovery via artificial intelligence," *Trends in Pharmacological Sciences,* vol. 40, no. 8, pp. 592–604, 2019.
20. C. J. Fluke and C. Jacobs, "Surveying the reach and maturity of machine learning and artificial intelligence in astronomy," *Wiley Interdisciplinary Reviews: Data Mining and Knowledge Discovery,* vol. 10, no. 2, p. e1349, 2020.
21. J. W. Goodell, S. Kumar, W. M. Lim, and D. Pattnaik, "Artificial intelligence and machine learning in finance: Identifying foundations, themes, and research clusters from bibliometric analysis," *Journal of Behavioral and Experimental Finance,* vol. 32, p. 100577, 2021.
22. K. N. Qureshi, A. Alhudhaif, M. A. Qureshi, and G. Jeon, "Nature-inspired solution for coronavirus disease detection and its impact on existing healthcare systems," *Computers and Electrical Engineering,* vol. 95, p. 107411, 2021.
23. C. Antoniades, P. Patel, and A. S. Antonopoulos, *Using artificial intelligence to study atherosclerosis, predict risk and guide treatments in clinical practice,* ed: Oxford University Press US, 2023.
24. I. Rani, K. Munjal, R. K. Singla, and R. K. Gautam, "Artificial Intelligence and Machine Learning-Based New Drug Discovery Process with Molecular Modelling," *Bioinformatics Tools for Pharmaceutical Drug Product Development,* pp. 19–35, 2023.
25. M. Subramanian, A. Wojtusciszyn, L. Favre, S. Boughorbel, J. Shan, K. B. Letaief, N. Pitteloud, and L. Chouchane, "Precision medicine in the era of artificial intelligence: implications in chronic disease management," *Journal of translational medicine,* vol. 18, no. 1, pp. 1–12, 2020.
26. F. McKay, B. J. Williams, G. Prestwich, D. Bansal, N. Hallowell, and D. Treanor, "The ethical challenges of artificial intelligence-driven digital pathology," *The Journal of Pathology: Clinical Research,* vol. 8, no. 3, pp. 209–216, 2022.
27. M. Corporation. *Microsoft Teams,* 2017. Available: https://teams.microsoft.com/
28. M. h. Mataoui, F. Sebbak, A. H. Sidhoum, T. E. Harbi, M. R. Senouci, and K. Belmessous, "A hybrid recommendation system for researchgate academic social network," *Social Network Analysis and Mining,* vol. 13, no. 1, p. 53, 2023.
29. K. Haresamudram, S. Larsson, and F. Heintz, "Three levels of AI transparency," *Computer,* vol. 56, no. 2, pp. 93–100, 2023.
30. Y. Moslem, R. Haque, and A. Way, "Adaptive machine translation with large language models," *arXiv preprint arXiv:2301.13294,* 2023.
31. V. Thakur, "Unveiling gender bias in terms of profession across LLMs: Analyzing and addressing sociological implications," *arXiv preprint arXiv:2307.09162,* 2023.

32. T. P. Pagano, R. B. Loureiro, F. V. Lisboa, R. M. Peixoto, G. A. Guimarães, G. O. Cruz, M. M. Araujo, L. L. Santos, M. A. Cruz, and E. L. Oliveira, "Bias and unfairness in machine learning models: a systematic review on datasets, tools, fairness metrics, and identification and mitigation methods," *Big data and cognitive computing,* vol. 7, no. 1, p. 15, 2023.
33. C. Wang, S. Liu, H. Yang, J. Guo, Y. Wu, and J. Liu, "Ethical considerations of using ChatGPT in health care," *Journal of Medical Internet Research,* vol. 25, p. e48009, 2023.

CHAPTER 3

AI Models and Data Analytics

Transforming Research Methods

Amna Khatoon, Asad Ullah, and
Kashif Naseer Qureshi

3.1 INTRODUCTION

A new era with the potential to change research procedures across numerous industries has just been ushered in by integrating Artificial intelligence (AI) models and data analytics [1]. This chapter examines the substantial effects of AI models and data analytics on research methods and how these tools have altered, collected, examined, and interpreted the data. By harnessing the powers of AI and advanced data analytics, researchers can now expose insights previously concealed within vast databases, encouraging innovation and growth in their respective fields. Data collection, analysis, and processing have long been labor-intensive components of research methodology. However, a paradigm shift has been compelled by the rapid expansion of digital data and the development of AI. Because they are based on intricate algorithms and Deep Learning (DL) approaches, AI models can process massive amounts of data with astonishing speed and precision [2]. These models have altered the study environment by identifying intricate patterns, anticipating outcomes,

DOI: 10.1201/9781032667911-3

and providing insights. The potential of AI to automate operating chores has streamlined research processes. For instance, in genomics research, AI-driven sequencing robots efficiently process genetic data, hastening the finding of disease-causing mutations. AI algorithms also help to clean data, minimize errors, and ensure data accuracy. When it comes to pattern recognition, AI is superior to humans. AI algorithms are used in astronomy to identify celestial objects and events in photos captured by telescopes, revealing exoplanets and distant galaxies. Detecting abnormalities like tumors in medical pictures using AI-driven algorithms is another application of the capacity to spot patterns in data [3]. Researchers utilize AI algorithms for predictive analytics, forecasting results based on historical data. AI-enhanced weather models process large meteorological datasets to increase forecast accuracy. Like scientists predict weather patterns, financial gurus utilize AI algorithms to support decision-making.

NLP techniques have revealed insights into textual data. Sentiment analysis is a tool used by social scientists to analyze user sentiment on social media platforms. By digitizing historical writings and identifying linguistic and cultural changes, NLP aids in historians' analysis of such materials [4]. AI-driven computer vision is revolutionizing image and video analysis. Ecologists employ image recognition to track the populations of wildlife using camera trap photographs. In archaeology, AI aids in the reconstruction of ancient artifacts from damaged remains. Despite the tremendous benefits of AI and data analytics, ethical concerns must be carefully explored. Issues like algorithmic bias, data privacy, and openness must be considered [5]. According to academics, the proper application of AI and technology development must coexist peacefully. The synergistic relationship between AI models and data analytics fundamentally alters research methodology. In many disciplines, including astronomy, finance, and linguistics, the incorporation of AI is accelerating research and creating new opportunities [6]. As they attempt to grasp the potential of AI and data analytics fully, researchers must manage the ethical problems to ensure that these paradigm-shifting tools are used responsibly to pursue the Role of AI and Data Analytics in Research. AI models and data analytics have ushered in a new era of transformative possibilities for research approaches across many fields. How researchers gather, handle, analyze, and interpret data has significantly changed due to the integration of these technologies. By leveraging AI and advanced data analytics, researchers may now extract insights from large datasets that were previously unattainable [7]. This transformative potential promotes innovation and progress in many different fields. In the past, research methods relied on lengthy analysis, manual data collection, and limited processing capacity. The exponential growth of digital data and the adoption of AI have sparked a paradigm shift. AI models based on these foundations can analyze vast volumes of data rapidly and accurately thanks to DL and sophisticated algorithms [8]. The research landscape has completely changed due to these models' capacity to recognize complex patterns, project future events, and provide insights.

One of AI's main benefits is its ability to automate data collection and preprocessing tasks, which speeds up research operations. For instance, AI-driven sequencing devices are helpful for genomics research because they efficiently process genetic data and hasten the finding of disease-causing mutations. AI algorithms also aid data sanitization, error reduction, and accuracy assurance. The ability of AI to recognize patterns outperforms human capabilities and has applications across numerous industries [9]. AI systems that identify celestial objects and events from telescope images enable astronomical discoveries such as exoplanets and distant galaxies [10]. Another field in which AI models excel is predictive analytics, which makes predictions based on past data. AI-enhanced models process a lot of meteorological data to improve forecast accuracy. With AI and data analytics, ethical concerns must be addressed. Issues like algorithmic bias, data privacy, and openness must be considered. Researchers must find a balance between legal AI use and technological advancement [11]. Integrating AI models and data analytics alters research approaches. The interaction brings up new possibilities in various disciplines, including astronomy, linguistics, and economics. Researchers must compromise on ethical considerations to fully explore the potential of AI and data analytics [12]. By effectively utilizing these revolutionary technologies, scientists may accelerate the speed of discovery and usher in a new era of knowledge.

3.2 EXISTING METHODS AND TECHNIQUES

As per the findings in [13], there is a connection between the emotion of social media and the state of the financial markets. By assessing the sentiment of tweets and applying that analysis to data collected from , the research aims to anticipate shifts in the level of trading activity on stock markets. It is accomplished by analyzing the data obtained from X. The method entails classifying the tweets of other users as either positive, negative, or neutral to examine the tone of the tweets that people have posted on X. This data source is beneficial for predicting how markets behave since the sentiments posted on social media can impact trading decisions and market patterns. Consequently, this makes the data source particularly important. In addition, market patterns are susceptible to being altered by news occurrences. Authors in [14] demonstrated how AI-driven picture analysis can be utilized to monitor the influence of urban growth on biodiversity. This monitoring process takes place during the conservation of biodiversity. The expansion of urbanization, which impacts ecosystems, may be tracked using AI algorithms that scan satellite pictures to look for changes in land cover. This research contributed to urban planning and conservation efforts by giving a quantitative assessment of the loss of habitat and the consequences of urbanization on ecosystems.

To bring attention [15] to the potential of AI in predicting gene expression based on Deoxyribonucleic Acid (DNA) sequences. This potential can be seen in examining vast datasets by AI algorithms, which provides the path for discovering previously undiscovered astronomical phenomena. This, in turn, expands the understanding of the cosmos. It is feasible to create predictions about the amounts of gene expression by utilizing Machine Learning (ML) models trained on DNA data. This has ramifications for the genomics study and elucidates the functional importance of genetic variation. Authors in [15] highlighted the potential role that AI might play in sentiment analysis to better comprehend public opinion amid political events using information obtained from various social media networks. This helped to understand public opinion in the context of political events. The emotions that are conveyed in social media posts are evaluated with the help of AI algorithms. The algorithms subsequently categorized the communications, providing information regarding how society reacts to political happenings. Authors in [16] illustrated how AI-driven language translation technology may support academics reviewing materials written in various languages to overcome language obstacles.

Authors in [17] discussed that the AI's enormous effect can revolutionize medical diagnosis and therapy by researching genetic data and medical imaging. This research was conducted in the context of the medical field. Authors in [18] investigated AI-powered predictive analytics in financial markets. Authors in [19] investigated the use of AI to analyze satellite pictures to track changes in the rate of deforestation. The literature findings show that AI models and data analytics have a transformative impact across various academic fields and subject areas. The literature also highlighted the importance of ethical considerations in AI-driven research procedures to evaluate data, predict discoveries, and drive innovation. Specifically, they highlighted the necessity of ethical problems in AI-driven research techniques. In particular, they bring attention to the necessity of ethical challenges in research methodologies powered by AI.

3.2.1 Data Automation with AI

The data needs different processes like data collection, preprocessing, and cleaning. AI models and data analytics have revolutionized these processes, allowing for efficient extraction of vast amounts of data from various sources. Traditional data collection methods can be time-consuming and resource-intensive [20]. Still, AI models can automate these processes by locating and substituting missing numbers, identifying and eliminating outliers, and standardizing data formats. It results in fewer errors and saves researcher's time. AI models and data analytics have significantly impacted research techniques in various disciplines. For example, in biomedical research, AI-driven automated data collection consolidates patient information and medical photographs from multiple sources [21], enabling investigation of disease progression and treatment effects. Data cleaning capabilities of AI improve the precision of study findings by identifying errors in clinical datasets and making necessary corrections. Automated data collection through satellite imaging and remote sensors allows researchers to track environmental changes in real time, ensuring data is ready for analysis. In affiliate marketing, AI and Big Data can automate data collection, enabling businesses to track affiliate performance, customer behavior, and market trends more effectively [22]. It allows for data-driven decisions, marketing strategy optimization, and identifying new opportunities in Affiliate Marketing. Figure 3.1 shows the connectivity among data, DSS, NLP, and anomalies for processing data.

3.2.2 Data Collection Processes

Efficient data collection is crucial for research success, and the integration of AI models and data analytics has significantly impacted research methodologies across various disciplines. AI models automate data collection by programmatically gathering information from various sources, accelerating data accumulation, and reducing the risk of human errors during manual data entry. This accuracy is vital in fields like healthcare, where precision in patient data is critical [23]. AI-powered automated data collection from remote sensors and satellite imagery provides information on ecosystems, enabling researchers to track

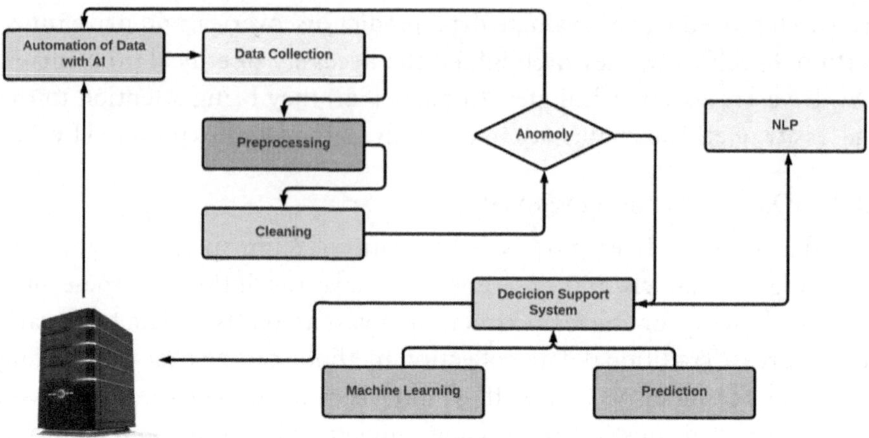

FIGURE 3.1 Connectivity among Data, DSS, NLP, and anomalies for processing.

changes in vegetation, analyze animal movement patterns, and monitor environmental shifts over time. It also analyzes online consumer behavior, providing insights from social media posts, reviews, and forums. AI models automate the collection of sentiment data from social media platforms, allowing real-time analysis of changing sentiments and attitudes [24]. Streamlining data collection processes through automation empowers researchers with automation, scalability, and accuracy, ultimately enhancing the quality and scope of research endeavors. The proliferation of digital data has brought the challenges of managing large and complex datasets, making AI models and data analytics crucial in addressing this challenge. AI models can handle vast amounts of data, from particle collisions to genomics, DNA sequences to finance, and market data to detect patterns and trends. Implementing AI technology in data handling processes can improve efficiency, accuracy, and effectiveness in Affiliate Marketing, enabling businesses to make data-driven decisions and optimize marketing strategies.

3.2.3 Data Cleaning Techniques

Accurate data is crucial for valid research outcomes, and manual data entry can lead to errors. Integrating AI models and data analytics has revolutionized data entry processes, reducing errors and enhancing the quality of research data. Optical Character Recognition (OCR) systems and speech recognition technologies can convert printed or handwritten text into digital data, reducing transcription errors [25]. AI models can validate and correct data during entry, using ML algorithms to predict valid data entries based on existing datasets. NLP techniques help to validate and standardize textual data entries, detecting anomalies, inconsistencies, and outliers. AI models and data analytics have transformed research methods across various fields, including health data, survey results, market data, and financial data. Automated data collection tools can gather data from various sources, reducing human error and enhancing data accuracy. AI algorithms can analyze and process data with high accuracy, minimizing the risk of errors during manual data entry [26]. Data cleaning is a critical step in ensuring the accuracy and reliability of research outcomes. AI models use statistical methods and ML algorithms to identify outliers automatically, impute missing values, and standardize data, simplifying data integration and analysis. This transformation enhances research quality and reliability across diverse fields, allowing researchers to focus on deriving meaningful insights from clean, accurate data.

3.3 STANDARDIZATION AND DATA TRANSFORMATION

Standardization and data transformation are crucial processes in data preparation, enabling consistent and meaningful analysis. AI models and data analytics have revolutionized research methodologies by introducing advanced techniques to standardize variables and transform data, leading to enhanced research outcomes. Standardization involves rescaling variables with a mean of zero and a standard deviation of one, ensuring fair comparisons and accurate analysis. Data transformation is used to achieve normality and stabilize variance in statistical analyses, ensuring reliable and valid results. Feature engineering involves creating or modifying new variables to enhance predictive model performance. These processes have revolutionized research methodologies across various disciplines, enhancing the accuracy and reliability of research outcomes. Data quality is paramount for accurate and reliable research outcomes [27]. AI models and data analytics have introduced advanced techniques to identify and rectify errors, inconsistencies, and anomalies in datasets, ensuring clean and accurate data. Automated data validation and cleaning techniques streamline research workflows, ensuring accurate electronic health records, reliable survey data, and reliable environmental data. The impact of enhanced data quality is evident in the integrity and validity of research findings, allowing researchers to analyze and interpret clean and accurate data confidently.

3.3.1 Pattern Recognition, Analysis, Anomaly Detection, and Outlier Identification

Pattern recognition, analysis, anomaly detection, and outlier identification are crucial to extracting meaningful insights from data. AI models and data analytics have introduced advanced techniques that enable researchers to identify hidden patterns, detect anomalies, and recognize outliers, leading to enhanced research outcomes. This transformation has significantly impacted research methodologies across diverse disciplines. AI models employ ML algorithms to identify patterns within datasets. These algorithms learn from historical data to recognize regularities and relationships, enabling researchers to uncover trends and insights that might be difficult to identify manually [28]. Pattern recognition and analysis are beneficial in fields like genetics, where AI identifies gene expression patterns.

AI-driven anomaly detection techniques automatically identify data points that deviate from the expected patterns. These anomalies might indicate errors, unusual events, or opportunities for further investigation. Anomaly detection is vital in cybersecurity, where AI identifies unusual network behavior that could signify a cyberattack [29]. Outliers are data points that differ significantly from the majority of the dataset. AI models utilize statistical methods and ML algorithms to identify outliers, ensuring researchers know data points that could skew analysis results. Outlier identification enhances the validity and reliability of research conclusions [30]. Through AI models and data analytics, pattern recognition, analysis, anomaly detection, and outlier identification have transformed research methods across various domains. Anomaly detection identifies irregular market behavior that might indicate financial fraud. Anomaly detection techniques identify unusual patient data that might require medical attention. Outlier identification in sensor data helps researchers detect sensor errors or environmental anomalies. These techniques ensure that accurate environmental insights are obtained. Pattern recognition, analysis, anomaly detection, and outlier identification through AI models and data analytics are transformative elements in research methodologies. These processes uncover hidden insights, identify anomalies, and enhance data validity. From economics and finance to medicine and environmental science, the impact of these techniques is evident in the accuracy and reliability of research outcomes. Researchers can now identify trends, anomalies, and outliers that inform their analyses and interpretations.

3.3.2 Uncovering Complex Patterns to Identify Non-Obvious Relationships

Research procedures have been changed by AI models and data analytics, which have made it possible for researchers to find complex connections, correlations, and interactions within datasets. These methods revolutionize research approaches across various fields by automatically identifying non-linear linkages and interactions using ML algorithms. DL and ML excel at identifying multi-level and hierarchical patterns in data. AI models are instrumental in finance and epidemiology because they are excellent at spotting temporal patterns and sequences within time-series data. These models have entirely changed how researchers find links in data, particularly those that might not be obvious at first glance. AI models can find tiny correlations and patterns by examining large datasets [31], assisting academics in determining potential research fields. Researchers can better understand the underlying causes of specific occurrences using AI-driven feature importance methodologies to identify the components that significantly impact an outcome. By helping researchers find hidden correlations and dependencies inside complicated datasets, AI models and data analytics have revolutionized research procedures and improved knowledge of complex phenomena.

3.3.3 Discovering Hidden Trends in Data

By helping academics find hidden trends inside datasets, AI models and data analytics have entirely transformed the study process. With these methods, researchers can spot temporal changes, new trends, and subtle patterns in time-series data. ML methods like recurrent neural networks and Auto-Regressive Integrated Moving Average (ARIMA) models capture these alterations over time. Clustering and association rule mining are advanced analytics approaches that combine comparable data sets to identify relationships that show changing preferences or behaviors. Through methods like heatmaps and trend lines, data visualization, made possible by AI models, aids researchers in uncovering trends that could otherwise go undetected [32]. Public health, market research, and climate science have all been revolutionized by this transformative characteristic of research methodology. AI-driven analysis of vast environmental data uncovers patterns in consumer behavior changes, disease outbreak trends, and climate change trends. These perceptions aid organizations in adjusting to shifting market dynamics and comprehending the effects of human activity on the environment. AI models and data analytics have also transformed multidimensional data analysis [33]. High-dimensional data can be simplified while retaining its fundamental structure using Principal Component Analysis (PCA) and t-distributed Stochastic Neighbor Embedding (t-SNE). AI models use clustering algorithms to group related data points within multidimensional datasets to obtain insight into complicated data structures. Research approaches across disciplines are changing due to the analysis of multidimensional data using AI models and data analytics, which improves researchers' abilities to derive insights from various sources.

3.3.4 Fraud Detection and Enhancing Quality Control

Research has been transformed by the early detection of anomalies in data made possible by AI models and data analytics, which enables scientists to stop, lessen, or react to abnormal events. These methods can be used in various fields, including manufacturing, cybersecurity, and finance. ML algorithms powered by AI may learn from previous data to spot typical trends and variations that might point to abnormalities. Researchers can find anomalies across various data sources using supervised and unsupervised techniques like isolation forests and autoencoders. Real-time data stream monitoring made possible by AI models enables researchers to spot anomalies as they happen. Predictive analytics enabled by AI predicted the upcoming patterns and behaviors, assisting in the early identification of anomalies [34]. Additionally, AI algorithms spot patterns that might indicate anomalies, enabling researchers to take preventative action.

AI models and data analytics have completely transformed fraud detection. They are excellent at finding intricate fraudulent patterns in massive databases, learning from the past, and examining user behavior to find anomalies. Real-time monitoring makes it possible to identify fraudulent acts as they occur, allowing institutions to act quickly and limit future harm. Accurate fraud detection lowers medical expenses and preserves the reliability of insurance programs. Quality control processes have also changed due to AI models and data analytics. These methods improve the ability to keep data consistent, accurate, and reliable. Datasets containing mistakes and abnormalities are automatically identified by AI-driven algorithms, increasing the accuracy and dependability of the data. Researchers can identify and correct quality problems in real-time monitoring of data streams. AI-powered prediction models minimize the possibility of data errors and guarantee the integrity of the research by foreseeing any quality concerns before they arise. An innovative feature of research methodology that improves data accuracy and dependability is improving quality control using AI models and data analytics. Now that they can access clean, precise data for their analysis, researchers may produce more reliable and legitimate findings.

3.3.5 Predictive Analytics and Decision Support Systems

Companies can accomplish their goals by making educated decisions, streamlining processes, and achieving targets with AI models and data analytics. A branch of advanced analytics called predictive analytics uses data from current and historical sources to predict future occurrences or patterns. Data from several sources, including databases, sensors, and social media, must be gathered and preprocessed in this procedure. Regression, decision trees, and neural networks are examples of statistical modeling techniques that can be used to build models [35]. After training on historical data, models are validated by using techniques like cross-validation. Assure the generalizability of the model. Once trained and validated, predictive models may forecast customer attrition, stock prices, equipment breakdowns, and disease outbreaks. Accuracy, precision, recall, and F1-score are used to assess performance. Decision Support Systems (DSS) are software, hardware, or mobile applications that help decision-makers by weighing their options and supplying pertinent data, as shown in Figure 3.2. DSS systems incorporate data from many sources, including databases, spreadsheets, and external data streams, to provide a comprehensive viewpoint.

Using various analytics and reporting platforms, they process and meaningfully present data through visualizations, dashboards, and reports. AI models like ML and DL algorithms can enhance predictive analytics and DSS by offering more precise forecasts and intelligent automation. Automating monotonous tasks, identifying intricate data patterns, and enabling prediction models to change in response to new circumstances are all possible with AI. Businesses can use data-driven insights to make wise decisions, optimize operations, and gain a competitive edge in various industries by deploying predictive analytics and decision support systems. By incorporating AI, these systems' capabilities are further improved, making them more potent and flexible in the always-changing data world.

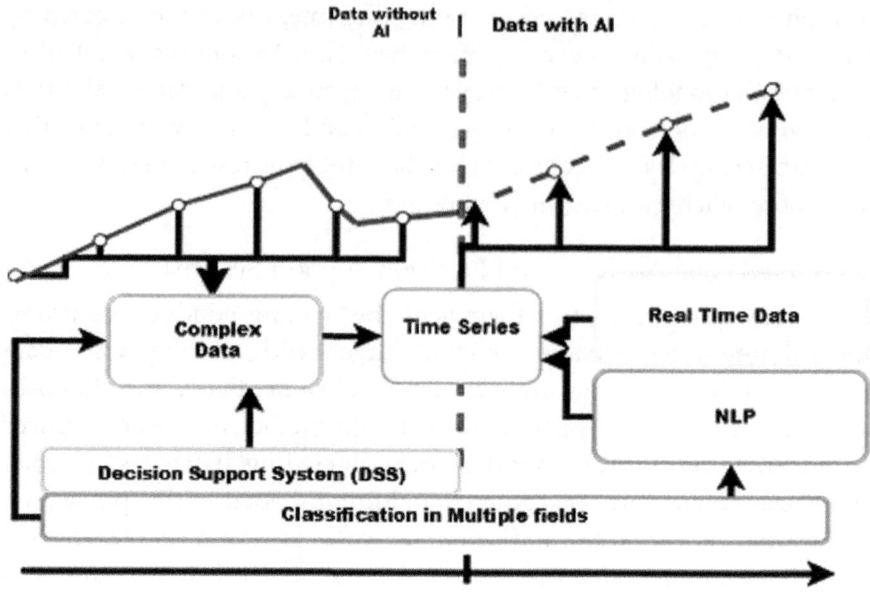

FIGURE 3.2 Data with and without AI in forecasting.

3.3.6 Predictive Models

Predictive model construction is a complex and time-consuming process. Data collection is the first step, which entails obtaining and checking all relevant information for errors or discrepancies. The following stage, data preparation, includes handling missing values, categorical data encoding, and feature scaling. The right model choice is crucial and is determined by the problem. One option is utilizing DL models, such as Convolutional Neural Networks (CNN) for picture recognition. A subset of the data is used to train the model after it has been selected, and the hyperparameters are changed to optimize performance. The model's effectiveness is assessed during the evaluation process using metrics including accuracy, precision, recall, and F1-score. With the aid of a distinct validation dataset, the evaluation is conducted. The model forecasts previously unseen data as soon as it meets the criteria. The algorithms used in ML cover a broad spectrum of unique techniques. In supervised learning, a model is developed using data labeled with a goal variable or related result for each sample. Data already gathered and categorized must be used for this kind of learning. Through unsupervised learning, models can detect patterns or clusters in data without explicit labeling. Reinforcement learning aims to teach agents to make sequential decisions that maximize cumulative rewards, typically in dynamic environments [36]. By instructing agents on how to maximize cumulative rewards, it is accomplished. The ML pipeline includes cross-validation, model evaluation, model selection, feature engineering, and hyperparameter tuning. One of these steps is method selection for example, NLP tasks requiring sentiment analysis may be carried out by using transformer-based models, such as BERT or Recurrent Neural Networks (RNNs).

3.3.7 Forecasting and Regression Models

A branch of statistics known as time series analysis is used to study data accumulated over a certain period. It comprises many elements: noise, seasonality, cyclical patterns, and trends. Untangling these elements through deconstruction allows for a deeper understanding of the fundamental patterns at work. Autocorrelation functions are responsible for demonstrating how data points are related to the values they previously possessed to make it simpler to spot patterns. Time series models are applied to the currently available data to provide predictions of the values that will occur in the future [37]. Models that fit into this category include state-space models and the ARIMA. By using "continuous" versions of both sets of data, regression analysis aims to determine the degree of connection between continuous independent variables (predictors) and continuous dependent variables (outcomes). According to the presumption that there is a linear relationship between the predictors and the outcome in linear regression, the coefficients representing each predictor's influence on the outcome are determined. Since linear regression requires a direct relationship between the predictors and the outcome, it is assumed that this linear relationship exists. It is expanded to accommodate many parameters using a method called multiple regression. When a result can only be expressed as true or false, yes or no, logical regression must be employed to evaluate the data. Regression analysis can be helpful in several ways, including producing predictions and providing in-depth information.

3.3.8 Decision-Making Processes

AI and data analytics use in decision-making has many advantages, including providing insights, recommendations, and automated decision support. The process starts with data collection and ends with data preparation for better quality. Following this, various data analysis techniques, including statistical analysis, ML, and data visualization, extract valuable insights from the data. Decision-makers can benefit from creating decision-support systems that can automate routine judgments or provide recommendations based on data-driven insights [38]. Data analysis from the past and the present is used to offer direction for decisions and plans that apply to the future. The strategies used by corporations today include trend analysis, risk assessment, scenario planning, and predictive modeling. Through analyzing trends, recurrent patterns from the past can be found and projected into the near future. The predictive modeling technique needs access to pertinent data to provide accurate predictions. Researching various outcomes and the ramifications of those outcomes is crucial in successful preparation for several circumstances. Analyzing potential risks that might have an impact on plans is part of the process of risk assessment. Either a positive or negative assessment may be made. Companies can develop a comprehensive strategy for ongoing success by incorporating the prior approaches. Real-time predictive analytics aims to process and analyze data as it comes in, frequently in near real-time or with little to no latency. Technologies that can handle data streams and highly efficient algorithms are needed. Real-time predictive models constantly revise their forecasts using the most recent information. Applications for this type of technology include fraud detection (spotting suspect transactions as they happen), recommendation systems (updating recommendations as users engage with a platform), and IoT devices (making immediate decisions based on sensor data). Organizations can respond quickly to changing circumstances thanks to real-time data.

3.3.9 Text Mining for Insights and Language Analysis

Text mining is extracting useful information and patterns from unstructured text using computer technologies to analyze. This process is known as data mining. Mining is the term used to describe this kind of activity. One may also hear the procedure of "mining" the text in specific contexts. The practice is frequently called "data mining" in common parlance. Data mining is one of the alternatives from which one might select one. One of the things that needs to be done to gain knowledge and insights from a textual source is called tokenization, and it's one of the activities that need to be performed. The task at hand is to analyze the piece of writing to determine its constituent pieces. The only text that needs to be processed to achieve this objective is the principal text. Tokenization refers to disassembling words into their parts one by one. This is responsible for performing analysis of the data and preprocessing the text. Finding recurring themes and terms in customer reviews can be helpful to businesses in determining which aspects of their services and amenities need to be improved. This is especially true in the case of hotel chains, which need to determine which aspects of their services and amenities need to be improved.

3.3.10 Summarization and Abstraction Techniques

Extracting knowledge from unstructured text is the process of turning unstructured, unorganized textual data into structured information that can be analyzed. It can be done using various techniques, such as text parsing, named entity recognition, and part-of-speech tagging. Executing legal document analysis and information extraction in the capacity of a legal services provider to identify key phrases, dates, and parties involved in contract negotiations. Condensed summaries of lengthy text can be created using various summarizing techniques. These summaries must be logical and uncomplicated and include crucial information. Additionally, the application of abstraction processes yields the creation of abstract representations, which are a byproduct of said application and maintain the significance of the concepts and relationships involved. These representations are still created even while the ideas and relationships involved remain relevant. Reducing a lengthy report of an investigation into a few lines that summarize the study's methodology, findings, and main hypotheses rather than leaving the report exactly as it was written.

3.3.11 Modeling and Trend Analysis for Text Data

Topic modeling is used to identify recurring topics or subjects present within a collection of documents. Performing a trend analysis entails keeping track of how certain subjects' popularity shifts over time to obtain new perspectives on developing arguments and interests. Finding topics in social media discussions pertinent to launching a new product and keeping track of those topics so that you may understand how customer sentiments and shifting trends affect your business. Sentiment analysis, or opinion mining, is used to evaluate the emotional tone communicated in text. The evaluation's results are categorized as positive, negative, or neutral. It is utilized as a tool for measuring the impression made by a brand on consumers as well as the general public. Analyzing people's remarks on social media about a politician to get a sense of how the general public thinks about their campaign.

3.3.12 NLP in Named Entity Recognition

Natural language processing, also known as NLP, is a subfield of AI that allows computers to comprehend, interpret, and produce human language. It encompasses a wide variety of tasks, some of which are language modeling, the production of text, and machine translation. This project aims to develop a Chabot that can hold conversations with customers in a natural language to respond to their questions, provide advice, or assist them in carrying out tasks [39]. The act of recognizing and categorizing named entities in text, such as names of people, places, organizations, dates, and other information is referred to as Named Entity Recognition (NER) and it falls under the purview of NLP. It is vital to have the ability to obtain structured information from unstructured text. The process of extracting information from news stories to identify specific individuals or organizations, such as firms and stock values in financial reports.

3.3.13 Text Classification and Categorization

Classifying texts requires arranging textual records into one of several separate categories or labels according to the information included within them. It is done as part of the process of categorizing texts. It can be accomplished in various ways depending on the situation. It is done following the information that can be found in the necessary texts which is used in this process. The great majority of the time, several different approaches from the field of ML are applied to complete this work correctly. These algorithms discover new information by studying training material that has been annotated. Emails received from customers looking for assistance are automatically categorized into various service request types, including requests for refunds and technical difficulties, to ensure they are distributed to the appropriate teams as quickly as possible.

3.3.14 Simulation and Modeling

Image and video analysis for research with simulation and modeling is a comprehensive technique that combines the skills of computer vision and ML skills with the insights generated from computational modeling. This method is established so that simulation and modeling may be used to research image and video analysis. Acquiring, processing, and analyzing data from pictures and videos to derive meaningful information from them are all included in this process. The goal of this process is to extract information from pictures and videos. In medical research, for example, this strategy may involve examining medical imaging to search for abnormalities or monitor disease progression, followed by applying computational models to simulate the effect of proposed treatments. In addition, this method may also involve monitoring the progression of diseases. In the field of robotics, conducting an analysis of video in real time may be necessary to guide a robot's actions while performing tasks in a simulated environment. Studying visually displayed data and using computational models together creates a synergy that helps us better comprehend complex processes and occurrences and improves our ability to foresee and maximize their effectiveness. This approach enables researchers to arrive at informed conclusions and propels innovation for the betterment of society. It holds tremendous promise in various industries, such as healthcare, environmental science, manufacturing, and autonomous systems, among others, and has the potential to benefit society immensely.

3.3.15 Computer Vision Applications for Object Detection and Recognition

Computer vision applications for object detection and recognition are a subset of AI and image processing that focuses on enabling machines to perceive and understand visual data such as photographs and videos. These applications include computer vision applications for object detection and recognition. Object detection recognizes and localizes specific items within an image or video frame. Object recognition, on the other hand, is the process of giving labels or categories to those specific objects that have been detected. These applications gather features from visual input using advanced algorithms and DL approaches, such as CNNs, to make informed choices. For instance, computer vision allows autonomous vehicles to see pedestrians, other vehicles, and traffic signs, contributing to their ability to navigate safely. In the retail industry, it makes cashierless checkout systems possible by identifying objects as they are being placed in customers' shopping baskets [36]. The ability to accurately detect and recognize things has transformative consequences in various sectors, including healthcare (identifying tumors in medical images), surveillance (tracking suspicious behaviors), and other fields. These applications allow machines to interpret their surroundings and engage with the outside world. They promise increased productivity, safety, and innovation across various industries, and they will ultimately shape the future of how humans and machines interact with one another.

3.3.16 Image Classification

Computer vision applications for object detection and recognition integrate AI and image processing to help machines interpret images like photos and movies. Computer vision applications are used for object identification and recognition. Computer vision applications like cameras and barcode scanners detect and recognize items. Object detection is detecting and localizing objects in an image or video frame. After spotting an item, object recognition identifies or categorizes it. These apps use cutting-edge algorithms and DL techniques like CNNs to extract data from visual input and make intelligent decisions. Computer vision lets autonomous vehicles see pedestrians, other vehicles, and traffic signals [38]. It enhances vehicle maneuverability and safety. This technology detects objects in clients' shopping baskets, enabling cashierless checkout systems in retail. The ability to effectively detect and recognize things transforms healthcare (identifying cancers in medical imaging), surveillance (monitoring suspicious behavior), and other fields. These ramifications go beyond healthcare and monitoring. These applications enable computing systems to understand and communicate with their surroundings. They promise to boost productivity, safety, and creativity across industries and change how humans and machines interact. They will shape human-machine interaction in the future.

3.3.17 Recognition and Monitoring for Visualizing Dynamic Processes

Identifying a person or object's movements and actions within visual data, such as films or image sequences, is the goal of computer vision and data analytics. The technology automatically recognizes and categorizes events or behaviors to offer insights into challenging real-world situations. For instance, by examining the motion patterns of people or objects in a security camera feed, surveillance can spot suspect behavior, such as trespassing or theft. By observing player placements and activities during a game, sports analysis can assist coaches and analysts in understanding player movements and strategy. Activity recognition can also enhance quality control in manufacturing by keeping an eye out for deviations from established protocols on the assembly line. The technology improves safety, efficiency, and decision-making across various industries by visualizing dynamic processes and providing real-time insights. Finally, activity recognition and monitoring for visualizing dynamic processes uses computer vision to automatically recognize, monitor, and categorize activities or actions in visual data. From security and sports analysis to manufacturing and beyond, this skill offers various uses that provide insightful information and aid in better decision-making. The potential for this subject to spur innovation and advancements in various fields remains significant as technology develops.

3.3.18 AI-Enhanced Simulation Techniques

Enhancing behavioral studies in AI-enhanced simulation techniques includes incorporating AI into simulation environments to make conducting in-depth analyses of human or agent behavior in various scenarios easier. The method combines the benefits of simulation, which offers controlled and repeatable experimental settings, with AI's strengths, enabling realistic and adaptable behavior modeling. For instance, AI-enhanced simulations can simulate various traffic circumstances in the development of autonomous vehicles, enabling researchers to examine how people respond to various driving situations. AI-driven simulations can be used in healthcare to simulate patient behavior, assisting in medical education and decision-making. This symbiotic relationship between AI and simulation enables researchers to understand human responses better, enhance training, and improve system designs through enhancing behavioral studies. As AI develops, its use in behavioral research in virtual settings has great promise to improve AI systems, enhance human-AI interactions, and advance various industries.

Applying cutting-edge computational techniques and technologies to improve the precision and speed of complicated simulation processes across various domains is called improving accuracy and efficiency for complex simulations in various fields. Computable models of complex real-world systems or events, such as weather patterns, nuclear processes, or financial markets, are known as complex simulations. These simulations frequently demand a lot of time and computing power. Researchers and practitioners can significantly decrease the time needed for these simulations by using high-performance computer clusters, parallel processing, and efficient methods. For instance, enhanced simulation techniques in climate research can result in weather forecasts that are more precise and timelier, assisting in disaster preparedness and agricultural planning. Engineering simulations can aid in designing and testing intricate systems, such as aircraft, to ensure their efficiency and safety. By tackling urgent problems across many fields, the quest for accuracy and efficiency through sophisticated simulation techniques stimulates creativity and speeds scientific discovery.

Integrating AI methods are used in product design and testing to increase effectiveness and creativity in real-time simulation. Engineers and designers may quickly iterate and improve their products by using this method, which uses AI algorithms to build dynamic and flexible simulations that imitate real-world scenarios. For instance, AI-driven simulations can simulate different driving scenarios in the automotive industry, enabling manufacturers to evaluate vehicle performance, safety, and fuel efficiency before producing prototypes. Similarly, AI-driven simulations in aerospace help with designing and fine-tuning aircraft components, improving performance, and speeding up development. Product development teams can discover and address problems early in the design process by utilizing AI and real-time simulation, cutting costs and time-to-market while promoting innovation. A revolutionary shift in product creation is being brought about by AI and simulation, boosting competition and expediting growth in various industries.

The use of computer-based simulations to evaluate and validate safety and risk management-related hypotheses is covered in the safety testing and risk assessment. Researchers and analysts can investigate fictitious scenarios using this method without putting real people or assets in danger since it uses mathematical models and algorithms to simulate real-world circumstances. For instance, the behavior of a reactor in the event of a failure can be predicted by simulations in nuclear power plant safety, allowing engineers to test safety procedures and emergency response plans. Simulations in the pharmaceutical sector help identify potential medication interactions and adverse effects before clinical trials. Organizations that rely on simulations can lower actual risk, cut down on trial-and-error costs, and make better decisions to improve safety and effectively mitigate risk. This strategy is critical for businesses and organizations seeking to put safety first and reduce unexpected difficulties in their operations.

3.4 AI IN HEALTHCARE

Drug discovery, healthcare medical imaging, and diagnostics form a vital intersection of modern medical technologies and academic fields. Researchers use computer modeling, virtual screening, and AI-driven techniques to speed up the discovery and development of new medicinal compounds in drug discovery. The time and expense involved in developing new drugs can be significantly reduced, for example, by using

AI algorithms to forecast possible therapeutic candidates from massive chemical datasets. In the meantime, medical imaging is crucial for identifying illnesses and tracking how treatments work. Medical diagnosis has been transformed by advanced imaging methods like MRI, CT scans, and ultrasound, which allow clinicians to visualize inside structures precisely [40]. AI-powered algorithms in diagnostics can examine medical images, spot anomalies, and help diagnose disorders like cancer or neurological diseases early. These technologies foster medical innovation by providing better patient outcomes, early disease identification, and quicker drug development procedures. The intersection of drug research, medical imaging, and diagnostics highlights how technology can revolutionize healthcare, offering improved patient care and therapies.

An important area of pharmaceutical research is accelerating drug discovery and predicting drug-target interactions which uses sophisticated algorithms and computational techniques to speed up the discovery and development of novel medications. Researchers employed computational models to forecast how a medication molecule interacts with specific biological targets like proteins or enzymes. Scientists can evaluate a drug candidate's potential efficacy and safety in silico by modeling these interactions, obviating the need for time-consuming and expensive laboratory tests. For example, ML algorithms may examine enormous biological and chemical datasets to spot possible drug-target interactions, which helps researchers find brand-new treatments for various ailments. Structural biology methods and molecular dynamics simulations also help researchers understand the complex atomic-level mechanisms behind these interactions. Ultimately, more prosperous and cost-effective drug development procedures and improved patient healthcare outcomes are anticipated due to the novel approach, which aims to accelerate drug discovery and reduce the likelihood of clinical trial failures. Integrating computational tools for predicting drug-target interactions in the pharmaceutical business is a game-changer, significantly improving drug development efforts' efficacy and success rate.

3.4.1 Compound Screening and Optimization for Personalized Medicine Approaches

Compound screening and optimization for personalized medicine approaches is a crucial component of contemporary healthcare that systematically assesses possible therapeutic compounds to customize medical treatments for specific individuals. High-throughput screening methods, computer modeling, and genomic analysis are used in this procedure to find medicines most likely suitable for a particular patient's genetic makeup or disease profile. Personalized medicine, for instance, examines a patient's tumor genetics during cancer treatment to identify tailored medicines that enhance effectiveness while minimizing adverse effects [35]. The strategy also includes customizing medicine dosages and combinations for specific patients to make sure that therapies are effective and safe. Compound screening and optimization represent a paradigm shift in medicine, moving away from a generalized approach to treatments highly customized to meet each patient's unique needs, thereby enhancing treatment outcomes and overall health.

3.4.2 AI in Medical Imaging Analysis and Automated Disease Detection

An innovative use of AI in healthcare is in medical imaging analysis and automated disease detection. The method analyzes medical pictures such as X-rays, MRIs, CT scans, and histopathology slides using ML algorithms, deep neural networks, and computer vision techniques to identify and diagnose diseases and anomalies effectively. These photos can teach AI systems to spot patterns, anomalies, and specific biomarkers, allowing for quick and precise disease identification. For instance, radiologists can use AI-powered image analysis to help find early-stage tumors in mammograms or diabetic retinopathy in eye scans. This technology accelerates and streamlines diagnosis while lowering the risk of human error. It is priceless in locations with limited resources and access to highly qualified medical personnel. AI in medical imaging is a game-changing development in healthcare, with the potential to transform patient care by enhancing patient outcomes, lowering healthcare costs, and delivering prompt and accurate diagnoses.

The extraction of numerous quantitative information from medical pictures, including CT, MRI, and PET scans, is a critical component of the developing area of radiology known as radionics. These characteristics contain complex information on the traits of tumors, the makeup of tissues, and disease patterns that might not be visible to the human eye [32]. Radiomics' primary objective is to make it possible for early and accurate diagnosis, treatment planning, and prognosis prediction for various illnesses, including cancer. For instance, radionics can identify minute alterations in tissue composition or tumor growth in cancer, offering crucial information for prompt intervention and individualized treatment regimens. Radionics helps understand the concealed information in medical images by utilizing cutting-edge ML and AI algorithms, improving patient outcomes through early diagnosis and customized therapy approaches. It also highlights the revolutionary potential of data-driven techniques in contemporary medicine, ultimately enhancing patient care and quality of life. This strategy not only improves the skills of healthcare personnel.

3.5 SOCIAL SCIENCES, HUMAN BEHAVIOR, AND CROSS-DISCIPLINARY COLLABORATION

Understanding complex societal issues and coming up with workable solutions to these issues requires a multifaceted approach that draws on the social sciences, studies of human behavior, and collaboration across academic domains. The investigation of human behavior and the relationships between various civilizations is the primary focus of the social sciences, including fields such as psychology, sociology, anthropology, economics, and political science. The social sciences also include fields such as political science. In this context, "interdisciplinary collaboration" refers to bridging the gap between several fields to obtain a deeper grasp of a topic and produce comprehensive answers to pressing problems. For instance, collaboration between epidemiologists, sociologists, and economists in public health can result in a complete understanding of health disparities and effective policy responses. In a manner comparable to this, combining the expertise of urban planners, psychologists, and economists can produce designs that are both more centered on humans and more environmentally friendly. This collaborative method acknowledges that several issues, ranging from protecting the natural environment to providing adequate medical care, call for various solutions considering the situation's scientific and human dimensions. These issues range from the preservation of the natural environment to the provision of adequate medical care. It recognizes that fresh information gleaned from various fields broadens our perspective and boosts the efficacy of treatments, and it recognizes the benefits these benefits provide. Cross-disciplinary collaboration, the study of human behavior, and the synergy that results from the three of these together are potent tools that may be utilized to address society's complex problems, encourage innovation, and bring about positive change in various fields as shown in Figure 3.3.

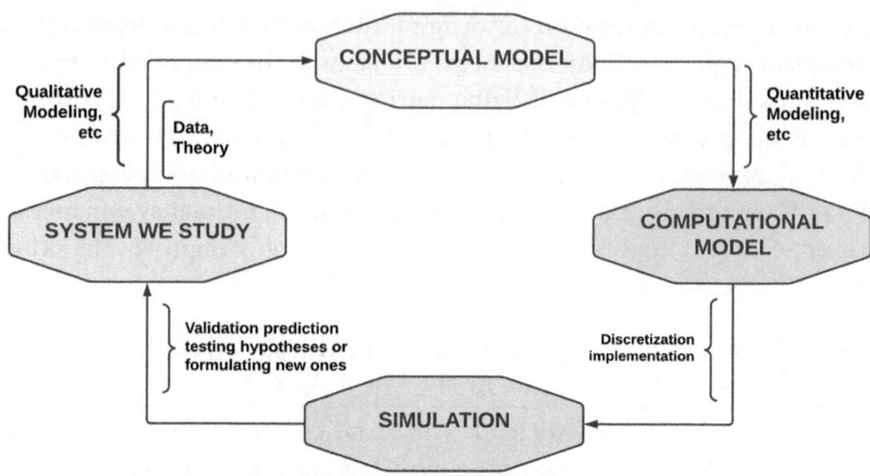

FIGURE 3.3 Multi-disciplinary Implementation of AI and research methods.

3.5.1 Analyzing Human Behavior in Social Data Analysis for Insights

In social data analysis for insights, human behavior analysis entails the methodical examination of enormous datasets collected from various social platforms and interactions to gain valuable insights into human behavior and the workings of social systems. During this step, advanced data analytics, ML, and natural language processing techniques are utilized to discover hidden patterns, trends, and correlations within the data. For instance, social media platforms produce user-generated content in large quantities. This content may be studied to understand public opinion better, track emerging trends, or identify future problems. When it comes to marketing, analyzing customers' behavior on e-commerce websites can reveal helpful information about their purchasing patterns and preferences. The insights gained from these analyses can inform decision-making in various contexts, including healthcare, public policy, and commercial strategy. This demonstrates the importance of using methods that are driven by data to comprehend and forecast human behavior. It opens the door for interventions that are more precisely targeted, enhanced products and services, and methods that are more efficient for a variety of industries.

3.5.2 Understanding Opinion Dynamics for Studying Online Communities

Understanding the dynamics of opinions for the study of online communities is an area of research investigating the formation, diffusion, and development of opinions inside digital platforms and social networks. This field focuses on how individuals express, share, and influence one another's thoughts in virtual environments, which incorporates approaches from network science, computational social science, and data analysis. For instance, academics in social media investigate how thoughts and feelings spread through retweets, likes, and comments, ultimately resulting in online trends and echo chambers [32]. It is essential to have a solid understanding of these dynamics for various applications, including but not limited to marketing and political campaigning, spotting false information, and improving participation in online communities. It sheds light on the intricate dynamic in the digital age between human psychology, technological advancements, and social institutions. Researchers can get significant insights into the behavior of online communities by delving into opinion dynamics. This can help facilitate better-informed decisions, create healthier online dialogue, and eventually contribute to our understanding of our digital society.

3.5.3 Bridging Gaps Between Fields, Data Scientists and Domain Experts

Bridge Building an initiative called Between Fields, Data Scientists, and Domain Experts seeks to bring together the knowledge of data scientists and experts from other fields or sectors to utilize data-driven insights successfully. This method combines many skill sets, with data scientists using their data analytics, ML, and computational abilities to tackle particular problems in industries like healthcare, finance, or environmental research. For instance, data scientists collaborate with medical professionals in the healthcare industry to evaluate patient data and create prediction models for early disease identification. They work with economists to develop financial models for calculating risk and formulating investment plans [41]. Domain experts contribute important context and practical knowledge, making these interdisciplinary collaborations crucial for solving complicated challenges. Data scientists also bring ahead-of-the-curve analytical methods and tools. By bridging disciplines, decision-making and problem-solving are improved, innovation is stimulated, and valuable solutions are developed for various industries.

3.5.4 Interdisciplinary Breakthroughs

When specialists from several professions work together to address complex problems, interdisciplinary breakthroughs reflect significant improvements in knowledge and problem-solving. To tackle complex issues more successfully than any one profession could alone, this collaborative approach uses several disciplines' distinctive views, approaches, and insights. In healthcare, for instance, the integration of biology, engineering, and computer science has resulted in advancements in the creation of medical equipment and therapies. Climate research also benefits from cooperation with meteorologists, ecologists, and data scientists to model and comprehend the complex dynamics of our planet's climate system. Innovation is sparked by interdisciplinary discoveries, advancing science and frequently leading to societally significant solutions. Interdisciplinary collaboration is a cornerstone of contemporary problem-solving and advancement because it is advantageous and frequently required to address the complex issues facing humanity, from climate change and healthcare to technology and sustainability, in a world increasingly shaped by interconnected challenges [39]. When specialists from several professions work together to address complex problems, interdisciplinary breakthroughs reflect significant improvements in knowledge and problem-solving. To tackle complex issues more successfully than any one profession could alone, this collaborative approach uses the several disciplines' distinctive views, approaches, and insights. In healthcare, for instance, the integration of biology, engineering, and computer science has resulted in advancements in the creation of medical equipment and therapies. Climate research also benefits from cooperation with meteorologists, ecologists, and data scientists to model and comprehend the complex dynamics of our planet's climate system. Innovation is sparked by interdisciplinary discoveries, advancing science, and frequently leading to societally significant solutions. Interdisciplinary collaboration is a cornerstone of contemporary problem-solving and advancement because it is advantageous and frequently required to address the complex issues facing humanity, from climate change and healthcare to technology and sustainability, in a world increasingly shaped by interconnected challenges.

3.6 COMPLEXITIES AND CHALLENGES

AI model complexity and data analytics challenges present substantial barriers and potential for changing research methodologies in various fields. The processing and analysis of enormous datasets made possible by AI models and data analytics have transformed research, but these tools also introduce challenges that must be overcome such as data bias and quality [30]. The accuracy and representativeness of the training data are crucial for AI models. Skewed AI models may reinforce injustices or produce incorrect conclusions if datasets are biased or inadequate. These biases must be addressed by researchers, who must also make sure their data sources are inclusive and varied. Model interpretability is another area of complexity. It can be challenging to comprehend why a model predicts anything specific because many AI algorithms, such as profound learning models, are frequently viewed as "black boxes." This lack of openness might be problematic regarding essential applications like healthcare or banking. To solve this problem, researchers are creating more interpretable AI models and methodologies [42]. The ethical aspects are also essential. The use of AI in research raises concerns concerning privacy, permission, and the possibility of unforeseen repercussions. Researchers must overcome these ethical difficulties to create rules for ethical AI use.

Additionally, interdisciplinary cooperation is required. Both technical and domain-specific expertise are needed for AI and data analytics. Researchers from many fields must work together to ensure that AI models and data analytics are implemented efficiently and appropriately. In conclusion, even if AI models and data analytics have the potential to alter research, they also bring with them new complexities and difficulties that researchers must work to overcome. Researchers can use the potential of these technologies to improve knowledge and solve challenging issues in a variety of sectors by giving priority to data quality, interpretability, ethics, and multidisciplinary collaboration.

3.7 ETHICAL CONSIDERATIONS, PRIVACY AND SECURITY

The transformation of research methodologies across several areas depends critically on ethical considerations and challenges for data privacy and security in AI models and data analytics. Using AI and data analytics in research raises several moral problems and security issues that need careful consideration. Data privacy is of utmost importance. Large amounts of sensitive and personal data are frequently needed for AI and data analytics, which raises questions about how this data is gathered, maintained, and used. Researchers must abide by strict data protection laws and respect people's privacy rights. Data encryption and anonymization methods are essential for protecting sensitive information. Fairness and bias are severe ethical problems. Discrimination and inequality can be sustained through biased information or biased algorithms. Researchers must actively uncover and reduce biases in their data sources and algorithmic decision-making processes to maintain fairness and justice in study outputs. Building trust in data analytics and AI models requires transparency and accountability [43]. To help stakeholders understand how decisions are made, researchers should work to make their processes and algorithms accessible and understandable. Setting up accountability systems is also crucial for dealing with problems as they arise. Security issues must not be disregarded. In light of the growing sophistication of cyber threats, researchers must implement robust security measures to safeguard sensitive research data and AI models from unauthorized access or cyberattacks [44]. Access controls, encryption, and recurring security audits are all part of it. When using human subjects in research, informed consent is crucial. When consent is required, researchers should ensure people are informed about how their data will be used. The rights and autonomy of study participants must be respected at all costs.

Effectively resolving these ethical and security challenges requires interdisciplinary collaboration. Develop best practices and recommendations for ethical AI and data analytics research by collaborating with researchers and authorities in data ethics, cybersecurity, and legal compliance. While AI models and data analytics are promising for revolutionizing research methodologies, they also present ethical dilemmas and security risks that scientists must carefully negotiate. Researchers can ensure that these technologies are used responsibly and ethically in their work, advancing knowledge and assisting society while protecting individual rights and data by prioritizing data privacy, fairness, openness, security, and informed consent.

3.8 PROTECTING SENSITIVE RESEARCH DATA AND ENSURING DATA ANONYMIZATION

Transforming research methodologies across multiple areas requires safeguarding sensitive research data and ensuring data anonymization in AI models and analytics. Data security and privacy are subject to new issues and obligations due to using AI and data analytics in research. Researchers must put the security of private research data first and foremost. It comprises information gathered from human subjects, confidential information, and any data that, if revealed, can have adverse effects. Strong security measures should be implemented to protect this data from illegal access or breaches, such as encryption, access limits, and secure storage [45]. When working with sensitive or personal data, data anonymization is crucial. While still enabling researchers to gain valuable insights, anonymization techniques, deleting direct identifiers, and adding noise to data help protect individual privacy. To avoid re-identification, anonymization techniques must be effective. It is necessary to abide by data protection laws, such as the General Data Protection Regulation (GDPR) in Europe. Data privacy rules and regulations should be familiar to researchers, who should ensure their study procedures comply with these legal frameworks. Data anonymization requires openness and informed consent, which are essential components. Participants must be informed how researchers will handle and anonymize their data. When necessary, obtaining informed permission ensures that people know and concur with the data management procedures. Data anonymization systems must undergo routine audits and evaluations to ensure compliance and efficiency. These audits can aid in locating potential hazards or gaps in data security procedures and enable prompt corrections.

To effectively handle data privacy and anonymization concerns, interdisciplinary collaboration involving data scientists, ethicists, legal experts, and domain specialists is essential. These partnerships let researchers use AI and data analytics to do transformational research while navigating tricky ethical and legal issues. In the era of AI and data analytics, safeguarding sensitive research data and ensuring it is anonymized are essential components of ethical and responsible research methods. Researchers may use these technologies to their full potential while protecting data privacy and security by implementing robust security measures, adhering to appropriate laws, and encouraging interdisciplinary collaboration. This will eventually advance knowledge and benefit society.

The GDPR and the ethical guidelines for bias and fairness in AI significantly impact how data analytics and AI models are developed, particularly regarding how research methodologies are changed. These legal and moral frameworks include crucial topics, including bias reduction, data protection, and justice in AI research. GDPR Any research involving the personal information of people living in the European Union must comply. The GDPR's guiding principles, which include getting express consent for data processing, ensuring data minimization, and putting strong data protection measures in place, must be followed by researchers. GDPR non-compliance can result in serious legal repercussions and reputational harm.

The ethical guidelines for bias and fairness successfully address the widespread bias problem in AI models. Researchers must know and follow these rules to prevent their AI models from prejudice or discrimination towards specific groups. Auditing and evaluating bias models, openly disclosing the data sources, and implementing countermeasures like fairness-aware algorithms and balanced dataset representation are all part of ethical AI practices [46]. Research methodologies considering GDPR and ethical principles provide legal compliance and advance responsible and moral AI development. It promotes trust among study participants and the general public and helps to avoid prejudiced outcomes. To successfully traverse the complexity of GDPR and ethical AI requirements, interdisciplinary collaboration between data scientists, ethicists, and legal professionals is crucial. By collaborating, researchers can create AI models and data analytics methods that adhere to moral and legal requirements, are more reliable, and are advantageous to society. Modern AI and data analytics research methodologies are heavily influenced by GDPR and ethical standards for bias and fairness. Following these guidelines is morally and legally required, ensuring that research advancements are done ethically and responsibly while protecting data privacy and advancing justice in AI systems.

The GDPR and the Ethical Guidelines for Bias and Fairness in AI significantly impact how data analytics and AI models are developed, particularly regarding how research methodologies are changed. Any research involving personal information of people living in the European Union must comply. The GDPR's guiding principles, which include getting express consent for data processing, ensuring data minimization, and putting strong data protection measures in place, must be followed by researchers. GDPR non-compliance can result in serious legal repercussions and reputational harm. The ethical guidelines for bias and fairness successfully address the widespread bias problem in AI models. Researchers must know and follow these rules to prevent their AI models from prejudice or discrimination toward specific groups. Auditing and evaluating bias models, openly disclosing the data sources, and implementing countermeasures like fairness-aware algorithms and balanced dataset representation are all part of ethical AI practices. Research methodologies considering GDPR and ethical principles provide legal compliance and advance responsible and moral AI development. It promotes trust among study participants and the general public and helps to avoid prejudiced outcomes. To successfully traverse the complexity of GDPR and ethical AI requirements, interdisciplinary collaboration between data scientists, ethicists, and legal professionals is crucial. By collaborating, researchers can create AI models and data analytics methods that adhere to moral and legal requirements, are more reliable, and are advantageous to society.

Modern AI and data analytics research methodologies are heavily influenced by GDPR and ethical standards for bias and fairness.

3.9 ALGORITHMIC BIAS FOR MITIGATING DISCRIMINATION IN RESEARCH

Addressing algorithmic bias for mitigating discrimination is of utmost importance in research incorporating AI models and data analytics, especially as these technologies become crucial to changing research techniques. "Algorithmic bias" describes how ML algorithms may produce unwanted, unjust, or discriminating results. Taking on this problem is essential to maintaining the morality, diversity, and objectivity of AI-driven research. The first step for researchers is acknowledging that bias can affect each step of the research procedure from data collection and preprocessing to model creation and application. It takes diligence and a proactive attitude to recognize and eliminate these biases [47]. To ensure that datasets are representative and inclusive, researchers should carefully review their data sources to spot potential biases. Model results can be skewed as a result of biased data. Justice-aware ML algorithms and debasing approaches are just some of the tools researchers should use to encourage fairness and equity. These methods seek to eliminate discriminatory trends in model predictions. Transparency in research methodologies is crucial. The decision-making procedures, model parameters, and data sources should all be documented by researchers. External oversight and accountability are made possible by this transparency. Recognizing and addressing bias when diverse teams are involved in the research process is more straightforward. Substantial bias mitigation requires teamwork between ethicists, subject matter specialists, and data scientists. Ethical standards and rules should govern research practices. It entails upholding pertinent legal and moral requirements, protecting people's privacy, and ensuring informed consent. As data sources or algorithms are updated, bias may alter. To recognize and address bias in real-world situations, researchers must put constant monitoring and auditing procedures in place. Understanding the possible effects of AI research and ensuring it complies with social values and expectations need interaction with affected populations and requesting their input. AI research is morally necessary and a technological matter to address algorithmic bias and reduce discrimination. Researchers can change research methods to be more ethical, inclusive, and equitable, advancing knowledge responsibly and helping society by adopting a holistic strategy that includes various views, transparency, fairness-aware approaches, and ethical considerations. Algorithmic Bias for Mitigating Discrimination is of utmost importance in research incorporating AI models and data analytics, especially as these technologies become crucial to revolutionizing research methodologies. "Algorithmic bias" describes how ML algorithms may produce unwanted, unjust, or discriminating results. Taking on this problem is essential to maintaining the morality, diversity, and objectivity of AI-driven research. The first step for researchers is acknowledging that bias can affect each step of the research procedure, from data collection and preprocessing to model creation and application. It takes diligence and a proactive attitude to recognize and eliminate these biases.

3.10 TRANSPARENCY AND ACCOUNTABILITY FOR RESPONSIBLE AI USE

Transparency and accountability are essential pillars for fostering responsible AI use in research within AI models and data analytics, thereby catalyzing the transformation of research methods toward ethical and trustworthy practices. Transparency refers to the openness and clarity with which AI research processes and outcomes are communicated. Researchers must provide comprehensive documentation of data sources, preprocessing steps, model architectures, and decision-making criteria [48]. Transparent practices enable scrutiny, validation, and the identification of potential biases or errors, fostering trust and credibility in AI research. Accountability entails taking responsibility for the ethical and societal implications of AI research. Researchers must adhere to ethical guidelines and legal regulations while actively monitoring and evaluating the performance of AI models in real-world applications. When adverse consequences or biases emerge, researchers should promptly rectify them and be prepared to adjust models or algorithms as necessary. Implement robust data governance policies to ensure data privacy, security, and responsible data handling throughout the research lifecycle. Adhere to ethical frameworks prioritizing fairness, equity, and societal welfare, ensuring that AI research aligns with these principles. Collaborate with experts from diverse fields, including ethicists, legal experts, and domain specialists, to provide holistic oversight and guidance on ethical and accountable AI research practices. Design AI systems and models to be auditable, enabling post hoc analysis and evaluation to detect issues or biases that may arise during deployment. Strive for model explain ability, allowing stakeholders to comprehend how AI decisions are reached and enabling accountability [49]. Encourage external review and audits of AI research practices to ensure impartiality and adherence to ethical and accountable standards. Embedding transparency and accountability principles into AI research practices is imperative for promoting responsible AI use in research. These principles drive the transformation of research methods in AI models and data analytics and ensure that research outcomes are ethical, reliable, and aligned with societal values, ultimately benefiting society.

3.11 ETHICAL AI DEPLOYMENT IN MONITORING AND AUDITING AI SYSTEMS

The transformation of research methodologies while upholding responsible and ethical standards depends on ensuring Ethical AI Deployment in Monitoring and Auditing AI Systems inside AI Models and Data Analytics. AI systems must be monitored and audited throughout their lifespan to ensure they operate impartially and ethically [50]. The constant assessment, validation, and accountability processes used in this process help to reduce potential risks and guarantee adherence to moral standards. Keep a close eye on AI systems used in real-world settings to spot any ethical transgressions or unforeseen effects. Implement systems that send out alerts when anomalies or biases are found. Utilize methods and technologies to identify and reduce bias in data analytics and AI models. Any discrepancies or discriminatory outcomes should be addressed right away. Make sure AI systems are transparent and understandable in their design. To increase responsibility, stakeholders should comprehend the reasoning behind AI decisions and actions. Conduct routine audits and evaluations of data analytics and AI models. External audits conducted by impartial professionals can offer unbiased assessments of ethical compliance [51]. Create and uphold ethical frameworks that guide the use of AI, focusing on justice, equity, and respect for people's privacy and rights. Implement robust data governance policies to manage data responsibly and ensure privacy, security, and compliance with applicable laws. Create feedback loops with stakeholders, including the impacted communities, to get their thoughts on AI systems' ethical implications and effects. Work with ethicists, lawyers, subject experts, and various stakeholders to thoroughly assess the ethical implications of AI deployment. Roles and responsibilities must be clearly defined to ensure the ethical use of AI. Determine who is responsible for addressing ethical issues and implementing remedies. Transparently share monitoring and audit findings with the organization and, where needed, the general public to foster accountability. The transformation of research methodologies in AI models and data analytics depends on ensuring ethical AI deployment through monitoring and auditing. Researchers may exploit the potential of AI while respecting responsible, ethical, and accountable standards, assisting society and reducing possible harm by incorporating these practices into research and deployment processes.

3.12 STRIKING A BALANCE BETWEEN AUTOMATION AND HUMAN EXPERTISE

A significant difficulty and opportunity in modernizing research methodologies is finding a balance between automation and human expertise in AI models and data analytics. A delicate balance must be struck to harness the power of AI while assuring its ethical and responsible use. Automation, powered by ML and AI, delivers efficiency, scalability, and the capacity to process massive amounts of data quickly. It can reveal trends, forecast outcomes, and support decision-making. However, relying only on automation can have unforeseen consequences, bias, and moral dilemmas [47]. Make sure that human specialists continue to participate actively in AI research. They should establish ethical standards, specify the research goals, and interpret the findings. Making morally and intelligent decisions requires human scrutiny. Establish specific ethical guidelines that will direct the creation and application of AI models. Fairness, accountability, transparency, and respect for personal privacy and rights should be prioritized in these frameworks.

Encourage cooperation between data scientists, subject-matter specialists, ethicists, and legal professionals. Multidisciplinary teams can offer a variety of viewpoints and tackle complex ethical and technological problems. Make AI systems understandable and transparent. Make sure that stakeholders can comprehend the decision-making processes used by AI models and that they have access to the justifications for these conclusions. Implement methods for identifying and reducing bias in AI models. Designing fair algorithms and spotting biased results both require human expertise. In detail, define roles and duties for people and teams working on AI research. Accountability guarantees that moral standards are upheld and that mistakes or prejudices are quickly corrected [50]. Watch for performance, prejudice, and ethical compliance in AI systems. Human professionals can interpret results and can also choose what adjustments are required. Create feedback loops with end users and affected communities to get feedback and ideas on AI systems' ethical implications and practical effects. Give the ethical and prudent use of AI models a top priority. The societal, legal, and ethical ramifications of using AI in various circumstances should be evaluated by human specialists [49]. It is crucial to balance automation and human expertise for research methods in AI models and data analytics to change. Researchers may use automation while keeping moral norms and assuring responsible AI use by fusing the advantages of AI with human oversight, ethical frameworks, interdisciplinary collaboration, and accountability.

3.13 CONCLUSION

A balanced and careful strategy is needed to fully realize AI promise for research in the context of the debates about how AI models and data analytics are changing research methodologies. However, AI must be appropriately used and ethically if it is to revolutionize science. Aim for high-caliber, diversified, and representative research datasets. AI models are powered by high-quality data, and incomplete or biased datasets might produce distorted findings. Throughout your research, abide by ethical standards and values. Give justice, accountability, and transparency top priority when conducting AI research. Collaboration between data scientists, subject matter specialists, ethicists, and legal scholars should be encouraged. Research findings are enriched by diverse viewpoints, which also aid in tackling complex moral and technological issues. Make data analytics and AI models clear and understandable. To foster responsibility and trust, stakeholders should be able to comprehend how AI choices are made. Continue to oversee AI research with humans actively. Humans should establish research goals, evaluate the findings, and guarantee ethical behavior. Put techniques for spotting and reducing bias in AI models into practice. Awareness of unjust or discriminating outcomes is crucial for ethical AI development. Watch for ethical behavior, performance, and potential prejudice in AI systems. The upkeep of ethical standards requires regular audits and evaluations. Emphasize the ethical and responsible use of AI models. The use of AI in various situations should be evaluated regarding its social, legal, and ethical ramifications. To get feedback and an understanding of AI systems' ethical ramifications and practical effects, consult with the affected communities and stakeholders. Identify their roles and responsibilities for people and teams working on AI research. Accountability procedures guarantee that moral principles are upheld, and problems are quickly resolved. The potential of AI to speed up research, reveal insights, and solve complicated problems may be unlocked by researchers by embracing these ideas and methods, which will also protect moral standards and ensure responsible AI use. An approach changes how research is conducted, advances knowledge, and benefits society.

In the context of AI models and data analytics, the core of responsible and transformational research methodologies is aligning scientific advancement with ethical commitments. AI has the potential to transform research, but for the sake of people's rights and the wellness of society as a whole, this development must be governed by moral standards. Researchers must establish clear ethical frameworks to prioritize justice, openness, accountability, and privacy in AI research. These frameworks are guiding principles for creating and implementing AI models and data analytics. Join forces with subject-matter experts from various disciplines, such as data scientists, ethicists, legal academics, and domain specialists. Interdisciplinary teams may offer comprehensive supervision and direction, ensuring technology advancements align with moral issues. Throughout the study process, maintain transparency. Be responsible for the outcomes of the deployment of AI by documenting data sources, preprocessing techniques, and algorithmic conclusions. Be proactive in addressing bias in data analytics and AI models. Utilize techniques to identify and reduce bias and closely monitor equity and fairness. Human specialists should actively participate in AI research to ensure ethical compliance and determine research goals. Instead of substituting for human skill, AI should enhance it. Obtain feedback and insights from communities and stakeholders regarding AI systems' moral ramifications and practical effects. Include their input in your research and deployment procedures. Conduct routine audits and evaluations of AI research methodologies. Objective assessments of ethical compliance can be made by impartial specialists conducting external reviews. Encourage education and knowledge regarding AI ethics among scientists, professionals, and the general public. To ensure ethical AI use, ethical literacy is essential. Emphasize the ethical and responsible use of AI models. Researchers should evaluate how diverse uses of AI will affect society, the law, and ethics. While conducting AI research, adhere to pertinent rules and regulations, such as data protection and privacy laws. Legal requirements provide the basis for ethical obligations. Researchers may use AI models and data analytics to their full potential to advance knowledge, tackle complex problems, and responsibly change research methodologies by balancing technological advancement with these moral commitments. This strategy supports the moral application of AI and fosters fairness, trust, and societal well-being.

REFERENCES

1. J. Sheng, J. Amankwah-Amoah, Z. Khan, and X. Wang, "COVID-19 pandemic in the new era of big data analytics: Methodological innovations and future research directions," *British Journal of Management,* vol. 32, no. 4, pp. 1164–1183, 2021. https://doi.org/10.1111/1467-8551.12441.
2. A. Husnain, S. Rasool, A. Saeed, H. K. Hussain, and A. c. Edu, "International journal of multidisciplinary sciences and arts revolutionizing pharmaceutical research: Harnessing machine learning for a paradigm shift in drug discovery," vol. 2, no. 2, 2023. https://doi.org/10.47709/ijmdsa.v2i2.2897.
3. M. A. Qureshi, K. N. Qureshi, G. Jeon, and F. Piccialli, "Deep learning-based ambient assisted living for self-management of cardiovascular conditions," *Neural Computing and Applications*, 2021. https://doi.org/10.1007/s00521-020-05678-w.
4. A. R. Moreno, "*Discourses Across Periods of Time Discourses Across Periods of Time Part of the Art and Design Commons, and the Artificial Intelligence and Robotics Commons*," 1932. Available: https://digitalcommons.lindenwood.edu/theses.
5. A. Ahmad, A. Tariq, H. K. Hussain, and A. Y. Gill, "Equity and artificial intelligence in surgical care: A comprehensive review of current challenges and promising solutions," *BULLET: Jurnal Multidisiplin Ilmu,* vol. 2, no. 2, pp. 443–455, 2023.
6. J. E. Hallsworth, Z. Udaondo, C. Pedrós-Alió, J. Höfer, K. C. Benison, K. G. Lloyd, R. J. B. Cordero, C. B. L. de Campos, M. M. Yakimov, and R. Amils, "Scientific novelty beyond the experiment," *Microbial Biotechnology,* vol. 16, no. 6, pp. 1131–1173. https://doi.org/10.1111/1751-7915.14222.
7. N. Pettorelli, W. F. Laurance, T. G. O'Brien, M. Wegmann, H. Nagendra, and W. Turner, "Satellite remote sensing for applied ecologists: opportunities and challenges," *Journal of Applied Ecology,* vol. 51, no. 4, pp. 839–848, 2014.
8. A. Kowalska and H. Ashraf, "*Advances in Deep Learning Algorithms for Agricultural Monitoring and Management.*" https://doi.org/0000-0002-2419-4626.
9. K. O'Leary, M. Lohman, F. Culver, A. Killarney, G. Smith, and D. Liebovitz, "The effect of tablet computers with a mobile patient portal application on hospitalized patients' knowledge and activation," *Journal of the American Medical Informatics Association: JAMIA,* vol. 23, 2015. https://doi.org/10.1093/jamia/ocv058.
10. M. Faaique, "Overview of big data analytics in modern astronomy," *International Journal of Mathematics, Statistics, and Computer Science,* vol. 2, pp. 96–113, 2013. https://doi.org/2023.10.59543/ijmscs.v2i.8561.

11. J. Moody, "Race, school integration, and friendship segregation in America," *American Journal of Sociology*, vol. 107, no. 3, pp. 679–716, 2001. https://doi.org/10.1086/338954.
12. F. Fui-Hoon Nah, R. Zheng, J. Cai, K. Siau, and L. Chen, "Generative AI and ChatGPT: Applications, challenges, and AI-human collaboration," in *Journal of Information Technology Case and Application Research*, vol. 25, ed: Routledge, 2023, pp. 277–304.
13. R. K. Mondol, E. K. A. Millar, P. H. Graham, L. Browne, A. Sowmya, and E. Meijering, "hist2RNA: An efficient deep learning architecture to predict gene expression from breast cancer histopathology images," *Cancers*, vol. 15, no. 9, 2023. https://doi.org/10.3390/cancers15092569.
14. A. Kowalska and H. Ashraf, "Advances in deep learning algorithms for agricultural monitoring and management," *Applied Research in Artificial Intelligence and Cloud Computing*, vol. 6, no. 1, pp. 68–88, 2023.
15. M. W. Libbrecht and W. S. Noble, "Machine learning applications in genetics and genomics," in *Nature Reviews Genetics*, vol. 16, ed: Nature Publishing Group, 2015, pp. 321–332.
16. L. N. Vieira, M. O'Hagan, and C. O'Sullivan, "Understanding the societal impacts of machine translation: a critical review of the literature on medical and legal use cases," in *Information Communication and Society*, vol. 24, ed: Routledge, 2021, pp. 1515–1532.
17. E. J. Topol, "High-performance medicine: the convergence of human and artificial intelligence," in *Nature Medicine*, vol. 25, ed: Nature Publishing Group, 2019, pp. 44–56.
18. S. Akter, S. Sultana, M. Mariani, S. F. Wamba, K. Spanaki, and Y. K. Dwivedi, "Advancing algorithmic bias management capabilities in AI-driven marketing analytics research," *Industrial Marketing Management*, vol. 114, pp. 243–261, 2023. https://doi.org/10.1016/j.indmarman.2023.08.013.
19. T. D. Gunawansa, K. Perera, A. Apan, N. K. Hettiarachchi, and D. Y. Bandara, "Greenery change and its impact on human-elephant conflict in Sri Lanka: a model-based assessment using Sentinel-2 imagery," *International Journal of Remote Sensing*, vol. 44, no. 16, pp. 5121–5146, 2023. https://doi.org/10.1080/01431161.2023.2244644.
20. A. Naveed, "Transforming clinical trials with informatics and AI/ML: A data-driven approach," *International Journal of Computer Science And Technology (IJCST)*, vol. 7, pp. 91–97, 2023.
21. B. El-Shweky, K. El-Kholy, M. Abdelghany, M. Salah, M. Wael, O. Alsherbini, Y. Ismail, K. Salah, and M. Abdelsalam, "Internet of Things: A Comparative Study," in *2018 IEEE 8th Annual Computing and Communication Workshop and Conference (CCWC)*, 2018, pp. 622–631.
22. M. K. Daoud, M. Alkhaffaf, M. Al-Qeed, and J. A. Al-Gasawneh, "Analyzing the impact of artificial intelligence in big data-driven marketing tool efficiency," no. S8, pp. 521–533, 2023. https://doi.org/10.59670/ml.v20iS8.4628.
23. K. N. Qureshi, A. H. Abdullah, and R. W. Anwar, "The Evolution in Health Care with Information and Communication Technologies," in Elsevier, *2nd International Conference of Applied Information and Communications Technology-2014, Oman*, 2014: Elsevier.

24. S. Mishra, M. Aggarwal, S. Yadav, and Y. Sharma, "An automated model for sentimental analysis using long short-term memory-based deep learning model," *International Journal of Engineering and Manufacturing*, vol. 13, no. 5, pp. 11–20, 2023. https://doi.org/10.5815/ijem.2023.05.02.
25. J. Bollen, H. Mao, and X. Zeng, "Twitter mood predicts the stock market," *Journal of Computational Science*, vol. 2, no. 1, pp. 1–8, 2011. https://doi.org/10.1016/j.jocs.2010.12.007.
26. C. Batini and M. Scannapieco, *Data Quality: Concepts, Methodologies and Techniques*, Springer, 2006.
27. L. M. Amugongo, A. Kriebitz, A. Boch, and C. Lütge, "Operationalising AI ethics through the agile software development lifecycle: A case study of AI-enabled mobile health applications," *AI and Ethics*, 2023. https://doi.org/10.1007/s43681-023-00331-3.
28. R. Deepa, S. Arunkumar, V. Jayaraj, and A. Sivasamy, "Healthcare's new Frontier: AI-driven early cancer detection for improved well-being," *AIP Advances*, vol. 13, no. 11, 2023. https://doi.org/10.1063/5.0177640.
29. K. N. Qureshi, G. Jeon, and F. Piccialli, "Anomaly detection and trust authority in artificial intelligence and cloud computing," *Computer Networks*, p. 107647, 2020. https://doi.org/10.1016/j.comnet.2020.107647.
30. S. Rasp, M. S. Pritchard, and P. Gentine, "Deep learning to represent subgrid processes in climate models," *Proceedings of the National Academy of Sciences of the United States of America*, vol. 115, no. 39, pp. 9684–9689, Sep 25 2018. https://doi.org/10.1073/pnas.1810286115.
31. X. Chen, G. Yu, G. Cheng, and T. Hao, "Research topics, author profiles, and collaboration networks in the top-ranked journal on educational technology over the past 40 years: a bibliometric analysis," *Journal of Computers in Education*, vol. 6, no. 4, pp. 563–585, 2019. https://doi.org/10.1007/s40692-019-00149-1.
32. H. Kharrazi, A. S. Lu, F. Gharghabi, and W. Coleman, "A scoping review of health game research: Past, present, and future," *Games for Health Journal*, vol. 1, no. 2, pp. 153–164, Apr 18 2012. https://doi.org/10.1089/g4h.2012.0011.
33. Y. LeCun, Y. Bengio, and G. Hinton, "Deep learning," *Nature*, vol. 521, no. 7553, pp. 436–444, 2015. https://doi.org/10.1038/nature14539.
34. M. W. Libbrecht and W. S. Noble, "Machine learning applications in genetics and genomics," *Nature Reviews Genetics*, vol. 16, no. 6, pp. 321–332, Jun 2015. https://doi.org/10.1038/nrg3920.
35. C. Angermueller, T. Parnamaa, L. Parts, and O. Stegle, "Deep learning for computational biology," *Molecular Systems Biology*, vol. 12, no. 7, p. 878, Jul 29 2016. https://doi.org/10.15252/msb.20156651.
36. O. B. Sezer, M. U. Gudelek, and A. M. Ozbayoglu, "Financial time series forecasting with deep learning : A systematic literature review: 2005–2019," *Applied Soft Computing*, vol. 90, p. 106181, 2020. https://doi.org/10.1016/j.asoc.2020.106181.
37. D. B. Rubin. *Multiple imputation for nonresponse in surveys*. vol. 81. John Wiley & Sons, 2004.

38. K. Cheng, X. Cheng, Y. Wang, H. Bi, and M. C. Benfield, "Enhanced convolutional neural network for plankton identification and enumeration," *PLoS One,* vol. 14, no. 7, p. e0219570, 2019. https://doi.org/10.1371/journal.pone.0219570.
39. A. Mittal, K. Manjunath, R. K. Ranjan, S. Kaushik, S. Kumar, and V. Verma, "COVID-19 pandemic: Insights into structure, function, and hACE2 receptor recognition by SARS-CoV-2," *PLoS Pathog,* vol. 16, no. 8, p. e1008762, Aug 2020. https://doi.org/10.1371/journal.ppat.1008762.
40. H. O. Alanazi, A. H. Abdullah, and K. N. Qureshi, "A critical review for developing accurate and dynamic predictive models using machine learning methods in medicine and health care," *Journal of Medical Systems,* vol. 41, no. 4, p. 69, 2017.
41. M. W. Libbrecht and W. S. Noble, "Machine learning applications in genetics and genomics," *Nature Reviews Genetics,* vol. 16, no. 6, pp. 321–332, 2015. https://doi.org/10.1038/nrg3920.
42. H. Sarma, M. A. Islam, J. R. Khan, K. I. A. Chowdhury, and R. Gazi, "Impact of teachers training on HIV/AIDS education program among secondary school students in Bangladesh: A cross-sectional survey," *PLoS One,* vol. 12, no. 7, p. e0181627, 2017. https://doi.org/10.1371/journal.pone.0181627.
43. A. Iftikhar, K. N. Qureshi, M. Shiraz, and S. Albahli, "Security, trust and privacy risks, responses, and solutions for high-speed smart cities networks: A systematic literature review," *Journal of King Saud University - Computer and Information Sciences,* p. 101788, 2023. https://doi.org/10.1016/j.jksuci.2023.101788.
44. D. J. Watts and S. H. Strogatz, "Collective dynamics of 'small-world' networks," *Nature,* vol. 393, no. 6684, pp. 440–442, 1998. https://doi.org/10.1038/30918.
45. A. Iftikhar and K. N. Qureshi, "Future privacy and trust challenges for IoE networks," in *Cybersecurity Vigilance and Security Engineering of Internet of Everything*: Springer Nature Switzerland, 2023, pp. 193–218.
46. Y. Guo, Y. Yang, and A. Abbasi, "Auto-debias: Debiasing masked language models with automated biased prompts," *Long Papers,* vol. 1 Available: https://github.
47. M. Śmietanka, A. Koshiyama, and P. Treleaven, "Review of algorithms in future insurance markets algorithms in future insurance markets," Available: https://ssrn.com/abstract=3641518.
48. H. Triastuti, K. Ningsih, L. Syafina, and I. Muda, "Digital accounting: The role of artificial intelligence and xbrl increasing transparency of accounting information. The overview based on financial data transparency act (FDTA)," *Russian Law Journal,* 2023, vol. XI Available: https://xbrl.us/home/government/legislation/.
49. "G. Edison, "Transforming medical decision-making: A comprehensive review of AI's impact on diagnostics and treatment." *Jurnal Multidisiplin Ilmu* vol. 2. no. 4, pp. 1121–1133, 2023.

50. Ahmad A, A. Tariq, H. K. Hussain, and A. Y. Gill, "Equity and artificial intelligence in surgical care: A comprehensive review of current challenges and promising solutions. *Jurnal Multidisiplin Ilmu*. vol. 2, no. 2, pp. 443–55, 2023.
51. P. Kowe, O. Mutanga, and T. Dube, "Advancements in the remote sensing of landscape pattern of urban green spaces and vegetation fragmentation," in *International Journal of Remote Sensing* vol. 42, ed: Taylor and Francis Ltd., 2021, pp. 3797–3832.

CHAPTER 4

Advanced AI-Based Healthcare Systems Applications and Services

Ayesha Aslam, Kashif Naseer Qureshi, and Adil Hussain

4.1 OVERVIEW

The healthcare field is witnessing rapid advancements in Artificial Intelligence (AI) since it holds immense potential in harnessing the capabilities of big data. This potential enables AI to provide valuable insights that assist evidence-based clinical decision-making and facilitate realizing value-based care. Health leaders must have a comprehensive knowledge of the current state of AI technologies and their prospective applications in improving the effectiveness, security, and accessibility of healthcare services. Such understanding is essential in enabling the digital transformation of healthcare. AI has a big effect on healthcare because it has spread to many areas, including helping doctors make decisions in hospitals, letting people with chronic illnesses take care of themselves at home, and doing real-life drug research [1]. However, the progress and use of AI

DOI: 10.1201/9781032667911-4

technologies present substantial challenges and cost constraints. Health organizations face numerous challenges that must be overcome to integrate AI technologies successfully. Multiple obstacles impede the implementation of AI technology into healthcare systems. First and foremost, there is a restricted understanding of the capacities and constraints of particular AI systems. In addition, there is a scarcity of clearly defined approaches for integrating diverse AI technologies into current healthcare systems to tackle the most pressing challenges health organizations face. Moreover, there is a shortage of proficient experts to support the integration of AI in the healthcare sector effectively. Also, the incompatibility between AI technologies and antiquated infrastructure is a substantial barrier. Furthermore, insufficient availability of comprehensive and varied medical data is necessary to train Machine Learning (ML) algorithms. The user's text lacks sufficient information to be rephrased academically.

4.2 APPLICATIONS OF AI IN HEALTHCARE

The healthcare industry utilizes the many applications of medical artificial intelligence, as shown in Figure 4.1.

4.2.1 AI for Drug Discovery

The application of AI technology within the healthcare sector has shown to be beneficial for pharmaceutical companies, as it has facilitated the acceleration of their drug discovery process. In contrast, it facilitates the automated detection of targets. Furthermore, the utilization of AI in the field of healthcare in 2021 facilitates the process of drug repurposing through the analysis of off-target molecules, as highlighted in [2]. Consequently, within the domains of AI and healthcare, AI drug discovery serves to streamline operational procedures and mitigate redundant tasks. According to [3], a variety of therapies have been identified and made accessible by prominent biopharmaceutical corporations. Pfizer uses IBM Watson, an AI system based on ML, to identify immuno-oncology therapies. Sanofi has consented to employ Exscientia's AI technology to identify pharmaceutical treatments for metabolic diseases. The search for anti-cancer drugs at Genentech, a subsidiary of Roche, is aided by a Cambridge, Massachusetts-based AI system created by GNS Healthcare. The vast majority of prominent biopharmaceutical companies have comparable relationships or internal programs.

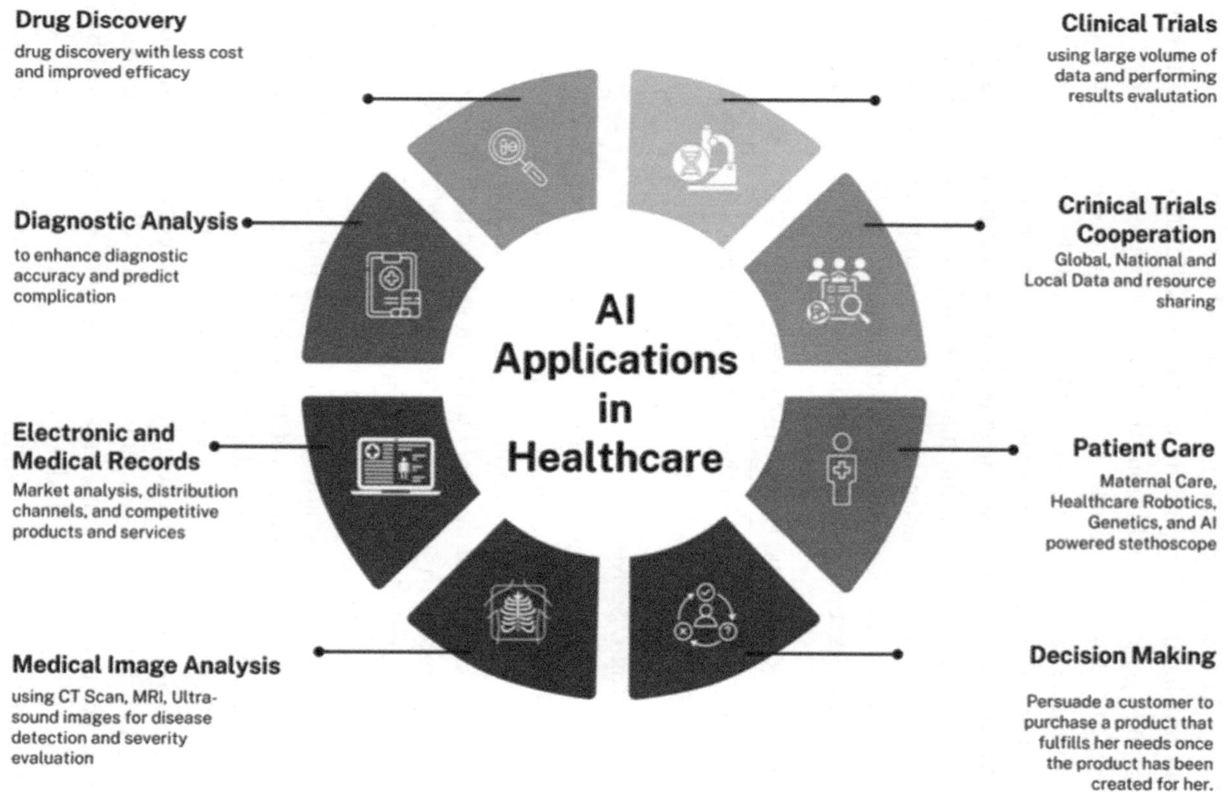

FIGURE 4.1 AI-based applications in healthcare.

Suppose the assertions made by advocates of these approaches are accurate. In that case, the utilization of AI and ML in drug development has the potential to usher in a novel era characterized by enhanced efficiency, reduced costs, and improved efficacy [4]. While there exists a degree of skepticism among certain individuals, the prevailing consensus among experts is that these instruments will assume an increasingly pivotal role in the future. Scientists encounter both challenges and prospects due to this shift, particularly when these methodologies are combined with automation.

4.2.2 AI for Clinical Trials

A clinical trial is a systematic process in which newly developed interventions are administered to individuals to evaluate their efficacy. This endeavor has required a substantial investment of both time and financial resources. However, the rate of success is rather modest. Implementing clinical trial automation has demonstrated its advantageous impact on both AI and the healthcare industry. Moreover, integrating AI in healthcare contributes to streamlining labor-intensive data monitoring protocols. Moreover, clinical studies that utilize AI assistance effectively manage substantial volumes of data and yield highly precise results. The above examples represent a selection of widely utilized AI applications within the healthcare sector, specifically in the context of clinical trials.

4.2.3 Intelligent Clinical Trials

Traditional clinical trials that follow a linear and sequential approach continue to be widely regarded as the benchmark for assessing the effectiveness and safety of novel pharmaceuticals. The established and comprehensive approach of sequential and well-defined phases in Randomized Controlled Trials (RCTs) was largely devised to assess widely available pharmaceutical products and has had minimal modifications recently. The use of AI can reduce the time of clinical trial cycles and improve productivity and outcomes in clinical development. In recent years, biopharmaceutical enterprises have experienced a notable surge in acquiring scientific and research data from various sources, commonly known as Real-World Data (RWD). However, there has been a recurring deficiency in their proficiency and access to the necessary resources for effectively using this data. Using predictive AI models and advanced analytics in analyzing RWD can enhance researchers' comprehension of diseases, facilitate the identification of pertinent patients and significant investigators, and facilitate the development of groundbreaking clinical study designs. In conjunction with a proficient digital infrastructure, AI algorithms have the potential to facilitate the cleansing, aggregation, coding, preservation, and maintenance of clinical trial data. In addition, enhanced Electronic Data Capture (EDC) can mitigate the influence of human error during data collection and facilitate seamless interaction with existing systems [5].

4.2.4 Clinical Trial Cooperation and Model Sharing

To assist in responding to the COVID-19 pandemic, scientists from various disciplines are engaged in a concerted effort of scholarly collaboration [6]. To have a worldwide influence through AI technologies, it is crucial to employ scalable methods for sharing data, models, and code, customize applications to fit local circumstances, and promote collaboration across borders. Collecting data is necessary for the creation and execution of AI systems. Data-sharing programs with an emphasis on COVID-19 are widespread across many levels of application, ranging from global to national and local contexts. The available resources comprise genetic sequences, genomic analyses, protein structures, clinical information regarding patients, medical imaging, event data, epidemiological data, movement data, social media comments, news headlines, and scientific literature. The excessive fragmentation of data-sharing efforts is concerning as it may hinder the progress of projects and communities. Utilizing scalable data, models, and code-sharing tools helps expedite the development and dissemination of innovative applications. Utilizing worldwide, transparent, inclusive, standardized, and auditable data-sharing programs can promote linkage and collaboration among diverse populations and geographical areas [7].

The advancement of capabilities within national health systems can be accelerated and information dissemination can be enhanced by fostering open science through the collaborative efforts of numerous stakeholders in AI. Utilizing the Epidemic Intelligence from Open Sources (EIOS) network exemplifies the application of open-source data in public health. This network utilizes open-source data to efficiently identify, validate, and assess public health concerns and hazards promptly. Entities within the healthcare intelligence network include governments, international organizations, and research institutions. These entities collaborate to assess and disseminate real-time information regarding epidemics. This collaboration is driven by collaboration rather than rivalry, focusing on early identification. Epidemiologists assert that the implementation of global standards and database interoperability has the potential to enhance global, national, and local response and decision-making. To acquire a better understanding of the epidemiological characteristics and risk factors associated with different demographic groups during the COVID-19 pandemic, it is crucial to take into account the capacity of healthcare systems, public health measures, environmental variables, and the societal ramifications of the virus.

Moreover, the exchange of trained AI models for the proposed applications in different initiatives is restricted, alongside data sharing. To overcome the special computational, design, and infrastructure constraints, several obstacles must be solved. The obstacles encompass inadequate documentation, verification and interpretability issues, and legal concerns about confidentiality and intellectual property. The distribution of pre-trained and certified AI models can accelerate the adjustment of solutions to various situations. Algorithms that encompass various applications, such as diagnosing illnesses through image analysis, predicting patient outcomes, identifying and mitigating the spread of misinformation on social media based on propagation patterns, and extracting knowledge graphs from extensive scholarly paper collections, have been identified as valuable models.

4.2.5 Patient Care

The influence of AI in healthcare has a significant impact on patient outcomes. Medical AI companies provide a comprehensive system that assists patients across various stages of healthcare delivery. Clinical intelligence is a process that involves the analysis of patient's medical data to provide valuable insights aimed at improving their overall quality of life. Several clinical intelligence systems have been identified as having a major impact on enhancing patient care.

A. Maternal Care

The resulting approach presents a feasible methodology for identifying high-risk mothers and mitigating maternal mortality and postpartum complications.

a) Utilizing electronic health data and AI to forecast the likelihood of pregnant moms encountering substantial challenges during childbirth.

b) Utilizing digital technology to enhance patient access to routine and high-intensity healthcare services, facilitating more advanced and frequent medical attention during pregnancy.

In contrast to delivering in clinics with higher acuity levels, greater resources, and extensive clinical experience, delivering infants by high-risk obstetric women in low-acuity clinics is associated with an increased likelihood of experiencing severe maternal morbidity.

B. Healthcare Robotics

In addition to healthcare professionals, specific medical robots assist patients. According to [8], exoskeleton robots can aid those with paralysis in regaining the ability to walk and achieve independence. A smart prosthesis serves as an exemplification of technology in operation. Bionic limbs are equipped with sensors that enhance their responsiveness and accuracy beyond that of natural body parts. Additionally, users can apply bionic skin to these limbs and establish connections with their muscles [9]. Rehabilitation and surgery have the potential to benefit significantly from robotic systems. Cyberdyne's Hybrid Assistive Limb (HAL) exoskeleton is designed to aid in rehabilitating patients with lower limb disorders resulting from spinal cord injuries and accidents. Electrical signals within the patient's body are effectively detected using sensors placed on the epidermis. Consequently, the exoskeleton responds by causing joint movement.

C. Genetics AI Data-Driven Medicine

The engagement of healthcare consumers in their medical treatment has witnessed a notable increase, ranging from the utilization of genome sequencing techniques to the generation of customized health profiles based on data obtained from fitness and activity trackers. The aggregation and correlation of extensive datasets generate a more prognostic representation of health or medical conditions. According to [10], using data-driven medicine can enhance the accuracy and efficiency of genetic illness identification and facilitate the development of personalized medical interventions.

D. AI-Powered Stethoscope

One significant advantage is that, in contrast to conventional stethoscopes, the readings can be obtained even in areas with high noise levels, enabling more precise diagnostic outcomes. According to [11], digital equipment requires no specialized training, allowing anyone to access and transmit the records to the doctor. Furthermore, this minimizes the likelihood of people contracting COVID-19 and allows for better healthcare services in distant areas and for people with chronic conditions. The emergence of AI and ML has enabled computers to efficiently detect disease patterns and anomalies by analyzing vast amounts of clinical data. The blood flow in healthy arteries displays discernible traits in contrast to the blood flow when a blood clot is present within the blood vessels. Therefore, the idea outlined earlier can be utilized in this particular situation.

The utilization of AI in the healthcare sector is steadily increasing due to its growing prominence in contemporary organizations and daily routines. AI can support healthcare providers across several domains, encompassing patient treatment and administrative functions. The healthcare industry benefits from various AI and healthcare breakthroughs; however, the tactics they support can exhibit notable differences. Although certain articles in the field of AI in healthcare assert that AI may match or surpass human performance in certain tasks, such as disease diagnosis, it will take considerable time before AI fully supplants human professionals across a broad spectrum of medical roles.

4.2.6 AI Application Areas in Healthcare

In contrast to conventional technologies that mostly operate within the physical domain, AI technology is pioneering novel advancements by extending its influence into psychological domains, encompassing areas such as experiential understanding, cognitive capabilities, and expert decision-making. Specifically, with the advent of deep learning technology, the efficacy of ML algorithms for pattern recognition has improved significantly. Consequently, AI technology now has a data pattern analysis capability comparable to a typical human's, especially in tasks such as image recognition and speech recognition. Due to their resemblance to artificial neural networks, which mimic the complex network of neurons in the human brain, deep learning algorithms are utilized in numerous medical data applications. These algorithms can acquire complex nonlinear correlations, making them highly effective for analyzing complex medical data. Several ongoing studies are investigating the use of AI-based technologies in healthcare, as shown in Table 4.1.

4.2.7 Medical Image Analysis

In recent years, the applications of ML algorithms to medical image processing have expanded significantly. This holds across other medical disciplines, not limited to radiology. This expansion encompasses other disciplines, including pathology, dermatology, cardiology, gastrointestinal, and ophthalmology, where images play a crucial role. ML algorithms

TABLE 4.1 AI Technologies in Healthcare

Technology	Application Description	Application Domain
Robotics	Improve the surgical procedures' precision and accuracy to provide high-quality care.	Health IT, Medical device
Digital Secretary	Find the best time to help by keeping an eye on the patient's state indicators and letting the nurse know when you need to.	Health IT, Medical device
Machine Learning (ML)	Predict and look for trends in the data that affect treatment outcomes. By processing a lot of diagnostic medical images and learning on your own, you can make medical care decisions with less uncertainty.	Health IT, Diagnostic medical image
Image Processing	Process several medical pictures quickly, then utilize the data to identify the type of disease and the status of positive and negative test results.	Health IT, Diagnostic medical image
Natural Language Processing (NLP)	Transform long unstructured text data (like medical charts) into a comprehensible and easily readable format.	Health IT, Medical device
Voice Recognition	Record the patient's words and facial expressions, then enter the relevant data into electronic medical records.	Health IT, Medical device
Statistical Analysis	Analyze a large amount of data from a patient's medical records quickly to predict the course of their treatment.	Health IT, Medicine
Big Data Analysis	Healthcare facilities can provide patients with individualized recommendations and prescriptions by analyzing the vast volumes of data they maintain.	Health IT, Medicine
Predictive Modeling	Make treatment outcome predictions using mathematical models, such as identifying high-risk diseases.	Health IT, Medicine

utilize several types of medical imaging data, including Computed Tomography (CT), Magnetic Resonance Imaging (MRI), ultrasound, pathology picture, fundus image, and endoscope data, to reliably detect diseases or assess their severity. Employing a machine learning algorithm for immediate colonoscopy resulted in a diagnostic accuracy of 94% and a negative predictive value of 96% while examining a subset of 466 tiny polyps. The Convolutional Neural Network (CNN) algorithm, renowned for its expertise in analyzing visual patterns, has proven highly effective in interpreting complex medical images.

Siemens Healthiness has developed the AI-Rad Companion Chest CT program, which employs AI in the healthcare sector to assist in diagnosing chest CT scans. In addition, GE Healthcare is now involved in advancing AI-driven technologies to analyze medical pictures. Furthermore, Philips Healthcare has successfully developed IntelliSpace Discovery, an accessible platform designed to conduct and apply AI research. Furthermore, Zebra Medical Vision and Aidoc are actively and assertively bringing AI-powered medical image-processing solutions to the commercial market. Currently, several firms, such as Vuno, Lunit, JLK Inspection, and Deepnoid, are bringing AI-driven analysis tools into the market. The Ministry of Food and Drug Safety has permitted these businesses to proceed.

A. Smart IoT Devices, Signal, and In Vitro Diagnostic Analysis

Tech companies such as Apple, IBM, Google, and Samsung are competing to develop and market products and services that enhance the health of their users. These initiatives utilize wearable devices and Internet of Things (IoT) technologies to gather health-related data regarding individuals based on their routine activities [12]. The FDA approved a deep learning algorithm that Apple integrated into its watch in 2017 because it could detect atrial fibrillation [13]. With the help of photoplethysmography and accelerometer monitors, the system learns the user's normal resting and active heart rates. If the actual numbers are very different from what is expected, the system sends a warning signal immediately. The deep learning method demonstrates a high degree of accuracy in analyzing Electrocardiogram (ECG) data. A recent empirical study employed a deep learning system to analyze 91,232 single-lead ECGs. The findings of this research demonstrated that the diagnostic accuracy is comparable to that of proficient cardiologists. Individuals afflicted with cardiac disease, chronic kidney disease, hyperkalemia, or both may find the application of ECG pattern analysis on smart devices exceedingly beneficial.

In addition to their global presence, multinational corporations are developing various cutting-edge medical technologies, such as non-invasive glucose meters. Bio-signal monitoring devices have the potential to be utilized in various medical settings, including critical care units, operating rooms, emergency rooms, and recovery rooms. These devices can aid in illness management and treatment by providing real-time information that can be utilized for forecasting and diagnosing diseases. An illustrative instance of this phenomenon may be observed in the partnership forged between Medtronic and IBM, wherein they jointly worked on creating SugarIQ. This innovative product integrates Medtronic's continuous glucose monitoring technology with IBM's AI system. The efficacy of the SugarIQ in mitigating symptoms of hypoglycemia and hyperglycemia has been demonstrated in diabetic individuals.

B. AI Using Electronic Medical Records (EMRs)

Much work is currently being done to develop AI systems that use EMRs. Several EMR companies in the US, including Allscripts, Athenahealth, Cerner, eClinicalWorks, and Epic, are doing research projects to improve hospital treatment procedures by leveraging AI technology [14]. In Korea, EvidNet is now developing technology for analyzing clinical big data across several hospitals. This technology is founded on the principles of observational health data sciences and informatics and the utilization of a Common Data Model (CDM).

C. AI And Decision-Making in Health Systems

For health systems, like public health or healthcare services, to work well, they must be run with a complex network of information processing jobs. Policymakers change how the health system is organized and run, how it pays for itself, and how it manages its resources to get the results and overall system goals they want, such as healthcare services and public health outcomes [15]. The process of screening and diagnosis, which involves looking at a patient's medical history, physical exam, and diagnostic tests, and the process of treatment and monitoring, which involves planning, carrying out, and constantly evaluating a series of steps meant to lead to a certain result in the future, are two of the most important information processing tasks in healthcare. The basic structure of these processes in managing the health system and giving care is to develop theories, test those hypotheses, and then act. ML could make it easier for a health system to develop and test hypotheses. ML can help to find previously unknown trends by revealing patterns in previously hidden data. So, this technology has a lot of potential to help the individual patient and the healthcare system in big ways.

ML is based on well-known statistical techniques and uses methods that don't make assumptions about how data is distributed. Its goal is to find patterns in the data, which can then be used to develop and test theories. Therefore, ML models pose challenges in terms of interpretability. Still, they can incorporate more variables, exhibit generalizability across diverse data types, and yield outcomes in intricate scenarios. These devices are implemented in various locations, usually within hospitals rather than in community environments. Additionally, it is worth noting that the data used to support these devices often comes from individual centers, which may have ramifications for the capacity to replicate the results and apply them to a wider population. Nevertheless, the exponential progress of ML persists in the field of healthcare and, more generally, in all domains involving information processing.

4.3 APPLICATIONS IN MEDICINE

The application of AI in medicine covers various fields, as shown in Figure 4.2.

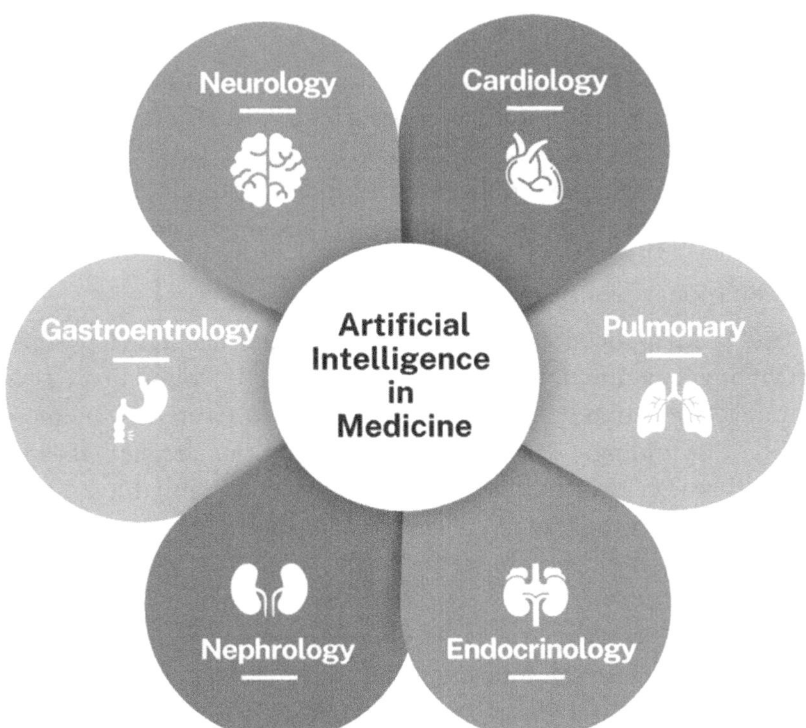

FIGURE 4.2 Artificial intelligence in medicine.

4.3.1 Cardiology

A. Atrial Fibrillation

The first time AI was used in medicine to find atrial fibrillation as soon as possible. In 2014, the Food and Drug Administration (FDA) gave AliveCor permission for its mobile app, Kardia. With this certification, the company could offer ECG tracking on smartphones and be able to spot atrial fibrillation. The results of the REHEARSE-AF [16] showed that remote ECG tracking through Kardia has a higher chance of finding atrial fibrillation in ambulatory patients than in standard care. The FDA has given Apple approval for its Apple Watch 4, which makes it easy to get ECG readings and diagnose atrial fibrillation. This information about the user's preference for a smartphone can be shared with a healthcare provider. Several criticisms have been acknowledged regarding wearable and portable ECG technologies. These critiques have shed light on the constraints associated with their utilization, including the issue of false positive results caused by movement artifacts, as well as the challenges in implementing wearable technology among elderly patients at a higher risk of experiencing atrial fibrillation.

B. Cardiovascular Risk

AI has been used in electronic patient data to make it easier to predict heart diseases like Acute Coronary Syndrome and Heart Failure, making the predictions more accurate than traditional scales. Recent literature studies have highlighted the influence of sample size on research outcomes.

4.3.2 Pulmonary Medicine

The field of AI applications in pulmonary medicine has shown promise in interpreting lung function testing. A recent study [17] found that AI-based software demonstrates enhanced precision in interpreting pulmonary function test findings, serving as a valuable tool for decision assistance. The study was criticized, with one critique which highlighted a notable disparity between the rate of accurate diagnosis among the participating pulmonologists and the national norm [18].

4.3.3 Endocrinology

Continuous Glucose Monitoring (CGM) technology lets people with diabetes get interstitial glucose levels immediately and all the time. This method gives important information about how quickly blood glucose levels change [19]. The U.S. FDA has approved the glucose monitoring device called Guardian, which Medtronic made. The goal of this method is to work with smartphones. In 2018, the organization collaborated with Watson, an AI system developed by IBM, to enhance its Sugar.IQ system. This collaboration's primary objective was to use repeated readings to assist individuals in lowering the frequency of bouts of low blood sugar. Patients may find it easier to manage their blood sugar levels, and hypoglycemia episodes are not as stigmatized in society with continuous blood glucose monitoring. Still, a study on how patients feel about glucose monitoring showed that, even though the participants said they trusted the system's alerts, they also felt like they weren't good enough at controlling their glucose levels [19].

4.3.4 Nephrology

AI has been employed in various contexts throughout the domain of clinical nephrology. For instance, this method has been utilized in prior research to forecast the decline in glomerular filtration rate among patients diagnosed with polycystic kidney disease. Additionally, it has effectively determined the likelihood of progression in IgA nephropathy. Nevertheless, limitations regarding the sample size required for making meaningful inferences.

4.3.5 Gastroenterology

The discipline of gastroenterology gains substantial benefits from the wide range of AI implementations in clinical settings. Gastroenterologists analyze images obtained from endoscopy and ultrasound by employing convolutional neural networks and other deep-learning models. Identifying and categorizing aberrant formations, such as colonic polyps, constituted the principal aim. The diagnosis of gastroesophageal reflux disease and atrophic gastritis, as well as the prognosis of gastrointestinal bleeding, survival rates in esophageal cancer, inflammatory bowel disease, and metastasis in colorectal cancer, and esophageal squamous cell carcinoma, all have been accomplished with the assistance of artificial neural networks.

4.3.6 Neurology

A. Epilepsy

Intelligent seizure detection devices exhibit considerable promise as technological advancements with the capacity to enhance seizure management via continuous ambulatory monitoring. In 2018, Empatica obtained approval from the FDA for their wearable device, Embrace. This device is equipped with electrodermal captors, which enable it to detect generalized epilepsy seizures. The Embrace device can also transmit this information to a mobile application. The patient's close family members and reliable doctors can then receive alerts from this program with further location-related details. In contrast to those who used heart-monitoring wearables, those with epilepsy found it easy to adopt seizure detection gadgets, according to a study focused on patient experiences. Moreover, these patients expressed significant interest in utilizing wearable devices.

B. Gait, Posture, and Tremor Assessment

Patients with MS, PD, tremors, and Huntington's disease can benefit from wearing sensors that objectively assess their gait, posture, and tremors.

4.3.7 Computational Diagnosis of Cancer in Histopathology

The FDA has granted Paige.ai breakthrough status for their AI-based algorithm, demonstrating remarkable accuracy in diagnosing cancer in computational histopathology. This advancement enables pathologists to allocate more time to analyzing critical slides.

4.3.8 Medical Imaging and Validation of AI-Based Technologies

The performance of radiologists and deep learning software in image-based diagnosis was compared in a meta-analysis that was eagerly anticipated [20]. Deep learning appears to have diagnostic efficacy on par with radiologists, according to the results. A trustworthy experimental design was required for 99.9% of the included studies, though, as pointed out by the authors. Furthermore, just 1% of the evaluated articles used algorithms to verify their findings in medical imaging from various demographics. To validate AI-based solutions, these results showed that extensive clinical trials are necessary.

4.4 APPLICATIONS IN MEDICAL IMAGING

4.4.1 Plain Film Radiography

The chest radiograph is often regarded as the most frequently conducted imaging examination on a global scale, with an estimated annual count of 2 billion [13]. Chest radiography's vast global availability and diagnostic versatility contribute to its significant appeal. Moreover, chest radiographs possess the highest abundance of labeled pictures, making them the most prevalent resource in the field of AI research. Due to these factors, chest radiography has attracted significant attention from researchers in the field of AI and remains a prominent topic of ongoing research.

An effective approach to commencing one's understanding of the subject is by examining the empirical basis for AI. Nevertheless, the ChestX-ray14 dataset has certain limitations, as extensively documented in many online sources. The dataset exhibits a pervasive presence of diagnostic uncertainty. Radiologists who are actively engaged in their practice would acknowledge the existence of ambiguity in numerous radiological diagnoses, as is apparent from the analysis of the ChestX-ray14 dataset. Authors in [21] employed text-mining techniques to extract the ground truth from radiology reports. The reports often included several potential diagnoses, possibly due to the inherent uncertainty in the radiological determination of the accurate diagnosis. There is a considerable degree of radiological overlap among several of the labels. For instance, pneumonia and atelectasis can exhibit comparable appearances. Moreover, the presence of conclusive data validating the accuracy of the radiological diagnosis remains uncertain.

In 2019, a notable advancement was made in the field of medical imaging with the introduction of Chester. Chester is a web-based disease prediction system that builds upon the existing ChestX-ray14 and CheXNet technologies. The primary objective of Chester is to provide the AI model known as CheXNet, which has undergone training using the ChestX-ray14 dataset, to a worldwide audience. Chester's underlying concept is around utilizing a web-based system to facilitate the broader dissemination of algorithms while simultaneously ensuring the preservation of patient confidentiality through local processing. The act of democratizing AI in this manner has the potential to contribute to the expansion of its accessibility inside nations that have limited resources. In a study [22], the ability of a deep-learning neural network to recognize tuberculosis-related chest radiographic features was evaluated. This study investigated AI applications in the analysis of diverse datasets. The authors proposed that resource-constrained developing nations could benefit from implementing such technology. Considering the increasing enthusiasm around the utilization of AI in radiology, the RSNA has taken the initiative to establish a dedicated platform known as Radiology: AI.

4.4.2 Advanced Imaging

AI exhibits numerous applications in the realm of enhanced imaging. One example of potential use is using AI models to augment Magnetic Resonance Imaging (MRI) techniques [23]. In this study, authors employed a deep learning algorithm to train on a dataset of 10 pre- and postcontrast brain MRIs. AI has also been employed as a comprehensive reader for screening examinations in controlled experimental settings. In [24], authors used data from the National Lung Cancer Screening Trial, which included 42,290 lung cancer screening CT scans available to the public. In the conducted study, the algorithm's performance is evaluated using a dataset. The results indicated that the algorithm attained an AUC value of 0.944. Notably, the algorithm demonstrated superior performance compared to six board-certified radiologists in cases where the prior imaging was not accessible.

A prevalent constraint in AI research is the insufficiency of suitably annotated data. Nevertheless, AI holds the potential to address this challenge. Authors in [25] conducted a study that employed deep learning techniques in brain MRI analysis. The researchers created an algorithm that helped to separate the brain by combining deep learning with a traditional probabilistic atlas. The deep learning algorithm did a better job than the baseline Gaussian likelihood functions. Notably, the algorithm was very good at separating deep brain regions, especially the hippocampus, with a Dice score of 81.1% compared to a baseline score of 73.1%. The significance of employing this technique lies in its potential to eliminate the laborious and resource-intensive task of annotating photographs, thereby enabling larger datasets for AI investigations.

4.4.3 Noninterpretive Tasks

The potential of AI may extend beyond visual interpretation, encompassing various other domains. AI has the potential to enhance workflow efficiency by providing support to radiologists in performing various activities. Protocoling the study is an initial step in the imaging approach. Several crucial factors need to be considered, including the clinical necessity of the research, the accuracy of the chosen acquisition parameters (such as the pulse sequence), and the potential need for contrast administration. Authors in [26] trained a deep learning network using a sample of musculoskeletal Magnetic Resonance (MR) requests. This network was made to tell apart different things, such as the language of the request, the use of contrast media, and the patient's personal information.

In addition, AI can enhance the process of acquiring scans. One primary limitation of Magnetic Resonance Imaging (MRI) as a modality is its extended duration for data collection. The extended duration of the acquisition process is a source of discomfort for patients and increases the likelihood of motion artifacts. In addition, extended acquisition durations harm the scanner's throughput. The technique created in [27] aimed to rebuild undersampled MRI data so that a high-quality picture could be made. In the future, AI has the potential to extract and utilize data from examination reports even after a significant period has elapsed. An ML method has been created to categorize unstructured radiological reports. Authors in [28] devised a model that utilizes the textual content of thoracic CT scan reports to accurately classify the occurrence, site, and duration of pulmonary emboli. Additionally, the research uncovers a potential alternate method for acquiring additional annotated data for future artificial intelligence investigations.

4.5 APPLICATIONS IN PATHOLOGY

In contrast to radiology, where the switch from lighted X-ray films to digital imaging happened quickly, pathology's shift to digital media has been a bit slower. This could affect how AI is used in this field. With the development of Whole-Slide Imaging (WSI), doctors can now look at histopathology slides with high-resolution images and the ability to change the depth. Despite WSI's widespread availability and advantages, converting glass slides into digital format is not commonly practiced. The rapid progression of technology has facilitated swift data transfers and provided extensive storage capacity for large quantities of information. This ability is especially helpful in the WSI area, where a lot of data needs to be processed quickly.

Radiology images that are shared using the Digital Imaging and Communications in Medicine (DICOM) protocol, on the other hand, have different numbers of pixels. For example, some nuclear medicine scans have about 4,000 pixels, while mammography images can have as many as 23 million pixels. The storage space needed for these photos depends on how many pictures are in each study. However, it has been shown in the online DICOM library that MRIs generally require approximately 3050 MB of storage capacity. Differences in file size and clarity between radiology and pathology studies could affect how AI will be used. The inquiry at hand pertains to the extent of AI proficiency in delivering crucial analytical capabilities within digital pathology. Despite the scarcity of published literature, a few studies have compared the performance of AI algorithms and pathologists in identifying and categorizing various cancer types.

Nevertheless, it is important to acknowledge that the pathologists in this study faced time limitations, with less than one minute per slide (although some flexibility was allowed). This aspect may not accurately represent the actual conditions of clinical practice. Furthermore, the pathologists employed glass slides instead of digital WSI, perhaps providing additional information but in an unfamiliar medium. However, this research emphasizes the capacity of AI to rapidly and accurately analyze tissue samples to identify instances of breast cancer metastases, matching the proficiency of an experienced pathologist indefinitely. In addition, the study [29] investigated how well deep learning algorithms could classify and find changes in histopathological images of non-small cell lung cancer.

4.6 APPLICATIONS IN OPHTHALMOLOGY

Diabetic retinopathy is a prevalent disorder encountered in everyday ophthalmology practice, representing a substantial and escalating public health concern. Diabetes-related Retinopathy (DR) is a common disease that causes working-age people to lose sight. In 2015, DR was thought to affect about 2.6 million people worldwide. By 2020, that number is expected to rise to 3.2 million [30]. The prevalence of diabetic retinopathy, a condition that poses a risk to vision, is decreasing in high-income nations due to improved diabetes management and advancements in ophthalmological treatments. However, this positive trend is counterbalanced globally by the rising prevalence of diabetes and the increasing occurrence of diabetic retinopathy in countries with limited resources. Since early detection leads to better outcomes and digital imaging techniques like high-resolution color retinal photography and optical coherence tomography are becoming more common, as well as the possibility of irreversible morbidity, DR is a good candidate for screening.

Nevertheless, the presence of stringent protocols and a scarcity of screening personnel have resulted in a need to streamline the process of referring patients to a specialized ophthalmologist for assessment. When doing prospective studies on AI systems in screening programmers, it is often necessary to include many people with no symptoms to find a small number of people with symptoms that can be detected. The fact that doctors can't agree on how to diagnose DR shows how hard it is to do so. It is imperative to establish processes to address these conflicts to develop a reliable reference standard, which can then be used to train the algorithms effectively. The establishment of ground truth is mostly achieved through a consensus reached by a group of independent experts or through resolving conflicts by an additional expert whose opinion serves as the reference for the DR grade. Authors in [31] tried to change the current way of doing things by improving a Google deep learning program to tell the difference between referable and non-referable DR.

However, it is important to acknowledge that this approach may unintentionally introduce complexities that could undermine the otherwise correct decision-making processes. AI algorithms have been successfully devised to analyze Optical Coherence Tomography (OCT) images, an additional diagnostic modality employed in the field of ophthalmology. A particular framework [32] employed a two-step procedure for identifying pathology. At first, a deep segmentation network is used to find differences concerning how the picture was taken. The segmentation map that is made is then looked at with a deep classification network to find any abnormal conditions. The referral suggestion is based on the Moorfields Eye Hospital's (London, UK) referral criteria, which put the most important diagnoses into four groups: immediate care, semi-urgent care, routine follow-up, or observation. Compared to a group of eight clinical experts, which included four retinal subspecialist ophthalmologists and four optometrists with additional medical retina training, the system correctly identified urgent referral cases with an AUC value of 0.9921. This performance was found to be on par with that of two ophthalmologists, surpassing the performance of all other experts in the group. The scientists attribute the accuracy of the results to the use of numerous neural networks, specifically five, for both segmentation and classification tasks. This approach can be likened to a panel of experts contributing to the overall precision of the system. Authors discovered that employing numerous neural networks yielded higher results than utilizing a single network. Remarkably, instances of uncertainty in which the networks exhibit divergent viewpoints and put out various conjectures can be effectively demonstrated through the medium of a movie. After that, this knowledge can be used to help the ophthalmologist make clinical decisions.

In ophthalmology, AI has been used to find more clinically important information and for screening and diagnosing. OCT-A is becoming increasingly known as a non-invasive alternative to fluorescence angiography (FA) for seeing the retinal blood vessels. OCT-A acts as a motion-contrast monitor that can tell the difference between moving blood and still nerve tissue. The extensive utilization of this modality is constrained by its high cost, the need for patient cooperation, its sensitivity to motion, and its narrow fields of view. When comparing the segmentation of vasculature from OCT by three physicians to the ground truth provided by OCT-A, the AI model demonstrated superior performance in terms of accuracy. After evaluating an OCT, clinicians usually proceed to OCT-A or FA to confirm a microvascular issue. Because of this, it's debatable if comparing AI to OCT-A or FA is fair. For these final diagnostic tests, the model may be useful. It could alleviate the problems with injecting contrast dye and with OCT-A's technical limitations if suggested as a replacement for the current diagnostic procedures. The ability of AI to accurately extract useful functional information from structural data is demonstrated in this work. The primary question of whether these new technologies will replace current investigative methods or improve diabetic retinal screening, diagnosis, and referral for other eye-threatening diseases is still unsolved. On the other hand, these innovations can potentially revolutionize ophthalmology as a whole.

4.7 APPLICATIONS IN DERMATOLOGY

Identifying visual patterns is a crucial diagnostic skill in dermatology, and AI can greatly enhance picture analysis and the accuracy of diagnoses in this domain. Recently, computational neural networks have emerged as a valuable tool for diagnosing skin disorders by leveraging visual image identification. Notably, these networks have exhibited equivalent, at times superior, sensitivity and specificity in classifying images compared to dermatologists with extensive clinical experience. Authors in [33] used a method called GoogleNet Inception, which was trained on 1.28 million pictures. After that, the algorithm was retrained using high-quality pictures of skin from a university dataset. The uploaded digital pictures were then looked at very closely, with each pixel being looked at individually to make a diagnosis. Compared to a group of 21 US board-certified doctors, the deep learning system did better, with an area under the curve (AUC) of 0.96 for carcinoma and 0.94 for melanoma.

In [34], the diagnostic accuracy of a convolutional neural network is compared with a group of 58 dermatologists, comprising 30 experts from various nations. Their research showed that the convolutional neural network outperformed the majority of dermatologists. Authors in [35] showed that convolutional neural networks outperformed when identifying melanoma dermatoscopic photos. Clinical images of a wide range of skin disorders have been successfully classified using convolutional neural networks, and this is in addition to skin cancers. In [36], authors used an algorithm to assess 12 skin conditions, including actinic keratoses, seborrheic keratoses, melanocytic nevi, pyogenic granulomas, hemangiomas, warts, and common skin cancers. AI can potentially contribute to advancing early detection and treatment methods for skin malignancies. Authors in [37] looked into the viability of providing laypeople with an early risk assessment tool for melanoma using smartphone technology. While making up a minor portion of all skin cancers, melanoma is the primary cause of skin cancer-related deaths. The assessed AI-based applications showed specificities ranging from 37.94% and sensitivities from 7.73% to specificities. Therefore, AI has not exhibited adequate diagnostic precision in smartphone technology. Due to the considerable probability of melanoma detection errors, the utilization of self-screening as a method among the general population may not now be deemed appropriate. Nevertheless, due to the limited number of studies and the quick advancements in this domain, conducting additional research may lead to more favorable outcomes in integrating smartphone technology and artificial intelligence for melanoma detection.

Significant challenges are associated with the implementation of AI in the field of dermatology. Despite the potential for AI to improve workflow efficacy, integrating AI into conventional clinical workflow systems has revealed obstacles. These challenges are expected to hinder the practical application of AI in enhancing diagnostics in dermatology, at least in the foreseeable future. The application of AI in dermatology poses significant challenges within the image analysis domain. In contrast to radiological imaging, the field of skin imaging exhibits a notable absence of standardization in various aspects, such as color representation, lighting conditions, imaging methodologies, and hardware configurations. The individual's skin tone can significantly influence the appearance of an image. In addition, it is necessary to have extensive collections of annotated, high-resolution imaging data that cover a wide range of diagnostic scenarios to establish a reliable reference point for the automatic advancement of algorithms in deep neural networks. Compiling big data sets can be characterized by a significant investment of time and financial resources.

Nevertheless, like other pragmatic situations, AI can offer a resolution to decrease the duration and expense of establishing the ground truth. In [38], authors demonstrated that their ML model, known as multi-instance multilabel, effectively performed algorithmic annotation on skin biopsy images. The efficacy of this model was validated using biopsy images from a clinical data set. Hence, computational neural networks may find the accurate and efficient annotation of skin imaging to be a dependable point of reference when classifying novel clinical data. Dermatologists have raised concerns regarding the implementation of AI in the field. Authors in [39] acknowledged the potential advantages that AI could bring in improving the accuracy and efficacy of diagnosis. Nevertheless, they advise against the careless application of this technology in place of taking into account other critical diagnostic resources, including patient history and clinical context.

4.8 CHALLENGES

Many challenges are likely to arise throughout the implementation of AI in healthcare. For the sake of clarity, we shall focus on three interrelated features of the research including overfitting, regulatory approval, and the black box problem. When discussing deep learning algorithms, the term "black box problem" describes their inability to explain their reasoning behind decisions. The traditional method has made it difficult to determine which imaging features were used, how they were assessed, and why the algorithm chose one result over another when interpreting a radiological discovery.

The majority of AI systems exhibit a similar problem. There have been multiple attempts made to deal with this issue, authors in [22] used a heat map to show regions of increased activation inside the deep learning network in their study on deep learning in tuberculosis. These sites showed higher activation levels, so it makes sense to assume they are very relevant for diagnosis. These techniques aid in the clarification of AI's inner workings. When AI systems are trained on one dataset and applied to another, they may show limited generalizability, a phenomenon known as overfitting. This behavior can be explained by the fact that the algorithm has learned only the statistical fluctuations in the training data instead of the general principles needed to solve problems. An algorithm's overuse of a certain dataset for training is the main cause of overfitting. Several factors, including the size of the dataset, the degree of heterogeneity within the dataset, and the distribution of the data within the dataset, impact the likelihood of overfitting. When there is a significant difference in the frequency and incidence of disease between the training and testing datasets, for instance, a model may show signs of overfitting. Overfitting can also occur if the training and testing datasets are generated with different parameters or equipment, and a limited sample size might exacerbate this problem. After training, algorithms can be assessed on several datasets to determine the presence of overfitting. In the case of an overfitting algorithm, it is expected that its accuracy, as evaluated by the AUC, will be much lower on datasets received from different sources than the training data. Obtaining regulatory permission will be a big barrier to developing AI systems.

The FDA will regulate medical AI like it regulates medications and medical equipment. The black box problem and overfitting are impediments to regulatory approval because assessors have difficulty understanding the inner workings of algorithms and determining the generalizability of their performance across different datasets. The FDA categorizes emerging AI tools based on three important factors: the potential risk to patient safety, the presence of an existing algorithm as a reference point, and the extent of human participation in their operation. High-risk algorithms, such as diagnostic tools with potentially serious repercussions for misdiagnosis and low human involvement, are evaluated through the premarket approval procedure. This approach requires considerable data from non-clinical and clinical trials to show the new tool's safety and efficacy. The De Novo technique is meant for reviewing and certifying novel technologies with low risks of safety and effectiveness. To delve deeper into the FDA regulation and approval of emerging AI technologies in medical imaging [40].

4.9 CONCLUSION

The significance of technology in contemporary society is paramount. Technology has a significant role in facilitating and enhancing human labor to increase efficiency. Furthermore, technology plays a crucial role in the healthcare industry by mitigating errors resulting from human incompetence. Without technological involvement in surgical procedures conducted by medical professionals, the potential risks and complications associated with the operation may increase, perhaps resulting in unfavorable outcomes. AI can be defined as a computational system designed to emulate human intellect, wherein robots are trained to exhibit cognitive processes and problem-solving abilities akin to those of humans. AI can enhance patient diagnoses, prevention, therapy, and clinical decision-making.

The utilization of AI within the realm of healthcare services presents both potential obstacles and benefits. The utilization of technology in the healthcare sector offers evident advantages, including enhanced decision-making and patient outcomes, alongside potential ancillary benefits such as decreased referrals, cost reduction, and time savings. Additionally, it has the potential to provide support to healthcare facilities located in remote rural regions and contribute to the recruitment and retention of healthcare professionals in these locations. In conclusion, this has the potential to enhance the fairness and inclusivity of the worldwide healthcare system. "The user's problems include assisting the early acceptance and sustainable deployment within the health system, a lack of respect for the user's perspective, and suboptimal utilization of technology," one possible method to reword the user's wording to be more academic. However, the adoption of AI in the public health sector remains critical. Implementing AI in the public health system provides a framework for investigating many aspects of AI adoption in the public sector.

The effectiveness of AI in improving healthcare administration has been established. AI will most likely be implemented into normal healthcare practice shortly. Nonetheless, concerns have been raised about the ethical and regulatory implications of applying AI in the healthcare sector. Concerns raised in this context include the possibility of bias, the lack of openness surrounding certain AI algorithms, privacy concerns with the data used to train AI models, security issues, and the division of duties when using AI in clinical settings. AI healthcare applications face several ethical issues. The major factors in this context include safety, efficacy, privacy, information and consent, decision-making authority, "the right to try," associated costs, and accessibility.

REFERENCES

1. F. Kiyani, K. N. Qureshi, K. Z. Ghafoor, and G. Jeon, "ISDA-BAN: interoperability and security based data authentication scheme for body area network," *Cluster Computing,* 2022. https://doi.org/10.1007/s10586-022-03823-9.
2. Ó. Díaz, J. A. Dalton, and J. Giraldo, "Artificial intelligence: a novel approach for drug discovery," *Trends in Pharmacological Sciences,* vol. 40, no. 8, pp. 550–551, 2019.
3. H. S. Chan, H. Shan, T. Dahoun, H. Vogel, and S. Yuan, "Advancing drug discovery via artificial intelligence," *Trends in Pharmacological Sciences,* vol. 40, no. 8, pp. 592–604, 2019.
4. S. Naseem, A. Alhudhaif, M. Anwar, K. N. Qureshi, and G. Jeon, "Artificial general intelligence-based rational behavior detection using cognitive correlates for tracking online harms," *Personal and Ubiquitous Computing,* 2022. https://doi.org/10.1007/s00779-022-01665-1.
5. I. Mayorga-Ruiz, A. Jiménez-Pastor, B. Fos-Guarinos, R. López-González, F. García-Castro, and Á. Alberich-Bayarri, "The role of AI in clinical trials," *Artificial Intelligence in Medical Imaging: Opportunities, Applications and Risks,* pp. 231–243, 2019.
6. M. A. Qureshi, K. N. Qureshi, G. Jeon, and F. Piccialli, "Deep learning-based ambient assisted living for self-management of cardiovascular conditions," *Neural Computing and Applications,* 2021. https://doi.org/10.1007/s00521-020-05678-w.

7. S. Lip, S. Visweswaran, and S. Padmanabhan, "Transforming Clinical Trials with Artificial Intelligence," in *Artificial Intelligence*: Productivity Press, 2020, pp. 297–306.
8. D. Shi, W. Zhang, W. Zhang, and X. Ding, "A review on lower limb rehabilitation exoskeleton robots," *Chinese Journal of Mechanical Engineering*, vol. 32, no. 1, pp. 1–11, 2019.
9. K. N. Qureshi, G. Jeon, and F. Piccialli, "Anomaly detection and trust authority in artificial intelligence and cloud computing," *Computer Networks*, p. 107647, 2020. https://doi.org/10.1016/j.comnet.2020.107647.
10. P. Hummel and M. Braun, "Just data? Solidarity and justice in data-driven medicine," *Life Sciences, Society and Policy*, vol. 16, no. 1, pp. 1–18, 2020.
11. A. Prabu, "SmartScope: An AI-powered digital auscultation device to detect cardiopulmonary diseases," *TechRxiv. Preprint*, vol. 14921268, p. v1, 2021, https://doi. org/10.36227/techrxiv.
12. N. Butt *et al.*, "Intelligent deep learning for anomaly-based intrusion detection in IoT smart home networks," *Mathematics*, vol. 10, no. 23, p. 4598, 2022. https://doi.org/10.3390/math10234598.
13. E. J. Topol, "High-performance medicine: the convergence of human and artificial intelligence," *Nature Medicine*, vol. 25, no. 1, pp. 44–56, 2019.
14. T. Sullivan, "Next up for EHRs: vendors adding artificial intelligence into the workflow. *Healthcare IT News*," ed, 2018.
15. R. Atun *et al.*, "Universal health coverage in Turkey: enhancement of equity," *The Lancet*, vol. 382, no. 9886, pp. 65–99, 2013.
16. J. P. Halcox *et al.*, "Assessment of remote heart rhythm sampling using the AliveCor heart monitor to screen for atrial fibrillation: the REHEARSE-AF study," *Circulation*, vol. 136, no. 19, pp. 1784–1794, 2017.
17. M. Topalovic *et al.*, "Artificial intelligence outperforms pulmonologists in the interpretation of pulmonary function tests," *European Respiratory Journal*, vol. 53, no. 4, 2019.
18. C. Delclaux, "*No need for pulmonologists to interpret pulmonary function tests*," vol. 54, ed: Eur Respiratory Soc, 2019.
19. J. Lawton *et al.*, "Patients' and caregivers' experiences of using continuous glucose monitoring to support diabetes self-management: qualitative study," *BMC Endocrine Disorders*, vol. 18, no. 1, pp. 1–10, 2018.
20. X. Liu *et al.*, "A comparison of deep learning performance against healthcare professionals in detecting diseases from medical imaging: a systematic review and meta-analysis," *The Lancet Digital Health*, vol. 1, no. 6, pp. e271–e297, 2019.
21. X. Wang, Y. Peng, L. Lu, Z. Lu, M. Bagheri, and R. M. Summers, "Chestx-ray8: hospital-scale chest x-ray database and benchmarks on weakly-supervised classification and localization of common thorax diseases," in *Proceedings of the IEEE Conference on Computer Vision and Pattern Recognition*, 2017, pp. 2097–2106.
22. P. Lakhani and B. Sundaram, "Deep learning at chest radiography: automated classification of pulmonary tuberculosis by using convolutional neural networks," *Radiology*, vol. 284, no. 2, pp. 574–582, 2017.

23. E. Gong, J. M. Pauly, M. Wintermark, and G. Zaharchuk, "Deep learning enables reduced gadolinium dose for contrast-enhanced brain MRI," *Journal of Magnetic Resonance Imaging,* vol. 48, no. 2, pp. 330–340, 2018.
24. D. Ardila *et al.*, "End-to-end lung cancer screening with three-dimensional deep learning on low-dose chest computed tomography," *Nature Medicine,* vol. 25, no. 6, pp. 954–961, 2019.
25. A. V. Dalca, E. Yu, P. Golland, B. Fischl, M. R. Sabuncu, and J. Eugenio Iglesias, "Unsupervised deep learning for Bayesian brain MRI segmentation," in *Medical Image Computing and Computer Assisted Intervention–MICCAI 2019: 22nd International Conference, Shenzhen, China, October 13–17, 2019, Proceedings, Part III 22,* 2019: Springer, pp. 356–365.
26. Y. H. Lee, "Efficiency improvement in a busy radiology practice: determination of musculoskeletal magnetic resonance imaging protocol using deep-learning convolutional neural networks," *Journal of Digital Imaging,* vol. 31, pp. 604–610, 2018.
27. C. M. Hyun, H. P. Kim, S. M. Lee, S. Lee, and J. K. Seo, "Deep learning for undersampled MRI reconstruction," *Physics in Medicine & Biology,* vol. 63, no. 13, p. 135007, 2018.
28. M. C. Chen *et al.*, "Deep learning to classify radiology free-text reports," *Radiology,* vol. 286, no. 3, pp. 845–852, 2018.
29. N. Coudray *et al.*, "Classification and mutation prediction from non–small cell lung cancer histopathology images using deep learning," *Nature Medicine,* vol. 24, no. 10, pp. 1559–1567, 2018.
30. S. R. Flaxman *et al.*, "Global causes of blindness and distance vision impairment 1990–2020: a systematic review and meta-analysis," *The Lancet Global Health,* vol. 5, no. 12, pp. e1221–e1234, 2017.
31. J. Krause *et al.*, "Grader variability and the importance of reference standards for evaluating machine learning models for diabetic retinopathy," *Ophthalmology,* vol. 125, no. 8, pp. 1264–1272, 2018.
32. J. De Fauw *et al.*, "Clinically applicable deep learning for diagnosis and referral in retinal disease," *Nature Medicine,* vol. 24, no. 9, pp. 1342–1350, 2018.
33. A. Esteva *et al.*, "Dermatologist-level classification of skin cancer with deep neural networks," *Nature,* vol. 542, no. 7639, pp. 115–118, 2017.
34. H. A. Haenssle *et al.*, "Man against machine: diagnostic performance of a deep learning convolutional neural network for dermoscopic melanoma recognition in comparison to 58 dermatologists," *Annals of Oncology,* vol. 29, no. 8, pp. 1836–1842, 2018.
35. T. J. Brinker *et al.*, "Deep learning outperformed 136 of 157 dermatologists in a head-to-head dermoscopic melanoma image classification task," *European Journal of Cancer,* vol. 113, pp. 47–54, 2019.
36. S. S. Han, M. S. Kim, W. Lim, G. H. Park, I. Park, and S. E. Chang, "Classification of the clinical images for benign and malignant cutaneous tumors using a deep learning algorithm," *Journal of Investigative Dermatology,* vol. 138, no. 7, pp. 1529–1538, 2018.

37. N. Chuchu *et al.*, "Smartphone applications for triaging adults with skin lesions that are suspicious for melanoma," *Cochrane Database of Systematic Reviews*, vol. 2018, no. 12, 1996.
38. G. Zhang *et al.*, "Augmenting multi-instance multilabel learning with sparse bayesian models for skin biopsy image analysis," *BioMed Research International*, vol. 2014, 2014.
39. B. Lim and G. Flaherty, "Artificial intelligence in dermatology: are we there yet?," *British Journal of Dermatology*, vol. 181, no. 1, pp. 190–191, 2019.
40. A. Kohli, V. Mahajan, K. Seals, A. Kohli, and S. Jha, "Concepts in US food and drug administration regulation of artificial intelligence for medical imaging," *American Journal of Roentgenology*, vol. 213, no. 4, pp. 886–888, 2019.

CHAPTER 5

AI Models in Environmental Social Science Research

Leveraging Technology for Sustainable Solutions

Cuong Pham-Quoc, Tran Vu Pham, Nguyen Cao Tri, Vu Hien Phan, and Kashif Naseer Qureshi

5.1 INTRODUCTION

In recent decades, the planet has faced an unprecedented array of environmental challenges, ranging from climate change and biodiversity loss to resource depletion and pollution. As the consequences of these issues become increasingly evident, urgent action is needed to safeguard ecosystems and secure a sustainable future for generations to come. In this critical quest for environmental preservation, technologies have emerged as powerful allies. Among these, the convergence of Artificial Intelligence (AI) and the Internet of Things (IoT) has revolutionized the landscape of environmental management, offering innovative solutions to tackle complex ecological problems with unprecedented efficiency and accuracy [1].

DOI: 10.1201/9781032667911-5

AI/IoT-powered environmental management systems represent a paradigm shift in how we approach and address environmental challenges. By harnessing the vast potential of AI algorithms and IoT-enabled devices, these smart systems can monitor, analyze, and optimize various aspects of the environment in real time, providing invaluable insights and predictive capabilities to guide informed decision-making. The integration of AI and IoT technologies presents a multifaceted approach to environmental management [2, 3]. Through a network of interconnected sensors and devices, IoT enables the collection of vast amounts of environmental data across diverse ecosystems. Simultaneously, AI algorithms leverage this data to recognize patterns, detect anomalies, and generate actionable knowledge, empowering stakeholders with the information required to design targeted interventions and adaptive strategies.

One of the primary strengths of AI/IoT-powered environmental management lies in its ability to facilitate proactive responses to environmental challenges. Unlike conventional systems that often rely on retrospective analyses, AI/IoT technologies enable real-time monitoring, early warning systems, and predictive modeling. These capabilities allow us to mitigate the impact of natural disasters, identify potential threats to wildlife, and optimize resource utilization in an ever-changing environment.

In this chapter, we delve into the vast potential of AI/IoT-powered environmental management systems, exploring their key components, functionalities, and applications across different domains. We shed light on how these transformative technologies can revolutionize energy management, waste reduction, air and water quality monitoring, wildlife conservation, and ecosystem preservation. Moreover, the successful implementation of AI/IoT-driven environmental management systems requires addressing critical challenges related to data privacy, security, and ethical considerations. As we explore the immense possibilities of these technologies, it is imperative to strike a balance between innovation and responsible use, ensuring equitable access to data and technology for all stakeholders [4].

This chapter aims to contribute to the growing body of knowledge surrounding AI/IoT-powered environmental management systems and serve as a guide for researchers, policymakers, environmentalists, and industry leaders alike. By understanding the potential and limitations of these cutting-edge solutions, we can collectively advance our commitment to sustainable development and cultivate a symbiotic relationship with our planet. As we navigate the path toward a greener and more resilient future, AI/IoT-powered environmental management stands as a beacon of hope, demonstrating the transformative potential of technology in the pursuit of environmental harmony.

5.2 RELATED WORK AND MATERIALS

In both academic literature and the industry, numerous air quality monitoring systems have been developed. These systems primarily utilize two different approaches: one relies on sensors to collect parameters of environmental management, while the other utilizes satellite images to estimate the parameters. In our work, we propose a novel system that combines both of these techniques, allowing them to complement each other effectively.

Regarding the former method, which involves using sensors for data collection, several systems have been developed worldwide. For instance, one of the earliest systems, GEMS/AIR – Global Environment Monitoring System, was established in Sweden to monitor and manage air pollution and climate change [5]. In North America, Kenneth L. Demerjian reported the presence of a consortium including the USA, Canada, and Mexico to operate over 4,000 monitoring stations for air quality measuring. Similar systems have also been implemented in Asian countries for collecting AQI-related parameters (Air Quality Index), as mentioned by studies in [6, 7].

The latter method utilizes images captured by satellites to estimate various values including LST (Land Surface Temperature) and Particulate Matter (PM10 and PM2.5) levels. Various studies reported and explored this method [8–13]. However, accurately estimating LST and PMx depends on the specific characteristics of the research areas, such as buildings, grasslands, forests, or water. Therefore, this research focuses on Ho Chi Minh City (HCMC) (representing the city center area) and Dong Thap province with the Tram Chim National Park (representing the rural area). In this book chapter, we introduce the comprehensive system, which seamlessly integrates the two aforementioned approaches.

5.3 STUDY AREAS

The proposed AIoT systems are used for environmental management systems and have been deployed in two areas: Ho Chi Minh City (city center) and Tram Chim National Park located in Dong Thap province (rural area). In this section, we briefly introduce the two deployment sites.

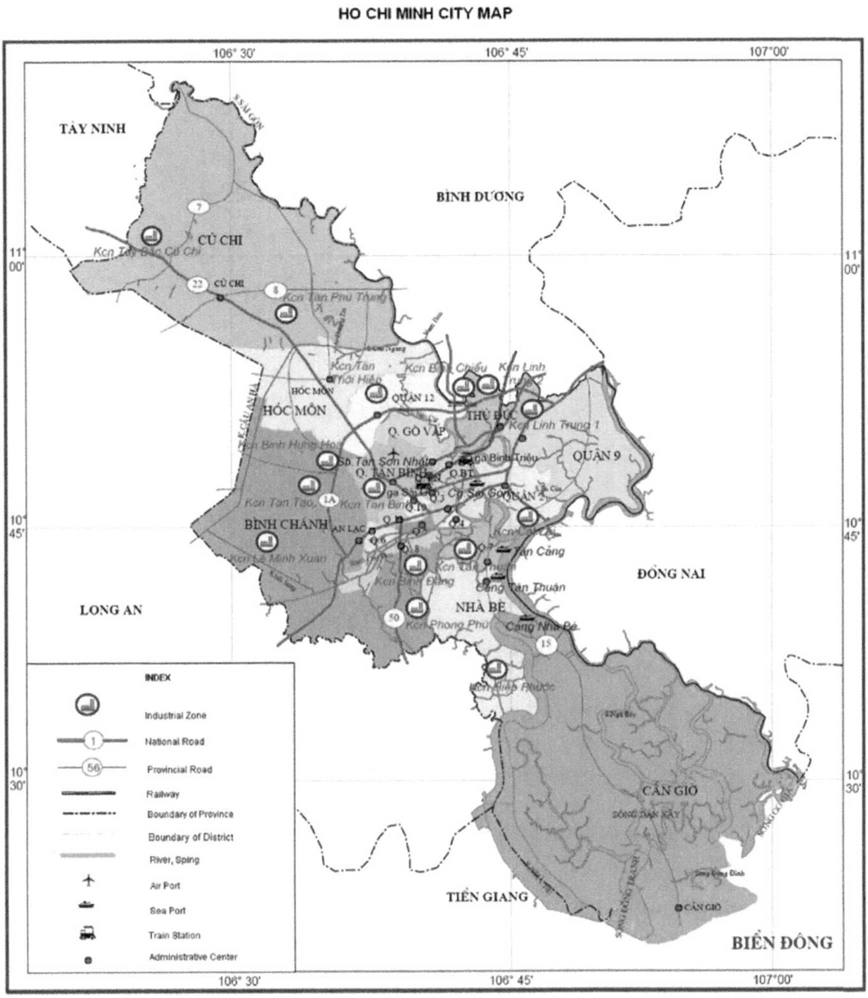

FIGURE 5.1 The map of Ho Chi Minh city.

This study focuses on investigating Ho Chi Minh City (HCMC), which is depicted in Figure 5.1 and located at coordinates 10°00'-10°380'N latitude and 106°220'-106°540'E longitude. HCMC is one of the most densely populated urban areas in Vietnam, covering approximately 2095.01 km^2 with a population of nearly 8.6 million people (estimated in 2020). The city is divided into 19 urban districts and 5 rural districts, bordered by Binh Duong, Tay Ninh, Dong Nai, Ba Ria-Vung Tau, Tien Giang, and Long An province as shown in Figure 5.1 [14].

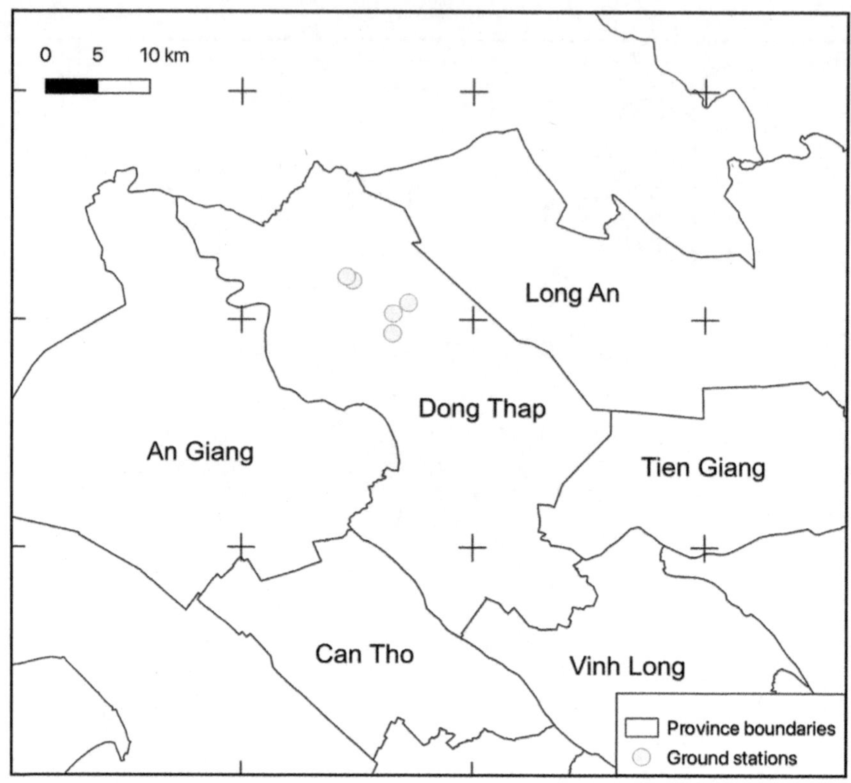

FIGURE 5.2 The map of Dong Thap Province and ground station positions.

HCMC experiences a tropical wet and dry climate characterized by two main seasons. The rainy southwest monsoon lasts from May through October, bringing approximately 159 rainy days per year. The dry season occurs from November to April, with an average of 2,490 hours of sunshine per year. The city maintains a relatively constant humidity of 75% and an average temperature of 28°C (82°F) throughout the year.

Figure 5.2 illustrates the positions of IoT ground stations in the Tram Chim National Park, Dong Thap Province. For this work, the Tram Chim National Park, situated in Dong Thap Province, Vietnam, is one of the study areas because this area is a representation of essential characteristics of the Mekong Delta region which is affected seriously by climate change.

The national park is a significant wetland and one of the last remaining areas of the threatened Plain of Reed's ecosystem. This park boasts remarkable biodiversity and holds both ecological and tourism value, housing more than 230 bird and 130 fish species, including the endangered Sarus Crane, included in the IUCN Red List. Recognized as a Ramsar site by UNESCO, Tram Chim's international significance is emphasized, reflecting its priority to society. Although it has ecological importance, Tram Chim faces challenges in environmental management and research due to limited and inconsistent survey data. These issues arise from difficulties in accessing faraway areas and the change of water level in the region. The lack of comprehensive data acquisition poses obstacles to preserving and protecting the ecosystem and wildlife of the park. To address these challenges, our intelligent environmental management system employs innovative data collection and processing techniques.

5.4 MATERIALS

In this research, besides telemetry data measured remotely by IoT ground stations, we also utilize satellite images for extracting values of areas where the stations cannot be deployed. Images from Landsat and MODIS satellites are collected and processed by AI computational models. In this section, we briefly introduce the images used for this work. Landsat-8, a satellite co-operated by the National Aeronautics and Space Administration (NASA) and the United States Geological Survey (USGS), represents one of the most recent additions to their satellite fleet. The primary objective of this satellite is to capture multispectral imagery at a medium spatial resolution, making it accessible to the public. This wide distribution of data brings numerous benefits to various sectors, including agriculture, science, and government applications, among others.

Remote sensing data of the Landsat-8 satellite is acquired through two sensors. One of them is the Operational Land Imager (OLI), offering 9 bands with a 30-meter spatial resolution. The second sensor is the Thermal InfraRed Sensor (TIRS), providing 2 bands with a 100-meter spatial resolution. As part of its public accessibility, Landsat-8 datasets can be downloaded free of charge from the USGS Earth Explorer [15]. Similar to Landsat, MODIS (Moderate Resolution Imaging Spectroradiometer) also offers daily land surface temperature and emissivity (LST&E) data with its MOD11A1 Version 6 product. However, it is important to note that the spatial resolution of MODIS images is lower than Landsat's, with a resolution of 1 km in a 1,200 × 1,200 grid.

To process MODIS images effectively, the MODIS Conversion Tool Kit can be employed. This toolkit enables the conversion of MODIS data from the HDF (Hierarchical Data Format) data format to the Geotiff image format, while also transforming the Sinusoidal datum into the WGS84 coordinate system. Fortunately, all the necessary data images and tools can be downloaded free of charge [16].

5.5 SYSTEM OVERVIEW

This section briefly depicts the proposed system for monitoring air quality in large cities, which utilizes multiple data sources. Figure 5.3 visually represents the proposed system, which incorporates data from four different sources as follows.

1. **Censoring systems:** These systems utilize sensors to collect precise values for various parameters at specific deployment positions. The collected data serves multiple purposes, including providing information to authorities and residents, as well as contributing to the development of AI-based models for satellite images.
2. **A cloud-based service for satellite image processing:** This component automatically gathers and applies the AI model to images from satellites like Landsat or MODIS. The results of this service are values for PM2.5 and LST at various locations within the city and the national park. This aspect is particularly beneficial in estimating harmful parameters at locations lacking physical sensors. Detailed information about the estimation models is available in our prior work [1].

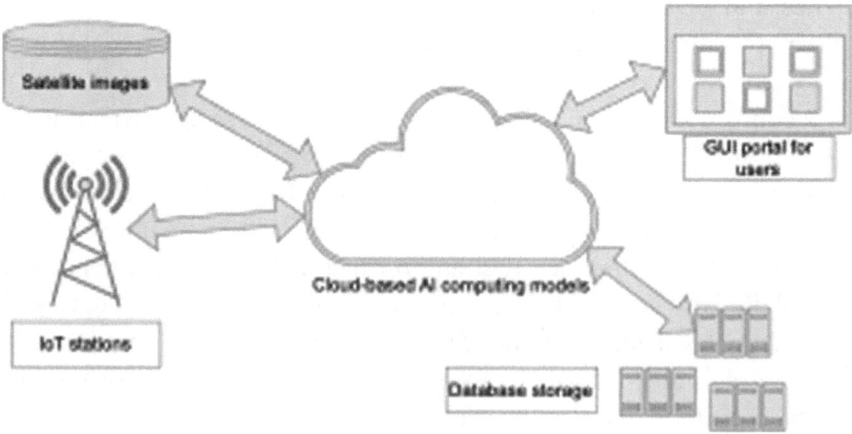

FIGURE 5.3 Overview of our AI/IoT-based system.

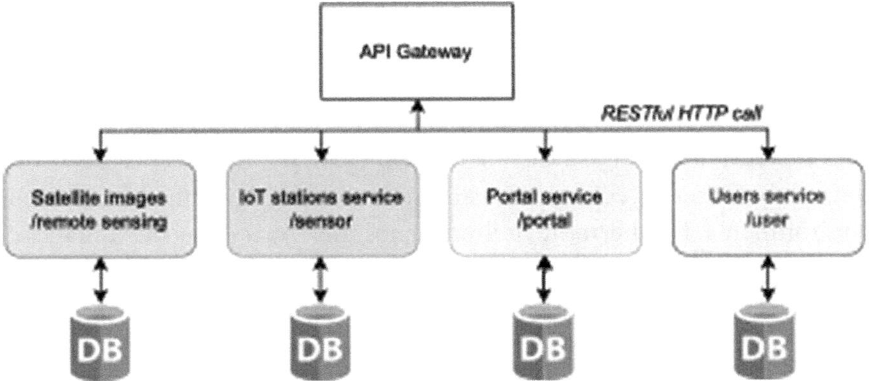

FIGURE 5.4 The hierarchy of APIs provided by our system.

3. **GUI and portal:** The system incorporates a user-friendly portal allowing users to access data and manually monitor potential hazardous situations. It is important to note that user feedback needs to undergo validation and approval by administrators or authorities before being made public.

4. **Database systems:** These systems are responsible for storing both raw and preprocessed data obtained from other sources. This data is preserved for future applications, such as AI-based prediction and forecasting.

Additionally, a cloud-based service is also responsible for processing data from different sources. However, due to the varied data structures of these sources in terms of size, types, and timestamps, the integration of multiple data sources requires the development of application programming interfaces (API). Figure 5.4 illustrates the architecture of the APIs designed in the system. These APIs follow the HTTP and REST standards, facilitating easy connectivity to the system for other developers who wish to develop related applications.

5.6 SYSTEM IMPLEMENTATION

In this section, we will provide a comprehensive overview and detailed explanation of the implementation of the proposed advanced system, which is built upon innovative architecture. The primary emphasis of the discussion will be centered around two crucial components: the cutting-edge IoT systems and the highly efficient cloud-based AI models for satellite image processing. By delving into the intricate workings and integration of these components, the aim is to offer a profound understanding of the seamless synergy that drives our system's exceptional performance and capabilities. As the study unravels the technical intricacies of the sensory systems, it highlights their state-of-the-art features, such as real-time data acquisition, precision measurements, and adaptability to various environments. Furthermore, it illuminates the unparalleled advantages of powerful cloud-based AI satellite image processing service, which enables seamless data storage, rapid processing, and secure accessibility for users. Together, these elements form the backbone of the proposed system's operational prowess, fostering innovative solutions and transforming the way to interact with technology.

5.7 IOT STATIONS FOR GROUND MEASUREMENTS

In this research endeavor, a highly efficient and robust IoT system is developed, ingeniously combining sensor nodes and gateway nodes. The sensor nodes play a pivotal role in directly acquiring a diverse range of critical monitored values, including but not limited to levels of carbon monoxide (CO), carbon dioxide (CO_2), temperature, PM2.5 particles, and humidity. Utilizing cutting-edge radio frequency (RF) technology, specifically Long-Range RF (LoRa), these sensor nodes seamlessly transmit the gathered data to their corresponding gateway node. The gateway node serves as a central hub, collecting and aggregating data from multiple sensor nodes, which is subsequently transmitted to cloud computing services via the internet, utilizing either 4G or WiFi connectivity. This ingenious architecture significantly reduces the system's reliance on Internet services compared to conventional monitoring systems in the vibrant city of Ho Chi Minh City (HCMC).

The integration of LoRa technology in the proposed system further amplifies its capabilities, allowing for an impressive communication range of several kilometers between sensor nodes and their respective gateway nodes. This expansive reach opens up new possibilities for deploying the system across larger geographical areas with enhanced efficiency. To ensure the utmost security and confidentiality during data transmission between the gateway node and sensor nodes, it employs AES128 bit encoding, a formidable encryption technique that safeguards sensitive information against unauthorized access or tampering.

A pivotal aspect of this work involves the design and meticulous implementation of sensor nodes, which are equipped with a sophisticated

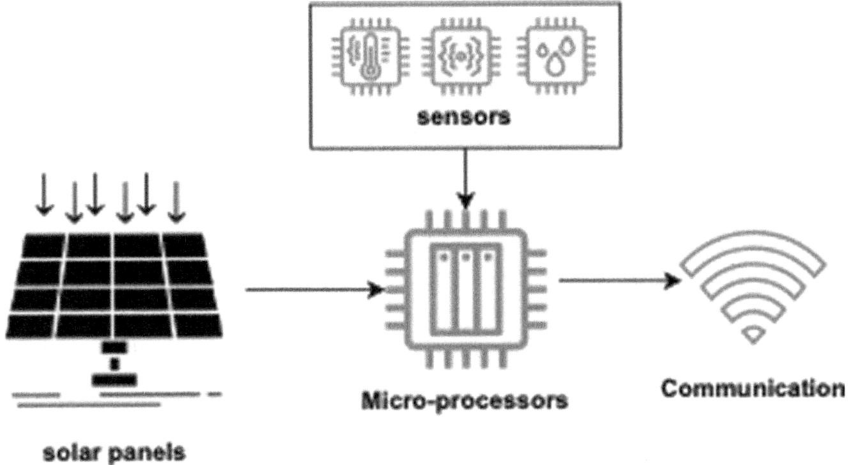

FIGURE 5.5 The architecture of used sensor nodes.

ensemble of components, including a micro-controller, a LoRa module, various sensors (CO, CO_2, temperature, PM2.5, and humidity), and a reliable power supply. Figure 5.5 presents an illustrative depiction of the architecture underpinning the sensor nodes, showcasing the seamless integration of these crucial elements. Through the amalgamation of advanced technologies, purposeful engineering, and secure communication protocols, the IoT system emerges as a trailblazing solution that promises to revolutionize environmental monitoring in HCMC and beyond. Its ability to collect and analyze diverse environmental data with minimal dependency on internet services, coupled with the remarkable reach of LoRa technology, solidifies its position as a cutting-edge solution with potential applications in various domains, such as smart cities, environmental conservation, and urban planning.

In addition to the proficiently crafted sensor nodes, equal strides are taken for developing gateway nodes, featuring an ingeniously designed mainboard tailored to meet unique requirements. The gateway nodes fulfill a pivotal role, acting as central hubs responsible for efficiently collecting data from the interconnected sensor nodes in their vicinity. This aggregated data is seamlessly relayed to cloud computing services through the internet, ensuring comprehensive and real-time monitoring capabilities. With an unwavering focus on data security and privacy, robust encryption techniques were adopted to fortify the communication channels within the system. Specifically, AES encoding plays a crucial role in safeguarding the data exchanged between sensor nodes and gateway nodes. Here, we utilize AES128-bit encryption, ensuring a high level of protection for the sensitive data transmitted between these critical components.

Moreover, recognizing the paramount importance of securing data transmission from the gateway nodes to the cloud services, we implement even stronger measures. For this purpose, we opt for the formidable AES256-bit encryption, which significantly bolsters the security of data communication between the gateway nodes and the cloud services. This additional layer of encryption assures that all data sent to the cloud remains invulnerable to unauthorized access, thereby instilling confidence in the integrity of the entire system. The diligent efforts in meticulously crafting both the sensor nodes and gateway nodes, coupled with the implementation of advanced AES encryption, culminate in an IoT system that not only excels in data acquisition and aggregation but also prioritizes the confidentiality and protection of the gathered information. This system not only showcases unparalleled capabilities in diverse application scenarios, ranging from environmental monitoring to industrial automation but also sets new standards for data-driven innovation in the burgeoning IoT landscape.

5.8 AI-BASED SATELLITE IMAGE PROCESSING

In this research endeavor, we have extended the scope of data acquisition beyond traditional sensor-based methods by incorporating the utilization of remote sensing images, specifically Landsat and MODIS images [17]. These images serve as valuable resources for estimating Air Quality Index (AQI) values, focusing on crucial parameters such as temperature and PM2.5 levels, particularly in regions where physical sensors are not present. By leveraging the powerful capabilities of remote sensing technology, the aim is to enhance the spatial coverage and accuracy of AQI estimation across a broader geographical area. To ensure precision and reliability in estimation models, a robust calibration process is established that involves leveraging the data collected by physical sensors. The information gleaned from these sensors is instrumental in fine-tuning and aligning the values extracted from the remote sensing images, allowing for more accurate and consistent AQI estimations. This integration of sensor data and remote sensing images serves as a vital cornerstone of this approach, enabling us to mitigate potential discrepancies and achieve a comprehensive understanding of air quality dynamics across diverse landscapes.

In previous work [1], we have elaborated on the intricate processing steps and mathematical models employed to estimate land surface temperature and PM2.5 levels from the remote sensing images. By presenting these methodologies, we lay the foundation for a transparent and replicable approach that fosters scientific rigor and credibility in the domain of air quality estimation using remote sensing technology. The commitment is to sharing these methodologies which allows fellow researchers and stakeholders to build the work and contribute to the collective advancement of air quality assessment techniques. The fusion of sensor data and remote sensing images exemplifies the dedication to exploring innovative avenues for environmental monitoring and highlights the transformative potential of interdisciplinary research. By bridging the gap between ground-based observations and satellite-based remote sensing, a comprehensive and multidimensional understanding of air quality patterns is used, which paves the way for evidence-based decision-making and targeted interventions to safeguard public health and the environment.

5.9 CLOUD SERVICES

The cloud services serve as the pivotal hub that seamlessly connects and synchronizes all subsystems within the system, making it the veritable heart of operations. In this subsection, an insightful depiction of the three-layer architecture comprising the cloud services illustrates the cohesive framework that drives our air quality monitoring system.

1. **Data Layer:** At the foundation of cloud services lies the data layer, encompassing two vital systems, the censoring system and the remote sensing images system. Through the censoring system, real-time air quality monitoring data is actively collected from sensor nodes, capturing essential parameters such as CO, CO_2, PM2.5, temperature, and humidity. Simultaneously, the remote sensing images system establishes a connection to the remote sensing server, efficiently retrieving crucial values of PM2.5 and land surface temperature, extracted from remote sensing images. This fusion of data sources ensures comprehensive and diverse data acquisition, enhancing the accuracy and coverage of air quality estimation.
2. **Cloud Layer:** Positioned as the central management entity, the cloud layer serves as the nucleus of cloud services. This layer undertakes the responsibility of processing the data collected from both sensors and

remote-sensing images. The processed information is then meticulously stored within databases for future reference and analysis. Moreover, the cloud layer efficiently disseminates the processed data to authorized users and governing authorities, empowering them with the insights required to make informed decisions and interventions. Furthermore, the cloud layer takes charge of managing users' access and privileges while actively monitoring the air quality values. In critical situations, the layer triggers timely alerts, ensuring proactive responses to potential air quality issues.

3. **Application Layer:** The apex of cloud services houses the application layer, replete with user-friendly graphical interfaces that offer seamless accessibility to the information generated by our system. These interfaces serve as the gateway for users to access and comprehend the analyzed air quality data. Additionally, the application layer equips administrators and governing bodies with powerful tools, allowing them to oversee and fine-tune the system through our intuitive configuration page. This facilitates customization options, enabling the specification of alert thresholds and contact details such as phone numbers. Furthermore, we extend the functionality of this layer by providing a dedicated portal that empowers users to submit warning information they observe at their respective locations. Such submissions undergo examination by administrators or governing authorities before being disseminated for public awareness.

Architecting the cloud services in this multi-layered approach ensures seamless integration of data from diverse sources, real-time analysis, and efficient user accessibility. The proposed system exemplifies an innovative fusion of cutting-edge technology and user-centric design, forging a robust foundation for addressing air quality challenges and fostering a healthier and more sustainable environment for all.

5.10 CONCLUSION

In this chapter, we introduce a pioneering and integrated system aimed at advancing environmental management through the seamless convergence of IoT stations and an AI-based model for satellite image processing. The escalating concerns regarding environmental degradation necessitate innovative approaches to effectively monitor and mitigate environmental factors. The proposed system harnesses the potential of IoT stations, equipped with an array of sensors, to gather real-time data on a diverse range of environmental parameters, including air quality, temperature, humidity, and more. Concurrently, it taps into the power of satellite imagery, incorporating data from reliable sources like Landsat and MODIS, to significantly enhance the spatial coverage and precision of our environmental monitoring efforts. By combining the capabilities of IoT stations and satellite images, we achieve a comprehensive and dynamic understanding of environmental conditions, empowering timely and well-informed decision-making for sustainable environmental management. Stakeholders, including environmental authorities, gain access to a centralized cloud-based platform that efficiently processes and analyzes the collected data. This platform serves as an invaluable resource, providing valuable insights and facilitating real-time alerts and notifications to address potential environmental hazards promptly.

Our research highlights the three-layered architecture of this cloud-based system, offering in-depth insights into the functionalities of each layer. The data layer adeptly synchronizes and consolidates information from both IoT stations and satellite images. Subsequently, the cloud layer takes charge of data processing, storage, and dissemination, while the application layer encompasses user-friendly interfaces catering to data access, system management, and public engagement. The integration of IoT and satellite-based technologies epitomizes a cutting-edge approach to environmental management, fostering sustainability and resilience in the face of the ever-growing environmental challenges. This seamless amalgamation opens new frontiers for precise and comprehensive environmental monitoring, heralding a new era of heightened environmental awareness and proactive action.

ACKNOWLEDGMENT

We acknowledge Ho Chi Minh City University of Technology (HCMUT), VNU-HCM, for supporting this study.

REFERENCES

1. V. H. Phan, D. P. H. Pham, T. V. Pham, K. N. Qureshi, and C. Pham-Quoc, "An IoT system and MODIS images enable smart environmental management for mekong delta," *Future Internet*, vol. 15, no. 7, p. 245, 2023.
2. Z. Ali *et al.*, "Edge based priority-aware dynamic resource allocation for internet of things networks," *Entropy*, vol. 24, no. 11, p. 1607, 2022.
3. A. Hasan and K. Qureshi, "Internet of things device authentication scheme using hardware serialization," in *2018 International Conference on Applied and Engineering Mathematics (ICAEM)*, 2018: IEEE, pp. 109–114.

4. Y. A. A. S. Aldeen and K. N. Qureshi, "New trends in internet of things, applications, challenges, and solutions," *Telkomnika,* vol. 16, no. 3, pp. 1114–1119, 2018.
5. M. D. Gwynne, "The global environment monitoring system (GEMS) of UNEP," *Environmental Conservation,* vol. 9, no. 1, pp. 35–41, 1982.
6. H. H. Duong, C. D. T. Nguyen, P. L. S. Nguyen, and H. T. To, "Fine particulate matter (PM2. 5) in Ho Chi Minh City: Analysis of the status and the temporal variation based on the continuous data from 2013–2017," *Science & Technology Development Journal: Natural Sciences,* vol. 2, no. 5, pp. 130–137, 2018.
7. Suzhou. "Air Quality Monitoring System." http://www.sinoitaenvironment.org/ReadNewsex1.asp?NewsID=1896.
8. J. Wang and S. A. Christopher, "Intercomparison between satellite-derived aerosol optical thickness and PM2. 5 mass: Implications for air quality studies," *Geophysical Research Letters,* vol. 30, no. 21, 2003. https://doi.org/10.1029/2003gl018174.
9. D. A. Chu *et al.*, "Global monitoring of air pollution over land from the earth observing system-terra moderate resolution imaging spectroradiometer (MODIS)," *Journal of Geophysical Research: Atmospheres,* vol. 108, no. D21, 2003. https://doi.org/10.1029/2002jd003179.
10. I. Kloog, B. Ridgway, P. Koutrakis, B. A. Coull, and J. D. Schwartz, "Long- and short-term exposure to PM2. 5 and mortality: using novel exposure models," *Epidemiology,* vol. 24, no. 4, pp. 555–561, 2013.
11. H. Lee, Y. Liu, B. Coull, J. Schwartz, and P. Koutrakis, "A novel calibration approach of MODIS AOD data to predict PM 2.5 concentrations," *Atmospheric Chemistry and Physics,* vol. 11, no. 15, pp. 7991–8002, 2011.
12. Y. Liu, M. Franklin, R. Kahn, and P. Koutrakis, "Using aerosol optical thickness to predict ground-level PM2. 5 concentrations in the St. Louis area: A comparison between MISR and MODIS," *Remote Sensing of Environment,* vol. 107, no. 1–2, pp. 33–44, 2007.
13. Y. Liu, P. Koutrakis, and R. Kahn, "Estimating fine particulate matter component concentrations and size distributions using satellite-retrieved fractional aerosol optical depth: Part 1—Method development," *Journal of the Air & Waste Management Association,* vol. 57, no. 11, pp. 1351–1359, 2007.
14. "Ho chi minh city." http://www.hochiminhcity.gov.vn/.
15. Z. Yang, C. Witharana, J. Hurd, K. Wang, R. Hao, and S. Tong, "Using landsat 8 data to compare percent impervious surface area and normalized difference vegetation index as indicators of urban heat island effects in Connecticut, USA," *Environmental Earth Sciences,* vol. 79, pp. 1–13, 2020.
16. "NASA official." https://earthdata.nasa.gov.
17. F. Gao *et al.*, "Fusing Landsat and MODIS data for vegetation monitoring," *IEEE Geoscience and Remote Sensing Magazine,* vol. 3, no. 3, pp. 47–60, 2015.

CHAPTER 6

Artificial Intelligence in Natural Science Research

Innovations and Limitations

Adil Hussain, Peng-Cheng Xu, Wang Shixin, and Kashif Naseer Qureshi

6.1 DATA ANALYSIS

Analytics 4.0 is the subsequent phase in advancing analytical capabilities within businesses, marking the advent of Artificial Intelligence (AI) and cognitive technology. The adoption of this approach has gained significant traction, encompassing not just the utilization of AI techniques but also a heightened emphasis on autonomy in method execution, namely in the realm of automated Machine Learning (ML).

6.1.1 Products and Services Development

AI can potentially enhance various products and services across several industries. The current embedded analytics strategies and methodology businesses use to pave the way to incorporate AI into a wide range of products and services. Industrial products may be improved

DOI: 10.1201/9781032667911-6

by using sensor data and ML analysis on "digital twins" of devices. There is potential for incorporating connectivity and smart features into various consumer devices, including autos and toothbrushes. The utilization of AI significantly enhances the process of incorporating analytics and expands the range of possibilities. In contemporary business practices, companies have the potential to integrate predictive or prescriptive analytics into their current products, such as a recommendation system, and subsequently progress toward the development of more advanced models over time. The use of AI enhances this endeavor by offering heightened levels of automation and complexity in the construction and upkeep of models. The acceleration of model generation and testing facilitates the exploration of novel markets, the creation of new products, the improvement of existing services and products, and the expeditious advancement of product development. Open-source AI services provided by major tech firms like Google, Amazon, Microsoft, and others make it possible for smaller organizations to affordably acquire AI capabilities that would be prohibitively expensive or impractical to create in-house. The cost of developing image recognition and Natural Language Processing (NLP) systems may be a significant barrier for certain industrial applications. However, integrating these capabilities into current products and services can be easily accomplished using recently developed AI services.

6.1.2 Internal Business Process Optimization

The most important effect of analytics is probably optimizing internal business processes. Many new applications would be impossible to develop without AI. Using ML techniques, such as incorporating them into internal business processes, can potentially enhance the effectiveness and efficiency of decision-making. Consequently, this can increase adaptability for organizations operating within dynamic and rapidly changing market environments. Firms can develop intricate propensity models, which enable salesforces to make informed decisions regarding the appropriate items to offer to specific customers. AI can offer ongoing, dynamic analysis that exhibits enhancement and refinement as time progresses. AI can provide significant advantages to commercial processes characterized by high-volume transactions and data transfer speeds. This encompasses both the supply chain and manufacturing aspects. As an illustration, Amazon employs AI significantly within its distribution centers to facilitate fulfillment operations and efficiently load its delivery vehicles [1, 2].

AI can restructure conventional business processes extensively. Instead of relying on traditional methods such as credit bureau reports and job information, this approach utilizes AI to assess numerous data points derived from social media and mobile activity. By leveraging this technology, credit applications can be approved within a matter of minutes, eliminating the need for a time-consuming approval process. Although the deployment of AI components in this application has proven successful, it is crucial to acknowledge and address the ethical and societal concerns associated with its use.

6.1.3 External Business Process Optimization

Companies born in the digital age, such as Google, Netflix, and Amazon, have revolutionized entire industries by integrating analytics and AI into their external business processes. Digital operations, sales strategies, marketing initiatives, and connections with customers are all part of this category. These innovations are advancing quickly and can be used in traditional workplaces as well. AI can also be utilized to offer comprehensive and dynamic analyses of client behavior. Applying this approach within the social media ecosystem allows for evaluating marketing initiatives in terms of their success and impact. When supplemented with proprietary customer data such as website, mobile, and healthcare CRM data, this data can estimate sales and develop novel marketing and customer success initiatives. These strategies aim to enhance customer satisfaction and broaden the client base by discovering innovative customer engagement approaches. Moreover, it has the potential to deliver precise digital advertisements and promotional offers to consumers by using their previous online activities. There has been a notable surge in the availability of commercially accessible "point" solutions that enable organizations to implement AI capabilities in novel manners. An illustration of this may be seen in the growing capabilities of Chatbot and intelligent assistants such as Google Home, Alexa, and similar technologies, which can now facilitate immediate and tailored client engagements. These proposed strategies can potentially yield cost savings for corporations while simultaneously improving brand image and fostering stronger consumer relationships. As the quality of these interfaces advances, they are expected to transition from being mostly utilized for customer support and retention purposes to being more employed for client acquisition endeavors.

6.1.4 Accelerating Analytics Capability

Numerous firms are currently encountering a talent deficit due to a scarcity of proficient analysts and data scientists. AI represents a viable solution for addressing the skill gap through the automation of processes, resulting in enhanced productivity and improved outcomes. AI has the potential to be integrated into conventional analytics data flow processes, including data purification and data input. This integration allows proficient analysts and data scientists to prioritize more advanced and strategic initiatives. In contemporary times, the size of data sets has escalated, so firms may find it imperative to employ AI and automation techniques to pre-process the data for subsequent analytical procedures adequately. Using ML in the form of "probabilistic matching" can potentially consolidate and combine divergent and isolated data sources into a cohesive collection of information.

According to authors in [3], the use of automated ML systems can significantly enhance the efficiency and effectiveness of data scientists and analytics professionals at the various stages of model building and deployment. In the past, the significance of data science experience and skill has been considerable in assessing suitable algorithms for addressing a given problem. The presence of erroneous selections or algorithmic bias can result in incorrect models and expensive corrective measures. Automated ML (AutoML) systems can compare with many learning algorithms and select the most appropriate one. After the models have been trained, AutoML can help other systems by creating APIs or writing code. That implies analysts won't have to know as much about the nuts and bolts of modeling to conclude. Consequently, seasoned data scientists can allocate their focus toward tackling the most challenges.

Prominent and niche cloud providers have recently consolidated numerous analytics and AI technologies into service offerings. Thanks to productized technologies like picture and text analysis, businesses can use new capabilities without learning the ins and outs of cutting-edge fields like AI. In addition, several companies that sell business software are beginning to include AI capabilities in their products. AI algorithms don't need to be integrated because they can readily get the required information themselves. For instance, the Einstein feature of Salesforce.com's customer relationship management features makes extensive use of ML approaches [4]. Due to AI's excitement, many businesses are creating AI efforts to take on projects that are above their capabilities. An aggressive approach can contribute to rapid growth in AI competencies, but any

AI strategy must consider the organization's current capabilities. Some organizations may not be prepared for AI or its more advanced variants. Firms can begin by evaluating their technology and data infrastructures to determine the disparity between their current analytical capabilities for successful AI implementation.

The utilization of data analytics in the healthcare sector is of utmost importance as it enables the conversion of vast quantities of data into meaningful insights, hence facilitating the enhancement of patient outcomes, operational efficiency, resource optimization, and the ability to predict disease outbreaks. Additionally, it facilitates improved patient health outcomes while enhancing operational efficiency within the industry. The utilization of data analytics plays a crucial role in the transformation of the healthcare sector, as it enables the conversion of raw and unprocessed health-related data into actionable and effective solutions [5]. This groundbreaking technology has substantially contributed to several healthcare domains, including clinical research, the advancement of novel therapies and pharmaceuticals, illness prognosis and prevention, clinical decision-making assistance, and enhanced diagnostic precision. Moreover, it provides automation for hospital administrative processes, optimizing operations and enhancing the efficacy of surgeries and medicine administration, ultimately resulting in direct patient benefits. Moreover, the software's capacity to accurately compute health insurance premiums guarantees precision in coverage estimations, becoming an essential instrument in the contemporary period.

6.2 IMAGE AND PATTERN RECOGNITION

Computer Vision (CV) is an academic discipline that uses computer-based simulations of human visual perception. Its primary goal is to analyze and comprehend images (or video) to make informed decisions or determinations. The field of CV has encountered significant growth and development in recent decades. Computer systems are used in the technological image recognition process, which entails processing, analyzing, and interpreting photographs to recognize and categorize various targets and objects. This is crucial for the advancement of computer vision. Data capture and processing via intelligent images are crucial and have far-reaching consequences. Image recognition technology has proven to be a proficient solution for detecting and identifying certain target items, such as faces, handwritten characters, and goods. Additionally, it can perform image categorization and subjective image quality rating, among other tasks.

Currently, image recognition technology exhibits significant commercial potential and promising prospects in several Internet applications, including but not limited to picture search, commodity suggestion, user behavior analysis, and face recognition. Simultaneously, advanced technologies include intelligent robotics, autonomous driving, and unmanned aerial vehicles [6]. Various fields, such as industry, biology, medicine, and geology, exhibit extensive potential for use in diverse contexts. The Scale-Invariant Feature Transforms (SIFT) [7] and the Histogram of Oriented Gradients (HOG) [8] are used as feature extraction techniques in the earliest iterations of picture recognition systems. These extracted features are subsequently utilized as input for the classifier, facilitating the process of classification and recognition. These traits are inherent to the design process that involves manual intervention. The extracted features directly influence the performance of a system in solving identification problems. Therefore, researchers must thoroughly investigate the problem regions to design adaptability to more effective features, ultimately enhancing the system's performance. The current phase of image recognition systems is primarily designed for specialized identification tasks, with limited data size and low generalization capacity. Consequently, achieving reliable identification results poses challenges in practical applications.

The authors in [9] initially contributed in the field of DL and primarily highlighted the utilization of Artificial Neural Networks (ANNs) for feature learning. The training model extracted the characteristics that are exhibited from the original data. Additionally, using an unsupervised learning algorithm enables the implementation of a "layer initialization" technique to establish a hierarchical expression of information within the input data. This approach effectively reduces the depth of the neural network and the level of difficulty associated with training. There has been a corresponding rise in the areas of speech recognition, picture identification, and natural language processing. As a result, the accuracy rate has improved from 20% to 30%. Merely one year later, the Convolutional Neural Networks (CNNs)-based DL model has exhibited significant advancements in the realm of large-scale image classification problems, thereby igniting a widespread surge of interest in DL. Several prominent Internet technology companies, like Google, Microsoft, and Facebook, compete intensely as they allocate substantial resources to research and develop extensive DL systems.

The CNNs have emerged as a prominent area of research within the domain of image recognition. This represents the initial implementation of a learning model that effectively trains multilayer neural networks, particularly when the network's input is multidimensional. The Conch neural network [10] has been utilized in various extensive ML applications, including speech recognition, image recognition, and NLP. This has been particularly significant as the recent surge in ML research has delved into these areas extensively. DL has emerged as a prominent area of research, with CNNs being widely employed. CNNs leverage fundamental components such as convolution, pooling, and fully connected layers to enable the network to learn and extract pertinent characteristics for subsequent utilization. This characteristic offers several advantages for various research endeavors, obviating the necessity for an excessively intricate modeling procedure.

Furthermore, DL has made significant advancements and achievements in other domains, such as image classification, object identification, attitude estimation, and image segmentation. On the one hand, the breadth of applicability in the field of learning is extensive, showcasing its versatility and potential for further expansion into several other domains. However, it is important to note that several opportunities for further learning warrant further investigation and exploration. In forthcoming times, even though numerous preceding deliberations are subject to supervision (evidenced by the final layer of the trained network computing a loss value grounded on the actual value and subsequently modifying the parameters), the supervised investigation has indeed attained considerable triumph. The utilization of DL within the context of unsupervised learning is anticipated to emerge as a prominent trajectory in the future. In the context of both humans and animals, it is often insufficient to rely solely on identifying an entity through its name to comprehend its nature. In the emerging domain of computer vision, it is anticipated that utilizing RNNs based on DL would gain significant traction as a network model. This will likely lead to notable advancements in several applied research areas, facilitating further growth in the field.

Furthermore, there is a growing possibility of employing robust chemical techniques to facilitate the training of an end-to-end learning system. This would enable the learning system to possess autonomous learning capabilities, proactively acquiring the relevant properties of representation and abstraction. The field of study involving deep and intensive learning integration is in its early stages. However, certain studies within this domain have demonstrated promising results in tasks such as multi-object identification and video game learning. Many researchers in various relevant fields are enthusiastic about one particular rationale. It is worth mentioning that natural language processing has the potential to advance to a stage where it can demonstrate its capabilities in tasks such as article writing or processing large texts. This can be achieved by employing neural network models, such as RNNs, and appropriate methods and strategies to comprehend textual content effectively. In contemporary times, individuals commonly employ profound learning techniques and rudimentary logical thinking to attain notable advancements in the domains of voice and image processing. There is a rationale to suggest that by enhancing the current network extraction feature to enable more flexible expression of characteristics and incorporating complex reasoning, the depth of learning in the application of AI can be advanced further.

6.2.1 Pattern Recognition

Within the realm of human intelligence, the capacity to identify and discern patterns emerges as a vital cognitive aptitude within the brain. In the context of machine intelligence, pattern recognition holds significant importance for both ML and AI. The successful and precise identification of patterns is crucial in addressing numerous complex, intelligent problems at a higher level. The levels of accuracy in various issue domains have exhibited substantial and quick improvements over time. For example, when utilizing the CNNs [11], on the MNIST data set, consisting of handwritten digits classified into ten classes, it is feasible to attain an accuracy beyond 99% without needing conventional manually engineered features [12]. Recent technological advancements have already significantly surpassed human performance. Accuracy improvements and new world records are common occurrences across various pattern recognition tasks, including face identification, speech recognition, and handwriting recognition. The accuracy problem of pattern recognition has been solved.

Nevertheless, the evaluation of performance encompasses more than just accuracy. When deploying a pattern recognition system with high accuracy in practical settings, it is common to encounter unsatisfactory and unforeseen outcomes, leading to a lack of robustness in the system. The underlying causes of these issues typically derive from a combination of many elements. Furthermore, introducing a carefully crafted minor alteration to the input sample can result in a significant perturbation, leading to an erroneous prediction in the output of a pattern recognition system. This creates a substantial threat from adversaries when using such a system in practical applications, necessitating strong safety measures. In addition, within the context of traditional pattern recognition, it is commonly presumed that the set of classes is closed. However, the issue of open sets with ever-changing class sets frequently occurs in the real world. Many previously solved problems become tough again when placed in the context of open-set challenges [13].

The presence of a distribution mismatch also leads to a notable decrease in performance for pattern recognition. It has been demonstrated that even a minor distribution shift can substantially impact the performance of high-precision pattern recognition systems. In practical situations, the importance of adaptability and transferability in a pattern recognition system is increased, in addition to the importance of accuracy. Many pattern recognition systems typically possess a singular input and output. Nevertheless, enhancing the system's resilience can be achieved by augmenting the input and output variety. Hence, incorporating multimodal learning and multitask learning constitutes significant considerations in enhancing the robustness of pattern recognition. Patterns are infrequently observed in isolation but tend to manifest alongside extensive contextual information. The enhancement of decision-making resilience through the analysis of inter-pattern dependencies is a significant issue within the field of pattern recognition.

The efficacy of pattern recognition systems heavily depends on the data quantity and quality for the training, as these systems tend to have a significant appetite for data. Any alterations to the quantity or categorization of the data significantly impact the ultimate performance. However, collecting extensive databases and generating precise manual labeling can pose significant challenges in practical scenarios. Hence, the few-shot or even non-shot learning capabilities of pattern recognition algorithms hold significant importance in practical contexts. The robustness of the pattern recognition system and its capacity to learn from noisy data are two ways to lessen the system's dependency on data quality. Learning from large amounts of unlabeled data and easily accessible surrogate supervisory signals is a major benefit of unsupervised, self-supervised, and semi-supervised. In addition to assuring accuracy, it is essential to emphasize strengthening pattern recognition's robustness.

There are two primary classifications for pattern recognition methods: two-stage and end-to-end. Most current approaches combine feature representation with pattern classification in two separate steps. The goal of feature representation is to reduce the original data to a smaller, more manageable space of features. The initial step in the data processing

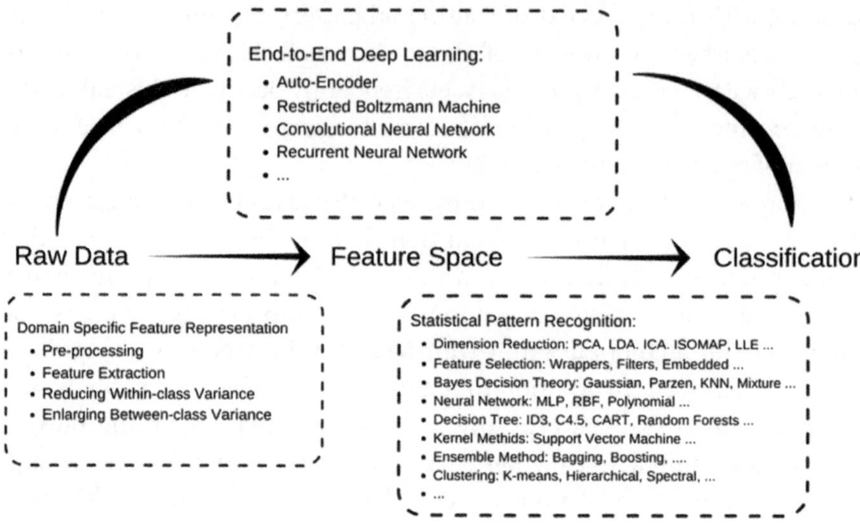

FIGURE 6.1 Brief overview of pattern recognition methods.

pipeline involves pre-processing techniques like noise removal and data normalization. This is done to decrease the variability within each class. Following this, feature extraction methods are employed to enhance the variability between different classes. It is important to note that this procedure is typically tailored to the specific domain being studied. The initial step in addressing novel pattern recognition challenges is creating a feature representation design. A well-designed feature may reduce the burden associated with classifier learning in the future. Various applications, including retinal recognition, gait recognition, and action recognition, exhibit these types of endeavors. Figure 6.1 shows the overview of pattern recognition methods.

Following the process of feature representation, the next step entails pattern classification, which presents a more extensive and complex challenge. This stage is commonly referred to as statistical pattern recognition, and it entails an exhaustive examination of multiple factors from diverse perspectives. To begin with, dimensionality reduction is a commonly employed technique to obtain a reduced-dimensional representation that may effectively support subsequent classification tasks. An alternative method for feature selection might be conceptualized as a discrete form of dimensionality reduction. Subsequently, a multitude of classical classification models can be employed. One of the most influential theories is Bayes' decision theory, which uses prior probability and class-conditional density estimation to get maximum posterior probability classification. Among the most popular ANNs for this purpose are the Multilayer Perceptron (MLP) network, the Radial Basis Function (RBF) network, and the polynomial network. Decision Tree (DT)-based methods employ a tree-like hierarchical structure to show the recognition rule better. Linear operations can be performed on higher-dimensional or infinite-dimensional spaces using a kernel mapping function. This has been frequently used to improve linear models. Support Vector Machine (SVM) is a popular method [14]. An ensemble method, which takes the predictions of several different models and mixes them, can boost performance. Clustering is a popular unsupervised strategy in pattern recognition.

In the context of two-stage approaches, it is common to encounter many options for both feature representation and classifier learning. Predicting the optimal combination for achieving the highest performance is a challenging task. In practice, it is seen that different pattern recognition challenges often require distinct configurations based on domain-specific knowledge and experiences. On the contrary, DL algorithms are characterized by their ability to simultaneously learn both the feature representation and classification tasks directly from the raw data. In this manner, the acquired features and classifiers exhibit more cooperation in addressing the assigned goal through a data-centric approach. Compared to two-stage methodologies, this approach is characterized by greater flexibility and discriminative capability.

Traditionally, DNNs have been commonly pre-trained layer-wise using unsupervised models such as auto-encoders and limited Boltzmann machines [9]. In contemporary times, there has been a notable advancement in the training of neural networks, wherein the depth of these networks has increased significantly. This progress may be attributed to the implementation of various enhanced techniques, including improved initialization, activation, optimization, normalization, and architecture. Image classification, detection, segmentation, and other visual recognition tasks have all made extensive use of CNN because of its efficient use of shared-weights architecture and local connection features [11]. The RNN's flexibility in dealing with sequences of varying lengths is one reason for its meteoric rise in sequence-based pattern recognition applications like speech recognition and scene text recognition. In addition, incorporating the attention mechanism has been shown to enhance the performance of DL models by directing their focus toward the most pertinent information. DL has emerged as a state-of-the-art approach for various pattern identification applications.

In addition, the vast class of statistical pattern recognition techniques has developed structural pattern recognition to exploit and comprehend the abundant structural information in patterns. In contrast to the fixed dimensionality of statistical feature representation, pattern structure lends itself to being understood in a non-Euclidean space. String matching and graph matching are two of the primary obstacles in structural pattern recognition. The learning capacity of structural pattern recognition issues has been improved by using various kernel methods, including but not limited to graph kernel, probabilistic graphical models, and graph neural networks. Structural pattern recognition research and practice are less common than statistical approaches.

6.3 NATURAL LANGUAGE PROCESSING

NLP is a challenging area of AI that attempts to solve the problem of automatically processing written human language. The objective can be accomplished by utilizing automated analysis, comprehension, and language creation processes. The execution of various tasks by machines is generally challenging, with Natural Language Understanding (NLU) being especially difficult. NLU encompasses both semantic and pragmatic aspects and is hindered by the inherent ambiguity of language and the nuanced variations in how humans interpret the meanings of words, phrases, and sentences. The primary objective of NLU is to facilitate language interpretation by empowering computers with the ability to read and comprehend textual information effectively. Hence, the primary inquiry revolves around the methodology for extracting significance from natural language while surmounting its intrinsic intricacies. NLU, in essence, is anticipated to facilitate the achievement of a longstanding objective in the field of AI, namely, Machine Reading [15]. New possibilities would arise in which machines would possess the capability to analyze, collect, and reason over vast quantities of data, enabling them to undertake activities that would be impractical for humans because of the sheer magnitude and time required.

The statistical techniques for natural language processing and supervised ML received a lot of attention, leading to the development of widely used methodologies based on probabilistic frameworks and high-performance classifiers. Tokenization, POS tagging, and syntactic parsing are just some of the text-processing tasks these approaches specifically handle. Among the many NLP tools available to the public, statistical machine translation is one of the most useful. This development progressed in the

2000s to integrate the core concept of translating phrases instead of only individual words, following the first word-based approaches proposed by IBM. Later, the target language version of these translated sentences is pieced back together. Recently, the introduction of DL has again transformed machine translation, leading to shockingly impressive outcomes. The widespread adoption of word embeddings has transformed the landscape of most lexical-semantic tasks dealing with language meaning, even as neural networks have delivered significant advancements to practically all areas of NLP.

6.3.1 To Supervise or Not to Supervise

***A. Supervised vs. Unsupervised*:** The supervised approach involves training an ML system using linguistically annotated items assigned to appropriate classes, demonstrating effective performance when enough annotated data is available. Regrettably, despite the ability to generate 17 comprehensive part-of-speech categories with abundant examples for each, the field of computational lexical semantics faces greater complexity due to the absence of consensus regarding the appropriate inventories and formalisms to employ. Additionally, each approach is accompanied by its own set of challenges. One potential solution is circumventing predetermined class labels, transforming the problem into an unsupervised task, and enabling the utilization of huge quantities of unprocessed textual data. Unsupervised approaches have been observed in all three tasks: Word Sense Disambiguation (WSD), semantic parsing, and Semantic Role Labeling (SRL). These methods include graph-based, probabilistic, and neural Word Sense Induction for WSD, unsupervised SRL methods, and semantic parsing techniques that do not rely on training data. However, the unsupervised assignment's performance is subpar and only recommended for languages with limited resources. The task at hand would provide significant difficulty, if not insurmountable, for individuals lacking both linguistic proficiency and familiarity with the external world.

***B. Supervision vs. Knowledge*:** One of the most important things to think about is whether to use knowledge, a framework that is more relevant to computational lexical semantics and NLU, or supervision, which refers to the use of traditional training data for an ML model. Here, you get to participate in a more interesting and challenging activity. When comparing the knowledge-based and supervised paradigms, it is clear that the two differ significantly in how data is presented to the classifier. In the knowledge-based paradigm, the classifier utilizes knowledge from external resources, such as WordNet, BabelNet, or other lexical-semantic resources, which are often delivered in a structured format. This approach is employed to carry out the task at hand. Significantly, the resource is not limited to a certain work and hence offers general information that may be applied to various other tasks as well. Current approaches in WSD encompass many knowledge-based methods. One such strategy utilizes Personalized PageRank [16], to effectively carry out the disambiguation task. Authors in [17] also suggested selecting the best sense candidates from a knowledge network expressing contextual information using the densest graph approximation algorithms. In contrast to supervised methodologies, knowledge-based approaches offer a significant benefit in terms of their extensive coverage while circumventing the need for resource-intensive and costly annotation tasks. The utilization of resources such as BabelNet facilitates the incorporation of several languages with minimal additional exertion, aside from the requirement to pre-process text, which involves tasks such as tokenization, part-of-speech tagging, and, occasionally, lemmatization. The assertion is supported by the observation that knowledge-based systems have demonstrated superior performance in WSD for non-English languages. Conversely, knowledge-based systems exhibit competitive performance for English unless a substantial amount of training data is available for each word. In the field of SRL, the possibility of learning semantic roles separately from the verb reduces the impact of sparsity, resulting in a state-of-the-art approach that is supervised in nature. In certain cases, semantic parsing is carried out using a knowledge- or graph-based approach, demonstrating satisfactory performance. This is mostly attributed to the intricate nature of the task and the limited availability of training data.

6.3.2 Scalability and Multilinguality

NLU systems must exhibit scalability across various dimensions to effectively process open text. Initially, it is vital to inquire whether the system can operate on the complete lexicon. The current trajectory of word embeddings suggests that extending their application to encompass the entire lexicon is feasible, mostly due to reduced data sparsity. The most recent improvements in bilingual and international embedding in a shared vector space of 5700 [18] make this goal more likely to be reached. In the given context, conducting training in one language and evaluating in another pose a significant challenge. For instance, it has been observed that this approach can yield competitive outcomes in WSD. Nevertheless, achieving a fair and equitable comparison across different languages remains unresolved. In addition, although the scalability issue appears to be resolved for individual words, it is not immediately clear whether the knowledge acquired by a supervised system about specific word senses or semantic roles can be easily extended to include other senses or roles within the same language. The issue can be alleviated by adopting the knowledge-based paradigm, which enables large-scale processing through the utilization of organized information and extensive resource coverage. Additionally, if the resource is designed to be bilingual, it can facilitate processing across different languages.

6.3.3 Language Independence and Universality

The universal Parts-of-Speech (POS) and syntactic dependency tagsets serve as a prominent illustration of how disparate efforts conducted in many languages, seemingly without clear interconnections, can be amalgamated into a cohesive framework. The task of semantics, on the other hand, presents greater challenges. Although it is generally accepted that a verb is a verb, there is some disagreement among linguists regarding this matter. Moreover, determining the various meanings associated with a specific verb is a complex task that is not easily accomplished. Additionally, there is no clear rationale for selecting one set of reasonable meanings over another unless granularity considerations are considered.

6.3.4 The Knowledge Acquisition Bottleneck

The phenomenon of limited availability of semantically annotated material is commonly called the knowledge acquisition bottleneck. The task of obtaining enough manually labeled data becomes unfeasible when transitioning toward a scale of hundreds of thousands of classes. The impact of this matter has had a long-standing influence on the domain of WSD, as evidenced by the existence of the most extensive manually curated dataset. Significant endeavors have been made to generate additional data, such as the Manually Annotated Sub-Corpus, OntoNotes, and certain CoNLL datasets. However, there is a need for larger datasets, particularly for languages other than English. One current approach involves the generation of annotated data in a semi-automated manner. This can be achieved through various methods, such as leveraging bilingual translations or utilizing knowledge-based systems. These systems can process many sentences with high precision but low recall.

Consequently, they can identify and select items that can be reliably utilized for training a supervised system at a later stage. For example, the TrainO-Matic4 system [19] has exhibited a notable level of performance in WSD, surpassing the current state-of-the-art in certain cases, all while operating without the need for any manual text annotation. As the transition to SRL is made, the difficulty may not increase significantly if the annotation is not dependent on the predicate. This means that the predicate specificity of PropBank arguments is disregarded. When it comes to semantic parsing, the difficulty is in requiring a significantly larger quantity of sentences that have been tagged with their corresponding semantic structures to carry out the work on unrestricted material effectively. Computers that can take in text and produce semantic representations must realize the goal, which requires a fully functional NLP architecture in multiple languages. However, despite significant advancements, current developments have only enabled text processing in numerous languages at the syntactic level. Semantic analysis remains a formidable obstacle, particularly when aiming to achieve computer comprehension of text in any given language. Current research efforts have focused on developing algorithms that can effectively analyze unstructured text and generate organized semantic representations. The ultimate goal is to ensure these representations are language-agnostic and not influenced by the specific linguistic expressions used.

NLU systems must exhibit scalability across various dimensions to effectively process open text. Initially, it is important to inquire whether the system can operate on the complete lexicon. The current trajectory of word embeddings suggests that extending their application to encompass the entire lexicon is feasible, mostly due to reduced data sparsity. However, achieving a fair and equitable comparison across different languages remains unresolved. Although the scalability of word-based models appears feasible, it does not appear that the knowledge acquired by a supervised system for specific words, in terms of either word senses or semantic roles, can be easily transferred to other senses or roles within the same language. The issue can be resolved by implementing the knowledge-based paradigm, which facilitates extensive processing through the use of organized information and comprehensive resources. Furthermore, if the resource is designed to be bilingual, it can support processing across different languages.

6.4 ROBOTICS

AI has seen significant advancements, as evidenced by the progress made in technologies such as SIRI and AlphaGo. Science fiction frequently depicts AI as humanoid robots, but in reality, AI encompasses a wide range of technologies, including e-commerce prediction algorithms and IBM's Watson computers. However, it is important to note that contemporary AI is commonly called weak AI, as it is specifically engineered to carry out specific tasks, such as facial recognition, internet searches, or driving vehicles. Although weak AI has demonstrated superior performance compared to humans in certain specialized tasks, such as chess playing or equation solving, general AI exhibits superior performance in practically all cognitive tasks compared to humans. In recent years, the United States government has provided backing for fundamental research in the field of AI, with a particular focus on robotics and the development of pattern recognition technologies, including many modalities such as speech and image analysis. Microsoft has recently made public advancements in the field of real-time translation robots and new picture recognition technology. Amazon employs AI technology to facilitate the operation of autonomous robots within its delivery systems. Facebook has also created a facial recognition technique, BDeepFace, which utilizes AI. In the United States, robots and AI are actively examined at universities. New technologies like enterprise-wide collaboration and DL are examples of creative thinking. The AI Laboratory at Stanford University has developed a robot vehicle that can complete a race lap faster than a human driver. The Massachusetts Institute of Technology's Computer Science and AI Laboratory has developed a robot that can clean and walk on four feet. Nevertheless, while conducting extensive research, it has been determined that contemporary AI technologies exhibit numerous limitations.

In recent times, there has been a significant advancement in AI technology, primarily attributed to the enhanced computational capabilities of computers and the amassing of vast amounts of data. Nevertheless, the existing capabilities of AI technologies are constrained to particular domains of intellectual inquiry, namely, image identification, speech recognition, and conversation response. AI is a specialized form of cognitive computing that operates inside a certain domain. Illustrative instances encompass methodologies such as CNN or Deep Residual Learning (ResNet) for visual recognition, RNN or DNN for audio recognition, and Represent Learning (RL) for conversation interpretation. These elements constitute the cognitive processes undertaken by distinct regions of the human brain, serving as mere surrogates that do not encompass the entirety of the brains functionalities. AI has yet to demonstrate the capacity to effectively engage in comprehensive cognitive processes, encompassing self-understanding, self-control, self-consciousness, and self-motivation. The limitations of the recent AI technologies for robotics are discussed in the next sub-sections.

6.4.1 Frame Problem

Given the multitude of occurrences in the physical realm, the time required for extensive data training restricts AI to a singular frame or problem category. For instance, if the algorithm is limited to the domains of chess, shogi, picture recognition, or speech recognition, it is anticipated that only specific outcomes will be obtained. Nevertheless, in attempting to address each occurrence present in the tangible realm, an extensive array of potentialities must be considered, resulting in an indefinite number of possibilities that necessitate anticipation. Consequently, the extraction duration becomes interminable due to the excessive burden placed on the database.

6.4.2 Problem with Association Function

ML and AI demonstrate exceptional proficiency in extracting certain patterns. Nevertheless, the outcomes of ML might be prone to misuse. Contemporary AI technology relies on extensive datasets and can generate outcomes solely based on numerical inputs. However, it lacks the associative capacity exhibited by the human brain. The intelligence of the entire brain surpasses the cognitive capabilities of any individual brain region.

6.4.3 Symbol Grounding Problem

Establishing a connection between symbols and their corresponding meanings is an essential undertaking, although it remains a persistent challenge within the field of AI. For instance, by possessing knowledge of the discrete definitions of the term "Bhorse^" and the term "Bstripes^", one may comprehend that the concept of "Bzebra" can be expressed as the combination of a horse and stripes, thereby signifying a horse adorned with stripes. Nevertheless, the computer cannot establish analogous associations among concepts.

6.4.4 Physical and Mental Problems

What is the link between the mind and the body? How can the mind influence the physical body if it is commonly conceptualized as immaterial? The feasibility of this matter has not been sufficiently clarified. In conclusion, numerous unresolved issues persist within contemporary AI. This study begins by providing a comprehensive overview of the latest algorithms utilized in the field of weak AI.

6.5 EVOLUTIONARY ALGORITHMS

Currently, the complexity of applications in the actual world has increased significantly. Robotics, operations research, decision-making, bioinformatics, ML, data mining, and other problems are distinguished by their intrinsic complexity and difficulty in solving them. Evolutionary Computing (EC) is one proposed approach for addressing intricate prob. EC encompasses a range of techniques, generally called the Evolutionary Algorithms (EAs) [20, 21]. EA employs simulated evolution to investigate potential solutions for intricate real-world situations. Utilizing the evolutionary process is most appropriate when heuristic solutions are not feasible and may yield unsatisfactory outcomes. There is a significant level of interest in the field of EA, notably in its use for addressing real problems. Over the past two decades, EAs have gained significant popularity as versatile tools for conducting searches, optimizing processes, and offering answers to intricate problems.

In the realm of natural phenomena, organisms need to undergo adaptive changes in response to their surroundings, a phenomenon commonly referred to as evolution. During the process of reproduction, individuals with advantageous traits for competition are selected, and their inferior characteristics are eliminated. Genes are the fundamental units that regulate these characteristics, and a collection of these genes constitutes chromosomes. In subsequent generations, only individuals with the highest fitness level can survive, resulting in the transmission of their most advantageous genes to their offspring through recombination. The term "crossover" is the phenomenon of natural selection and optimization that gives rise to the development and proliferation of EAs. EAs are inspired by the mechanisms observed in biological evolution, such as reproduction, mutation, recombination, and selection. A set of probable solutions is generated randomly to maximize the quality function. A quality function, specifically an abstract fitness function, is used within the domain. Depending on fitness function, individuals with superior qualities are identified as possible candidates in the following generation. This is performed by using the recombination and/or mutation methods on the entities listed above. The binary operator denotes recombination. The operator above can be employed on a set of two or more chosen individuals, referred to as parents, and subsequently produces one or more novel individuals, known as children. In contrast, the process of mutation is specifically applied to a single candidate, leading to the creation of a solitary offspring. Following the execution of this recombination or mutation process, a collection of novel candidates is generated, determined by their respective fitness functions. The procedure at hand is characterized by its iterative nature. The process can be extended until a candidate of satisfactory quality is identified.

6.5.1 Types of EAs

Evolutionary Algorithms' subfields are Evolutionary Programming (EP), Evolution Strategies (ES), Genetic Programming (GP), and Genetic Algorithm (GA). All effective altruists operate based on a shared principle that involves simulating individuals' evolution through selection, mutation, and reproduction. However, it is possible to distinguish between them based on their implementation and how they are applied to a specific situation.

A. Genetic Algorithm

The GA is often recognized as the most common type of EA due to its ability to effectively imitate the natural evolutionary process within a computational framework and is frequently used in the fields of ML, pattern recognition, and optimization [22]. Its principal application is in adaptive search and adaptive system design. To solve a given problem, GAs use recombination and mutation operators. The answer is represented as a bit-string, which is a sequence of integers, often in binary format, that represents the genes.

B. Genetic Programming

In contrast to utilizing universal binary coding to express attributes in GA, GP employs the representation of programs or sets of instructions as attributes. GP is a computational technique that produces solutions through computer programs. The effectiveness of GP in solving a given computational problem is evaluated based on its fitness function. The use of tree-based encoding in genetic programming has been demonstrated to be useful in a variety of fields, including arithmetic operations, mathematical functions, Boolean operations (e.g., AND, OR, NOT), and recursive functions [22].

C. Evolution Strategy

This technique aims to determine the precise representation of a characteristic by eliminating extraneous coding. In the context of evolutionary strategy, the answers are often encoded as vectors consisting of real numbers, and the approach incorporates the utilization of self-adaptive mutation rates. Several notable uses of evolution techniques include routing and networking [22], biochemistry, optics, and engineering design.

D. Evolutionary Programming

EP closely resembles that of ES. In contrast to ES, there are no limitations on utilizing data types for characteristics. The EP paradigm is characterized by a defined program structure that facilitates the evolution of numerical parameters. EP is applicable in various domains such as forecasting, generalization, games, and automatic control [22]. The most effective approach to choosing a specific strategy for problem-solving is first to choose an appropriate representation that is well-suited to the problem at hand. Following this, one should carefully select variation operators that align with the chosen representation. The independence of selection operators from representation is attributed to their exclusive reliance on the fitness function.

6.5.2 Extensions to EAs

Two modifications are proposed to enhance the effectiveness of EA approaches, specifically Memetic Algorithms (MA) and distributed EA [23].

A. Memetic Algorithms

The use of heuristics is frequently referred to as "local search," which describes the methods used by a person. The MA combines elements of local search with evolutionary algorithms. In the natural environment, individuals undergo postnatal modifications and endeavor to acclimate to their surroundings to ensure their survival. The phenomenon of adaptation is commonly referred to as plasticity. In the field of EA, it is observed that everyone possesses a certain degree of flexibility, allowing them to adapt to their environment. Most of the time, people are chosen for further growth based on random criteria. That chance may have a role in the relative importance given to local search and the EA technique. By incorporating the MA into the EA, a more powerful multi-start search of the local search heuristic is made possible [20].

B. Distributed Evolutionary Algorithms

Implementing EA approaches necessitates a substantial allocation of computational resources. There exist two fundamental rationales for this phenomenon. One potential explanation is the presence of a significant delay in the evaluation of the fitness functions. Additionally, the size of the population has the potential to become excessively huge. To address this issue, a viable approach would involve the distribution of the entire workload among multiple computers, thereby enabling parallel processing for all computations. In the natural world, it is common for species to tend specialization, enabling them to explore new ecological niches and ultimately evolve into distinct species that are not found elsewhere. In addition, EA periodically exchanges the most optimal person available, and each subpopulation tends to converge toward the globally optimal solution encompassing all suboptimal solutions. Distributed evolutionary algorithms employ parallelism to decrease computing time significantly. Additionally, it enhances the efficacy of the solution obtained within multi-modular search spaces.

While EA has a broad range of applications across several fields, its performance is generally limited to minor improvements. Therefore, the current endeavors in the field of evolutionary algorithms are concentrated on the application of supplementary algorithms. Using EA can potentially improve performance. One prevailing practice in the field involves the hybridization of many algorithms or the enhancement of pre-existing algorithms. The collective approach is expected to yield superior outcomes compared to the solo approach.

6.6 PREDICTIVE MODELING

Predictive modeling encompasses using mathematical or computer techniques to construct models capable of prognosticating forthcoming outcomes. The former makes use of equation-based models, whereas the latter requires the use of simulation techniques. Clinical decision-making and clinical trials are just two examples of the many medical contexts in which predictive modeling has proven useful. However, many restrictions currently prevent the widespread use of predictive modeling in this field. The dynamic nature of medicine and the wide range of patient populations seen in modern healthcare settings create special difficulties in applying such tactics. In addition, the successful development and implementation of accurate predictive models necessitate a comprehensive comprehension of the data being utilized and sufficient resources to facilitate the model creation and deployment process.

6.6.1 Key Principles of Predictive Modeling

Predictive modeling makes use of computational and mathematical methods to foresee potential outcomes. Regression analysis, DT, random forests, neural networks, and SVM are just a few of the many methods that fall under these umbrella terms. By recognizing patterns and links in the data, these techniques use algorithms to build predictive models. Two primary approaches are used to achieve this goal, both described in detail below.

A. Equation-Based Predictive Modeling

Equation-based models are specific models that use math equations to show how factors are connected. These models are widely used in the physical sciences, the chemical sciences, and the engineering disciplines to foretell the behavior of physical systems. These models rely on embedded parameters that elucidate the relationship between inputs and the examined outcome to forecast future results by employing mathematical projections of changes in input variables. Time-series regression frameworks have been shown to effectively demonstrate this method by employing linear regressions to anticipate aircraft traffic volume or fuel efficiency, considering engine speed and load variations. The simplicity and convenience of assessing this methodology make it particularly advantageous for projecting results related to a certain condition. Hence, it is frequently employed in the field of predictive analysis when there is a substantial volume of data accessible. However, problems arise when trying to establish a connection between the variables used to build a model and the particular outcome variable under investigation. The scarcity of available data may hinder a person's ability to establish a connection between independent and dependent variables. Moreover, the task of determining the most influential independent variables within a certain setting can provide difficulties. Hence, when constraints are employed to examine intricate systems encompassing several variables that contribute to the emergence of result variables. Consequently, the utilization of predictive modeling in medicine is limited and inappropriate for scenarios characterized by temporal variability of the dependent variable, such as monitoring the occurrence of diseases within a population over time. The concerns emerge due to the potential obsolescence of model parameters when utilized for predicting future instances of the mentioned diseases within populations. As a result, regular updates are required to reflect the changing dynamics within these communities throughout time accurately. To do so with equation-based models, it is crucial to have a solid grasp of how all of the input variables relate to the target variable. However, ML methods may provide more value in this context. In this method, the datasets are not used to generate equations. Instead, ML techniques such as DT and neural networks are used to make sense of the connections between things. The correlations between variables that are obtained using these methods are contingent upon the specific dataset utilized rather than being predetermined by the method itself. These characteristics make the predictive modeling approach flexible and applicable in various contexts.

B. Computational Predictive Modeling

Computational predictive modeling, in contrast to the mathematical method, uses models that formulae cannot easily explain. Simulation approaches are crucial for developing predictions utilizing the "black box" approach. In contrast to traditional techniques such as curve and surface fitting or time series regressions, computational modeling does not offer explanatory insights into the relationship between input components and results. ML methodologies, such as neural networks and bagged DT, can facilitate various tasks, such as assessing a borrower's creditworthiness [24]. Many industries have successfully employed these strategies to tackle various difficulties. Nevertheless, this approach exhibits several limitations, including its restricted suitability for handling datasets of small sizes and the heightened computational intricacy it poses for larger datasets. Using computational modeling has proven to be highly advantageous in the domains of insurance and finance. This is mostly due to its capacity to generate precise forecasts, enabling stakeholders to make well-informed decisions. However, implementing a computational model might present challenges, including high costs and difficulties in accurately interpreting the models. As a result, adopting a suitable modeling approach necessitates a thorough evaluation of the multiple aspects involved. Furthermore, integrating several methodologies and techniques can yield superior results by minimizing inaccuracies and simultaneously offering valuable perspectives on the matter under consideration.

Irrespective of the methodology, developing a predictive model necessitates adhering to a standardized procedure. The process involves cleaning and improving data with modeling in mind, zeroing in on an appropriate method for prediction, training and evaluating the model using subsets of data, gauging performance with goodness-of-fit tests, and finally verifying correctness on unused datasets. After arriving at a point of satisfaction with the results of these steps, models can be used confidently for forecasting. To optimize performance or boost the model's efficacy, it is possible to repeatedly conduct the operation under varying conditions in various predictive modeling applications. In addition, the model can be improved over time to anticipate future outcomes better. In conclusion, deploying on several devices allows predictions to be made at any given time. However, it is important to note that this entire process may necessitate substantial effort and time. However, the benefits of a proficient predictive model are substantial, suggesting its escalating usage in our everyday existence and its burgeoning appeal.

C. Applications and Limitations

Healthcare data includes health-related information acquired from medical records and surveys that pertain to individuals or groups. Hospitals, doctors, psychiatrists, and pharmacists are just a few healthcare professionals who rely heavily on healthcare analytics. This tool may also be utilized by stakeholders seeking to improve the quality of treatment through well-informed decisions backed by dependable insights derived from comprehensive datasets, including Electronic Health Records (EHRs) and registries based on claims. Predictive modeling, an integral component of data analytics, leverages AI, ML, and data mining methodologies to analyze historical and current data to generate future predictions. In the context of healthcare settings, predictive analytics plays a crucial role in examining both present and past medical records, thereby assisting healthcare practitioners in enhancing their decision-making processes. This analytical approach also enables the prediction of trends and facilitates more effective management of disease outbreaks.

The application of predictive analysis methods that consider medical records, age, socio-economic factors, and individual anatomy has facilitated healthcare institutions in assessing the likelihood of patients developing different lifestyle-related ailments. Using sophisticated techniques, such as simulation, variable modeling, and predictive analysis, has become crucial for healthcare organizations worldwide to augment their decision-making capabilities. These techniques are especially valuable in industries characterized by high speed and urgency, such as healthcare, where the efficient analysis of large volumes of data is essential, alongside properly managing risks. The integration of cutting-edge technologies has greatly enhanced the problem-solving capabilities of medical enterprises, consequently facilitating the identification of chances to improve global healthcare systems. Implementing novel approaches in day-to-day activities can significantly augment operational efficacy and alleviate stress levels for individuals tasked with addressing the healthcare requirements of a global population.

Despite the progress made in predictive modeling for drug-target interactions in the last two decades, a significant knowledge gap remains regarding understanding the mechanisms through which these associations translate into clinical outcomes. There is a specific requirement to predict the activities of receptors throughout the entire genome and the selectivity of their functions in response to novel substances, specifically in terms of distinguishing between agonist and antagonist effects. Nevertheless, attaining this goal presents a formidable challenge due to the limited availability of data about receptor activity and the requirement of training models using varied shifting distributions that are applicable in real-world scenarios [25]. Given the rapid growth of technology and research, predictive modeling is expected to continue to play an important role as an effective tool in improving patient outcomes across many disciplines. There appears to be no end to how this unique approach could be used, which bodes well for significant advances and discoveries in the field of medicine. Each new finding and improvement brings us that much closer to realizing the full potential of predictive modeling in the medical field. This has the potential to revolutionize our comprehension of disease diagnosis, treatment alternatives, and tactics for prevention, among other areas.

6.7 GENOMICS AND PROTEOMICS

Proteomics is not restricted to quantifying a given biological system's various proteins and proteoforms. In addition, it can be used as a "read-out" after any treatment that boosts certain classes of functionally distinct proteins. Using a bait molecule, which can be another protein of interest or a small molecule drug, proteins are separated by mass using a pull-down test in the field of interactomics [26]. Engaging in this process on a comprehensive scale results in a complex web of interconnections, the disruptions of which can be examined. In a similar vein, the cellular proteome has the potential to be fractionated by the process of centrifugation, followed by the utilization of mass spectrometry-based proteomics techniques on the resulting fractions. This analytical approach ultimately yields a comprehensive depiction of cellular organelles in the form of a map. PTMs play a crucial role in regulating the activity status, localization, and turnover of a significant proportion, if not the majority, of proteins. Mass Spectrometry (MS)-based proteomics has significantly transformed the field in the past 20 years by revealing an unprecedented level of complexity in Post-Translational Modifications (PTMs) and their control [27]. In addition to the several dimensions inherent to proteomics, integrating proteomics data with other omics or clinical data within the same project is crucial.

ML and DL techniques are in their infancy when applied to analyzing genomics' downstream metabolomics, lipidomics, and proteomics data. These methods are typically used in the preliminary phases of data analysis. This phenomenon could potentially be attributed to limited sample numbers, inadequate interpretability, and a general insufficiency of reference, training, and validation data. There have been instances where lipidomics datasets have been utilized as proof of concept in ML applications. For example, authors in [28] employed such datasets to predict the state of Type 2 Diabetes (T2D). Authors in [29] also used lipid profile data from dried blood spots to predict clinical lipid concentration.

Integrating diverse datasets is a crucial task in multi-omics, involving the potential integration of datasets across several omics hierarchies such as genomic, epigenomic, proteomic, and lipidomic levels. An enduring inquiry pertains to how alterations in transcriptome levels can indicate corresponding changes in the proteome. There is not a direct and exclusive relationship between a transcript and the protein or proteins it ultimately generates due to the intricate regulatory mechanisms involved in gene expression at the mRNA level and subsequent translation into proteins, as well as the complex processes of regulated protein degradation. Predictions for (phospho)proteins are made from DNA and RNA data in recent work [30] by using crowdsourcing via the NCI-CPTAC DREAM proteogenomic challenge. Tree-based and ensemble-based models are just two examples of the many ML methods that are employed. However, the predictive capabilities of these models fell short of the baseline methodology, and in several cases, they even exhibited performance levels comparable to random chance. Therefore, the researchers concluded that it is crucial to measure the proteome directly instead of using transcriptome or copy numbers as substitutes.

The field of genomics is widely recognized as a discipline that deals with large-scale datasets, commonly referred to as Big Data. The obstacles associated with this approach include the large volume of available datasets (measured in Gigabytes), their heterogeneity (including the sequencing of coding or non-coding genes and gene variations), and their diverse nature. DL necessitates a substantial volume of data. In certain instances, and concerning a particular issue, the existing data may be insufficient in achieving optimal performance. The phenomenon of imbalanced classes refers to a situation in which the distribution of instances across different classes in a dataset is significantly skewed, Typically, genomics data that is gathered tends to have an uneven distribution of examples across different classes. Therefore, the DL model can lack generalization regarding certain categories. The topic of interest pertains to the interpretation of a model. The DL technique is widely recognized as a significant issue. The comprehension and interpretation of acquired patterns might provide challenges for model designers. The issue at hand is commonly referred to as the black box.

However, many challenges still need to be overcome before genomics and personalized medicine can become standard practice in clinical settings. Among these difficulties is the call for more stringent regulation and guarantee of genetic testing's quality. Clinical usefulness and cost-effectiveness of genetic and genomic diagnostics are difficult to assess due to a lack of high-quality, prospectively gathered information [31]. The pharmaceutical industry has raised concerns about the possibility of market segmentation and a subsequent decrease in the number of patients for whom a certain prescription is approved and prescribed if drugs are instead tailored to specific genetic subpopulations. A bigger issue is that doctors and nurses aren't very familiar with genetics and genomics [32].

6.8 ENVIRONMENTAL MONITORING

The global population growth and human activities, including urbanization, the establishment of Hydroelectric Power (HEP) and nuclear power plants, deforestation for agricultural purposes, and the combustion of fossil fuels, contribute to a substantial environmental transformation [33]. Satellite images and GIS data are useful tools for environmental monitoring since they allow for investigating and forecasting human impacts on the natural world. For instance, the anticipation of dam building alterations can be approached from various academic viewpoints, encompassing the realms of environmental studies, forestry, hydrology, agriculture, and geology. In addition, the quantity of Remote Sensing (RS) data gathered by a wide range of airborne and space-based sensors is growing rapidly. Due to this meteoric rise, environmental monitoring systems need to use powerful AI-based algorithms, big data analytics, data fusion mechanisms, and ample computing resources to be the most effective. As a result, satellite RS has become a standard procedure for making detailed maps of crops. This enables the accurate identification of quick phenological transitions at the field level and facilitates the prediction of many environmental consequences.

6.8.1 Impacts on the Environment

Environmental changes occur for a variety of reasons, for instance, the construction of hydroelectric dams, the growth of metropolitan areas, the combustion of fossil fuels, and deforestation can have substantial effects on the environment, leading to serious issues such as land use and land cover changes, water pollution, greenhouse gas emissions, and loss of species diversity. To address this objective, numerous research papers have been offered to examine the identification and anticipation of the effects, employing AI and ML models. Land use and changes LULC undergoes substantial transformations in the context of urbanization, dam construction, and several other factors. Hence, a crucial objective in this scenario is to utilize DL-based methodologies to examine, categorize, and detect alterations in present and past LULC data derived from multispectral satellite imagery. Consequently, these approaches can be employed to forecast future LULC changes. A range of DL models have been examined within this particular setting. These include, among others, Long Short-Term Memory (LSTM), Random Forest (RF), Extreme Gradient Boosting (XGboost), CNN, DT, and Fully Convolutional Neural Network (FCN).

Waterlogging susceptibility, the building of canyons and reservoirs resulting from fast urbanization, dam construction, and other activities in forest areas, can give rise to significant wildlife habitat loss and fragmentation, leading to severe WLS concerns. Consequently, researchers have employed DL models, specifically CNNs, to perform long- and short-term analyses and predict dams' Water Level Series (WLS). For example, authors in [34] employed geographic features collected from RS photos and input them into an RF classifier to produce Coarse-Weighted Least Squares (CWLS) probability maps. The latter facilitates the estimation of hazard probabilities and the prediction of WLS events. Using satellite imagery, CNN models have been employed to detect reference objects and determine WLS depths. Gully Erosion Susceptibility (GES). The soil has a crucial role as a medium for supporting biodiversity, thereby facilitating the fulfillment of many human requirements, such as access to clean water, food, and clean air. In light of this objective, the prevention of soil deterioration becomes of paramount significance. One example is the GES, an entity accountable for significant soil degradation. Severe soil erosion is a common phenomenon that gives rise to several environmental concerns, such as sedimentation in dams, depletion of groundwater nutrients, and deterioration of surface runoff. One potential approach to accomplish this objective involves using innovative ML methodologies that may forecast and delineate the vulnerability of gully erosion. Such methodologies include ANN, naïve-Bayes tree, RF, and CNN.

ML tools are a set of computational methods and techniques that automate how computers learn and improve from experience. These resources can help to make sense of mountains of data and spot trends. A systematic classification of AI instruments used for environmental monitoring is carried out. Unsupervised, Semi-Supervised, and Supervised models are the three main types of ML models. The latter's subcategories include classical classification, clustering, DCNN, GAN, RNN, ensemble models, Structural Causal Models (SCM), and AI-based segmentation. RS photos receive a significant amount of attention excessively. Three essential components of these approaches include geographical image analysis and mapping, the extraction of multi-modal characteristics using ML, and the fusion of these heterogeneous features to enhance the informativeness of images. Nevertheless, using RS images for environmental monitoring possesses distinct and unparalleled attributes that significantly differentiate it from conventional image analysis methods. This encompasses a range of factors that must be anticipated, such as the various forms of environmental influences, the diverse and numerous devices/sensors used to capture images, the quality of these images, the metrics employed for evaluation, the suitable data fusion systems for multi-modal images, and numerous other considerations. Furthermore, the effectiveness of constructing reliable and efficient environmental monitoring technologies is contingent upon the quality of the ML models implemented and the accessibility of the data utilized for their training. The latter continues to pose a challenge because the majority of current frameworks have only been tested using proprietary datasets. Consequently, assessing the current state-of-the-art and accurately carrying out equitable comparisons becomes difficult. In this context, more endeavors are necessary to cultivate innovative datasets for environmental monitoring, particularly about GES, Water Quality and Quantity (WQQ), and Land Surface Imaging (LSI).

6.9 DRUG DISCOVERY

6.9.1 Predicting Drug Efficacy and Toxicity Using Machine Learning

The prediction of the effectiveness and potentially harmful effects of potential therapeutic molecules is a significant application of AI in the field of medicinal chemistry. The conventional methodologies employed in drug discovery frequently depend on arduous and protracted experimental procedures for evaluating the possible physiological impacts of a chemical on the human organism. The process can be characterized by its slow pace and significant expenses, while the outcomes frequently exhibit uncertainty and a notable level of unpredictability. AI approaches, such as ML, can surpass these constraints. ML algorithms can discern patterns and trends that may elude human researchers, as evidenced by the study of extensive data. This approach facilitates the generation of novel bioactive molecules with minimal adverse effects in a significantly expedited manner compared to conventional techniques. As an illustration, a DL algorithm has recently undergone training utilizing a dataset comprising identified pharmacological molecules alongside their respective biological activities. Subsequently, the program demonstrated a remarkable ability to forecast previously untested chemicals' activity accurately. Previous studies have made noteworthy advancements in mitigating the toxicity of possible medicinal molecules. These advancements involve using extensive training methods that rely on vast databases containing both toxic and non-toxic chemicals for ML purposes.

Another significant utilization of AI in the field of drug development pertains to the detection and analysis of drug-drug interactions. These interactions occur when multiple medications are administered concurrently to a patient for the same or separate medical conditions, potentially leading to modified therapeutic effects or undesirable consequences. The identification of this issue can be facilitated through the utilization of AI-based methodologies, which involve the analysis of extensive datasets about established medication interactions, hence enabling the recognition and interpretation of prevailing patterns and trends. The accurate prediction of interactions between novel drug pairings has recently been addressed using an ML system [35]. The utilization of AI in identifying potential drug-drug interactions within the framework of personalized medicine holds significance. This application of AI facilitates the creation of tailored treatment strategies that aim to mitigate the occurrence of unfavorable reactions. The objective of personalized medicine is to customize medical treatment based on the unique attributes of individual patients, encompassing factors such as their genetic makeup and specific reactions to pharmaceutical interventions.

The preceding instances from the scholarly literature illustrate that the application of AI in the field of pharmaceutical research has the potential to enhance the accuracy of forecasting the effectiveness and adverse effects of prospective therapeutic compounds. This has the potential to facilitate the advancement of more efficacious and less hazardous pharmaceuticals and speed up drug discovery.

6.9.2 Impact of AI on the Drug Discovery Process and Potential Cost Savings

One of the primary applications of AI in the field of drug discovery involves the development of innovative molecules that possess distinct features and activities. Conventional approaches frequently depend on the discernment and alteration of pre-existing substances, a procedure that might be characterized by its sluggishness and demanding nature. In contrast, AI-driven methodologies have the potential to facilitate the expedited and effective development of innovative molecules possessing favorable characteristics and functionalities. For example, a DL algorithm was recently used to train on a dataset of established medicinal compounds and their corresponding properties. This training aimed to develop novel therapeutic molecules with desirable properties such as solubility and activity [36]. This successful application demonstrates the potential of these methodologies to expedite and improve the development of new drug candidates.

DeepMind has recently made a significant contribution to AI research by developing AlphaFold, an innovative software platform that enhances our understanding of biological processes. This research employs a system that utilizes protein sequence data and AI techniques to predict the three-dimensional structures of proteins accurately. The anticipated impact of this structural biology innovation will likely result in a paradigm shift in personalized medicine and drug discovery. AlphaFold represents a significant advancement in the application of AI to structural biology and the broader life sciences. ML Strategies and Molecular Dynamics (MD) simulations are currently utilized in de novo drug design to improve precision and efficacy. To maximize the synergistic effects of their integration, the exploration of integrating diverse approaches is currently underway. Applying Interpretable ML (IML) and DL techniques substantially contributes to this endeavor. Using AI and ML, researchers have acquired an unprecedented ability to improve the efficacy and efficiency of drug design.

6.9.3 Limitations and Challenges of Using AI in Drug Discovery

Despite the prospective advantages of AI in drug discovery, there are several obstacles and restrictions to consider. One of the primary obstacles that researchers face is the limited accessibility of appropriate data. AI-based methodologies commonly necessitate a substantial amount of data for training. In numerous instances, the accessibility of data may be constrained, or the data itself may exhibit deficiencies in terms of quality or consistency. These factors might have a detrimental impact on the precision and dependability of the outcomes. Another problem arises from ethical considerations since AI-based systems may raise questions regarding fairness and bias. For instance, in the case when the data utilized for training an ML algorithm exhibits bias or lack of representativeness, the consequent predictions may exhibit inaccuracy or unfairness. AI's ethical and equitable utilization in advancing novel medicinal compounds is a crucial matter that necessitates attention. Various tactics and methodologies can be employed to surmount the challenges encountered by AI within the domain of chemical medicine. One such methodology uses data augmentation, which entails creating synthetic data to complement pre-existing datasets. Including additional data, both in terms of quantity and variety can potentially improve the training of ML algorithms. In turn, this can enhance the accuracy and dependability of the results. An alternative strategy employs explainable AI (XAI) techniques [37], which aim to provide interpretable and transparent justifications for ML algorithm predictions. This technique can effectively mitigate bias and fairness concerns in AI-based methodologies while augmenting understanding of the mechanisms and assumptions that drive predictive outcomes.

The existing AI-based methodologies do not serve as a viable alternative to conventional experimental techniques, and they are incapable of supplanting the knowledge and proficiency possessed by human researchers. AI is limited to generating predictions solely based on the data it can access. Consequently, the outcomes produced by AI necessitate subsequent validation and interpretation by human researchers. Nevertheless, incorporating AI in conjunction with conventional experimental approaches can potentially augment the drug development process. By integrating AI capabilities with the knowledge and proficiency of human researchers, it becomes feasible to enhance the efficiency of the drug discovery process and expedite the advancement of novel pharmaceuticals.

6.10 PERSONALIZED MEDICINE

Like any academic discipline, the field of evolution undergoes progress through significant research breakthroughs and developmental advancements that contribute to the improved usage and understanding of this specific area of study. AI systems have been effective in resolving challenges across diverse domains. The selection of AI algorithms for addressing issues in Personalized Medicine, such as disease detection, prediction, accurate diagnosis, and treatment optimization, depends on their respective capabilities and suitability. The phenomenon of certain individuals experiencing varying levels of efficacy or adverse reactions to a given medicine is consistently perplexing. Another issue that arises pertains to the inquiry into the factors contributing to the development of certain diseases, such as malignancies, in certain individuals while others remain unaffected. The presence of genetic variations and other distinct factors, such as age and lifestyle choices, may contribute to the occurrence of these issues. Accordingly, it is argued that the field of medicine should adopt an approach that recognizes the individuality of each patient's disease and develop medications that are specifically customized to their unique medical history and biological characteristics [38]. The medical practice referred to in this context is commonly recognized as Personalized or Precision Medicine. As a distinct discipline or extension of Medical Sciences, personalized medicine employs clinical practice and medical decision-making processes to provide individualized healthcare services to patients. According to the assertions made by [39], personalized medicine plays a significant role in forecasting an individual's susceptibility to a particular ailment, attaining precise diagnoses, and optimizing the most effective treatment options. The customization of medical therapy or administration is facilitated through the utilization of genetic information, which serves as a fundamental component of the baseline data. However, many reproducible results do not identify prevalent genes contributing to vulnerability or resilience against diseases. Instead, the current knowledge predominantly revolves around infrequent genetic variations, although some prevalent variations have also contributed to our understanding. The field of medicine has experienced substantial growth in recent years, with a heightened focus on disease prevention through the utilization of advanced technologies. This involves assessing the likelihood of an individual developing a particular disease and administering appropriate treatments, such as pharmaceutical interventions, to mitigate the anticipated occurrence of said disease.

Furthermore, using technology enables clinical workers, such as doctors and pharmacists, to provide healthcare services with much-enhanced efficiency compared to conventional methods. When used in the formation and development of personalized medicine, AI approaches can greatly improve the accuracy with which diseases are diagnosed, and treated, and drugs are administered. The management of adverse drug reactions and the metabolism of enzymes might contribute to variations in drug elimination among individuals, potentially leading to instances of drug overdose in some cases and rapid drug elimination in others. Electronic Health Record (EHR) systems and the use of computers to record medical activities in healthcare settings like hospitals and clinics already offer significant medical benefits. This information and data can serve as a reference point for improving the delivery of medical services. Numerous algorithms from the areas of ML and AI have found use in the field of Medicine, most notably in Personalized Medicine. Some examples of ML methods are the Nave Bayesian (NB), ANN, and SVM.

A. Naïve Bayesian

The NB algorithm is based on a probabilistic model to facilitate the incorporation of uncertainty into a model in a systematic manner by calculating the probability associated with various outcomes. NB is a commonly employed technique in several systems, encompassing spam filtering, recommender systems, and text classification. Additionally, it finds application in the field of medicine and meteorology for various purposes. This characteristic renders the method suitable for both classification and prediction tasks. One of the notable advantages of the NB algorithm is its robustness in handling noise present in input data. Additionally, NB demonstrates efficacy even when trained on a very small amount of data. One of the drawbacks of the NB algorithm is the potential decrease in accuracy due to the assumption of class conditional independence. One of the drawbacks of NB is its reliance on the availability of predictors in the training data. If a given predictor is not present in the training data, Naive Bayes assumes that there is no chance that a given record will contain a new predictor category.

B. Artificial Neural Network

Diagnostics, medical imaging, evaluation of back pain, dementia, pathology, and prognosis evaluation for conditions like appendicitis, myocardial infarction, acute pulmonary embolism, arrhythmias, and psychiatric disorders, as indicated by [40], are just some of the many areas in which ANNs are being used in medicine. Neural networks can acquire knowledge about both linear and non-linear models. Additionally, the statistical measurement of model accuracy can be applied to neural networks. Neural networks can tolerate incomplete input and noise to a certain extent. Neural network models possess a high degree of flexibility due to their capacity for updates, rendering them well-suited for dynamic environments such as the health sector. ANNs are considered black box algorithms, as they lack transparency in revealing their internal structure, hence limiting their ability to provide meaningful insights. Furthermore, while it can derive generalizations from a given collection of instances, the system in question is limited in its predictive abilities when exposed solely to cases falling within a specific range.

C. Support Vector Machines

SVM generates reliable classification results, even when the input data cannot be linearly separated. Additionally, the accuracy outcome is not dependent on the level of competence in human judgment when selecting the linearization function for non-linear input data. One drawback of SVM as a non-parametric method is its limited transparency in producing results. Ensuring that the settings are configured accurately is imperative to obtain precise outcomes for a given activity or problem. Selecting a kernel that yields precise outcomes for task “A” may result in subpar outcomes for task “B”.

6.11 PROBLEMS IN PERSONALIZED MEDICINE

The difficulties of personalized medicine vary widely depending on the nature of the condition being treated. Government policies, such as those governing access to public medical records and genetic research, healthcare workers’ mindsets, knowledge, and training, technological advances, and financial constraints, all play a role. Nevertheless, the primary objective of this research article is to examine the challenges associated with

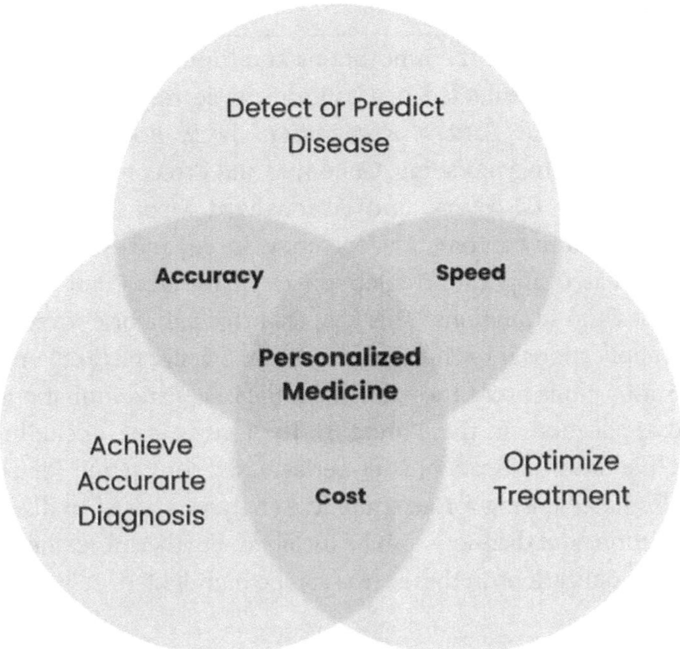

FIGURE 6.2 Problems in personalized medicine.

adopting Information technology (IT) in the healthcare sector. The specific problems under investigation, as illustrated in Figure 6.2, encompass disease detection or prediction, attainment of precise diagnosis, and the identification of appropriate treatment strategies. Figure 6.2 illustrates the challenges encountered in the field of personalized medicine.

The utilization of AI algorithms plays a significant role in implementing Personalized Medicine. Nevertheless, the current state of this phenomenon remains nascent and encounters various obstacles, a subset of which are intricately connected to AI. The other crucial difficulties, such as research and implementation costs and government restrictions, are also important for successfully applying for customized medicine. Nevertheless, Personalized Medicine encounters not only problems but also presents certain obstacles. One such obstacle involves the transformation of the medical profession and practice to the extent that some futurists speculate that algorithms and computers could potentially supplant a significant portion of the tasks currently performed by doctors. Ultimately, the successful integration of customized medicine has the potential to enhance patient outcomes and revolutionize the field of medicine significantly.

6.12 CONCLUSION

This chapter highlighted the innovations and limitations of AI-based applications that are implemented in Natural Science research, including Data Analysis, Image and Pattern Recognition, NLP, Robotics, Evolutionary Algorithms, Predictive modeling, Genomics and Proteomics, Environmental Monitoring, Drug Discovery, and Personalized Medicine. AI has several advantages, including automation, accuracy, speed, and cost benefits, which result in AI-based application adoption in many areas, but they still face several issues and limitations. This chapter provided a brief overview of the AI-based applications implemented in Natural Science research areas. It covered the innovations over time, along with the issues and limitations faced by AI-based applications in these areas. In the future, areas including natural disasters like flood mitigation, time series-based forecasting for power and weather, Remote Sensing for geographical analysis, and AI in disease detection, prediction, and diagnosis can be included. Furthermore, the limitations of AI-based applications in these areas can be highlighted.

REFERENCES

1. J. Dimon, "Chairman and CEO letter to shareholders," *jpmorganchase.com, 8th April,* available at: http://www.jpmorganchase.com/corporate/annual-report/2014/ar-introduction.htm, last accessed 2nd June, 2015.
2. S. Iqbal, H. Maryam, K. N. Qureshi, I. T. Javed, and N. Crespi, "Automised flow rule formation by using machine learning in software defined networks based edge computing," *Egyptian Informatics Journal,* vol. 23, pp. 149–157, 2021.
3. T. H. Davenport, "Realizing the benefits of automated machine learning," *(DataRobot Working Paper),* 2018.
4. J. Sinai, "Introducing salesforce einstein–AI for everyone," *Salesforce Blog,* 2016. https://www.salesforce.com/blog/2016/09/introducing-salesforce-einstein.html.
5. M. A. Qureshi, K. N. Qureshi, G. Jeon, and F. Piccialli, "Deep learning-based ambient assisted living for self-management of cardiovascular conditions," *Neural Computing and Applications,* 2021. https://doi.org/10.1007/s00521-020-05678-w.
6. K. N. Qureshi, A. Sikandar, and P. Dhawankar, "Next-generation connected traffic using UAVs/Drones," *Secure and Digitalized Future Mobility: Shaping the Ground and Air Vehicles Cooperation,* CRC Press, p. 65, 2022.
7. D. G. Lowe, "Distinctive image features from scale-invariant keypoints," *International Journal of Computer Vision,* vol. 60, pp. 91–110, 2004.
8. N. Dalal and B. Triggs, "Histograms of oriented gradients for human detection," in *2005 IEEE Computer Society Conference on Computer Vision and Pattern Recognition (CVPR'05),* 2005, vol. 1, pp. 886–893: IEEE.

9. G. E. Hinton and R. R. Salakhutdinov, "Reducing the dimensionality of data with neural networks," *Science,* vol. 313, no. 5786, pp. 504–507, 2006.
10. A. A. M. Al-Saffar, H. Tao, and M. A. Talab, "Review of deep convolution neural network in image classification," in *2017 International conference on radar, antenna, microwave, electronics, and telecommunications (ICRAMET),* 2017, pp. 26–31: IEEE.
11. Y. LeCun, L. Bottou, Y. Bengio, and P. Haffner, "Gradient-based learning applied to document recognition," *Proceedings of the IEEE,* vol. 86, no. 11, pp. 2278–2324, 1998.
12. C.-L. Liu, K. Nakashima, H. Sako, and H. Fujisawa, "Handwritten digit recognition: benchmarking of state-of-the-art techniques," *Pattern Recognition,* vol. 36, no. 10, pp. 2271–2285, 2003.
13. W. J. Scheirer, L. P. Jain, and T. E. Boult, "Probability models for open set recognition," *IEEE Transactions on Pattern Analysis and Machine Intelligence,* vol. 36, no. 11, pp. 2317–2324, 2014.
14. S. Naseem, A. Alhudhaif, M. Anwar, K. N. Qureshi, and G. Jeon, "Artificial general intelligence-based rational behavior detection using cognitive correlates for tracking online harms," *Personal and Ubiquitous Computing,* 2022. https://10.1007/s00779-022-01665-1.
15. O. Etzioni, M. Banko, and M. J. Cafarella, "Machine Reading," in *AAAI,* 2006, vol. 6, pp. 1517–1519.
16. E. Agirre, O. López de Lacalle, and A. Soroa, "Random walks for knowledge-based word sense disambiguation," *Computational Linguistics,* vol. 40, no. 1, pp. 57–84, 2014.
17. A. Moro, A. Raganato, and R. Navigli, "Entity linking meets word sense disambiguation: a unified approach," *Transactions of the Association for Computational Linguistics,* vol. 2, pp. 231–244, 2014.
18. A. Conneau, G. Lample, M. A. Ranzato, L. Denoyer, and H. Jégou, "Word translation without parallel data," *arXiv preprint arXiv:1710.04087,* 2017.
19. T. Pasini and R. Navigli, "Train-o-matic: Large-scale supervised word sense disambiguation in multiple languages without manual training data," in *Proceedings of the 2017 Conference on Empirical Methods in Natural Language Processing,* 2017, pp. 78–88.
20. A. E. Eiben and J. E. Smith, *Introduction to evolutionary computing.* Springer, 2015.
21. N. M. Al-Salami, "Evolutionary algorithm definition," *American Journal of Engineering and Applied Sciences,* vol. 2, no. 4, pp. 789–795, 2009.
22. C. A. Coello Coello, "An introduction to evolutionary algorithms and their applications," in *International Symposium and School on Advancex Distributed Systems,* 2005, pp. 425–442: Springer.
23. D. Sudholt, "Parallel evolutionary algorithms," *Springer Handbook of Computational Intelligence,* Springer, pp. 929–959, 2015.
24. Y. Jiang, "A Primer on Machine Learning Methods for Credit Rating Modeling," 2022.

25. T. Cai, K. A. Abbu, Y. Liu, and L. Xie, "DeepREAL: a deep learning powered multi-scale modeling framework for predicting out-of-distribution ligand-induced GPCR activity," *Bioinformatics,* vol. 38, no. 9, pp. 2561–2570, 2022.
26. I. Bludau and R. Aebersold, "Proteomic and interactomic insights into the molecular basis of cell functional diversity," *Nature Reviews Molecular Cell Biology,* vol. 21, no. 6, pp. 327–340, 2020.
27. R. Aebersold and M. Mann, "Mass spectrometry-based proteomics," *Nature,* vol. 422, no. 6928, pp. 198–207, 2003.
28. S. Y. Ho, K. Phua, L. Wong, and W. W. B. Goh, "Extensions of the external validation for checking learned model interpretability and generalizability," *Patterns,* vol. 1, no. 8, 2020.
29. S. G. Snowden, A. Korosi, S. R. de Rooij, and A. Koulman, "Combining lipidomics and machine learning to measure clinical lipids in dried blood spots," *Metabolomics,* vol. 16, pp. 1–10, 2020.
30. J. Yang, Z. Gao, X. Ren, J. Sheng, P. Xu, C. Chang, and Y. Fu, "DeepDigest: prediction of protein proteolytic digestion with deep learning," *Analytical Chemistry,* vol. 93, no. 15, pp. 6094–6103, 2021.
31. M. J. Khoury, M. Gwinn, P. W. Yoon, N. Dowling, C. A. Moore, and L. Bradley, "The continuum of translation research in genomic medicine: how can we accelerate the appropriate integration of human genome discoveries into health care and disease prevention?," *Genetics in Medicine,* vol. 9, no. 10, pp. 665–674, 2007.
32. K. Kawamoto, D. F. Lobach, H. F. Willard, and G. S. Ginsburg, "A national clinical decision support infrastructure to enable the widespread and consistent practice of genomic and personalized medicine," *BMC medical informatics and decision making,* vol. 9, no. 1, pp. 1–14, 2009.
33. S. Weichenthal, M. Hatzopoulou, and M. Brauer, "A picture tells a thousand… exposures: opportunities and challenges of deep learning image analyses in exposure science and environmental epidemiology," *Environment International,* vol. 122, pp. 3–10, 2019.
34. L. Xu and A. Ma, "Coarse-to-fine waterlogging probability assessment based on remote sensing image and social media data," *Geo-spatial Information Science,* vol. 24, no. 2, pp. 279–301, 2021.
35. H. Y. Jang, J. Song, J. H. Kim, H. Lee, I.-W. Kim, B. Moon, and J. M. Oh, "Machine learning-based quantitative prediction of drug exposure in drug-drug interactions using drug label information," *NPJ Digital Medicine,* vol. 5, no. 1, p. 88, 2022.
36. R. Gómez-Bombarelli, J. N. Wei, D. Duvenaud, J. M. Hernández-Lobato, B. Sánchez-Lengeling, D. Sheberla, J. Aguilera-Iparraguirre, T. D. Hirzel, R. P. Adams, and A. Aspuru-Guzik, "Automatic chemical design using a data-driven continuous representation of molecules," *ACS Central Science,* vol. 4, no. 2, pp. 268–276, 2018.
37. D. Minh, H. X. Wang, Y. F. Li, and T. N. Nguyen, "Explainable artificial intelligence: a comprehensive review," *Artificial Intelligence Review,* vol. 55, no. 5, pp. 1–66, 2022.

38. W. Karen, "*GNS aims to help MDs know which treatment will work the best for each patient*," The Boston Globe," ed.
39. U. Ozomaro, C. B. Nemeroff, and C. Wahlestedt, "Personalized medicine and psychiatry: Dream or reality?," *Psychiatric Times*, vol. 30, no. 10, pp. 26–26, 2013.
40. I. Y. Khan, P. Zope, and S. Suralkar, "Importance of artificial neural network in medical diagnosis disease like acute nephritis disease and heart disease," *International Journal of Engineering Science and Innovative Technology (IJESIT)*, vol. 2, no. 2, pp. 210–217, 2013.

CHAPTER 7

AI Models and Ethical Considerations in Research

Amreen Shafique, Muhammad Shahid Saeed, Danish Javeed, and Kashif Naseer Qureshi

7.1 AI AND ETHICAL STANDARDS

Artificial Intelligence (AI) is remarkably promoted as a core pillar to transform the traditional paradigm of scientific innovations, leading the world toward a better and more flourishing living standard. The unmatched potential of AI ranging from sophisticated data analysis, pattern recognition, and predictive modeling, to even creative content generation, has expanded the horizons of research in disciplines as diverse as medicine, social sciences, engineering, and the humanities [1]. As AI-driven research continues to advance along with the expanding domain of its applications, it is imperative to concurrently undertake a comprehensive analysis of the ethical implications associated with these developments [2]. Throughout history, the field of research ethics has been structured based on a set of principles aimed at safeguarding the rights, integrity, and dignity of those who participate as subjects in research studies. In the realm of AI-driven research, ethical considerations are strongly interlaced with AI models. The subject matter encompasses several aspects such as data

DOI: 10.1201/9781032667911-7

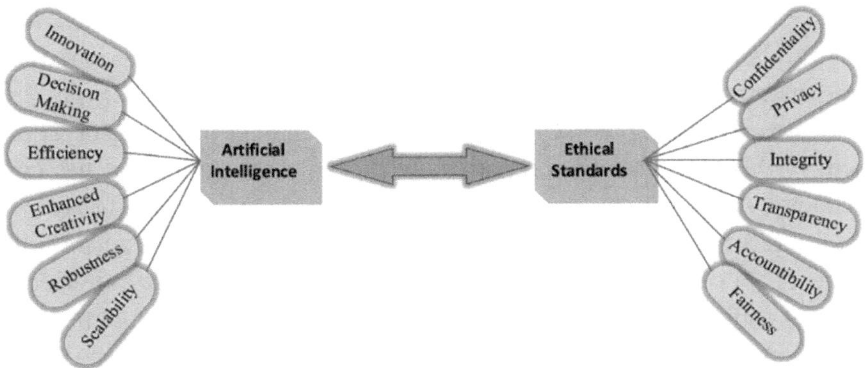

FIGURE 7.1 The interaction between AI and ethical standards.

privacy, algorithmic fairness, transparency, accountability, and the epistemological foundations of insights created by AI [3]. Figure 7.1 projects the interaction between AI and ethical standards.

In recent years, it has been observed that ethical considerations play a pivotal role in shaping the direction and influence of AI-empowered research. Various studies indicate that the lack of transparency and replicability in numerous sophisticated Machine Learning (ML) models presents significant problems to the fundamentals of research ethics. The prioritization of human rights is upheld by ethical values [4]. AI systems are entrusted with the management of sensitive information, such as healthcare records, financial transactions, etc. To safeguard privacy and prevent unauthorized access to data, ethical rules are implemented. These norms serve to uphold the principles of privacy protection and prevent any potential breaches of data security [5]. Furthermore, the establishment of ethical principles contributes to the development of confidence in AI-driven approaches.

Several researchers suggest that the trustworthiness of AI-driven models can be established among consumers, lawmakers, and industry stakeholders by considering common research ethics [6]. These objectives can be achieved through the implementation of transparent, interpretable, and fair AI models. Ethical considerations have an essential role in mitigating biases, hence promoting principles of fairness and diversity. It is an undeniable fact that AI systems that have been trained on biased data have the potential to perpetuate societal assumptions and provide biased outcomes

[7]. Researchers are obligated by ethical guidelines to thoroughly examine data and models to guarantee that AI solutions cater to diverse populations and accurately represent the entirety of human experiences. Researchers provide useful insights regarding the integration of ethical considerations within AI research and claim that AI-driven research approaches contribute to the cultivation of sustainable innovation in the long run [8]. Moral technologies exhibit a higher likelihood of being adopted and integrated, hence assuring their longevity and utility. The incorporation of ethical considerations within AI research contributes to the enhancement of societal well-being, consequently resulting in a more reliable array of innovations [9].

As an uprising paradigm of transformational innovations, AI possesses the capabilities to address enduring challenges, such as enhancing accessibility for individuals with disabilities, optimizing resource allocation among vulnerable populations, etc. AI plays a crucial role in promoting justice, equity, and positive societal transformation while also enhancing efficiency [10]. It is observed that the advantages of conducting ethical research in the field of AI are extensively widespread. It is observed that prioritizing concepts like respect, justice, and transparency is essential for navigating this age of technological development [11]. By adhering to ethical norms, AI models can be built to minimize bias, make good use of training data, and generate findings that humans can comprehend [12]. AI models offer the appropriate means to go beyond the limitations of traditional research, while ethical concerns provide the groundwork for doing so ethically and innovatively. Their interdependence and importance in conducting ethical and fruitful research cannot be emphasized enough [13].

7.1.1 Major Research Contributions

1. This research study aims to explore the chronological interaction between AI models and ethical considerations in advanced research practices.
2. The study investigates the potential impact of AI-driven research approaches on the most common data-related ethical principles.
3. We have addressed the common concerns associated with AI-enabled research approaches and also provided feasible solutions to combat such challenges.

4. We have provided real-life case studies to explain the positive societal impact of ethical considerations and also discussed the possible issues that may arise due to the negligence of modern research ethics.
5. We have figured out the significance and influence of AI ethics on general research attributes.

7.2 LITERATURE REVIEW

The integration of AI in modern research practices is obtaining considerable attention. Literature environs a plethora of contributions made to interpret the interconnection of AI-driven researchers and modern ethical considerations. In this section, we have discussed some of the research studies addressing this domain. Researchers in [14] present a comprehensive examination of the multifaceted aspects of AI, subsequently followed by an analysis of the diverse applications of AI across various contexts. They examine the matters associated with constructing and implementing AI models, including ethical difficulties, privacy concerns, and potential implications for work. The primary objective of this research is to address the ethical considerations that emerge during the development of AI models. Researchers suggest the implications of Human-in-the-Loop (HITL) systems to establish reliable coordination among the data owners and the AI models supposed to train on their data. Researchers in [15] focus on acquiring Explainable Artificial Intelligence (XAI) that interprets the complex computations involved in decision-making by AI models and makes them easy to understand by AI analysts and researchers. The authors in [16] provide an analysis of the present status of constructing XAI models while acknowledging that AI systems possess an opaque nature with limited understanding of their internal workings. The utilization of Trusted Artificial Intelligence (TAI) serves the purpose of fostering confidence in data security, accuracy, cost management, credibility, and decision-making protocols within different sectors and to diverse extents. Researchers in [17] discuss the current challenges to implementing AI approaches. Thus, XAI must be used to integrate more explainable models with high accuracy that allow humans to understand, trust, and manage next-generation innovations. Through the prism of ethics and law, researchers explore certain issues and concerns posed by artificial intelligence [18]. They examine the current legislation that is being discussed in the European Parliament about law and AI.

This will involve looking at a variety of ethical and legal considerations. The authors further recommend a cross-case-analysis technique that enables AI analysts to draw logical relationships between the diversified data streams. This process aims to protect the confidentiality and integrity of sensitive data. Researchers in [19] provide a comprehensive examination of the ethical considerations in AI research, as well as an analysis of the many techniques presented in existing research to address these concerns. They demonstrate that organizations possess a strong awareness of the ongoing debate surrounding AI ethics and have a proactive inclination to aggressively address ethical concerns. They further suggest appropriate data labeling while training complex AI models on segregated users' data. The scheme provides a clear understanding regarding various aspects of large volumes of data during the entire training mechanism of AI frameworks.

The scholars of [20] engaged in a discourse regarding the optimal utilization of AI technology and the establishment of a sound legal framework and regulatory structure. As previously mentioned, the ongoing advancement of suitable systems, particularly the establishment of ethical regulations and standards by various authorities, has laid a solid groundwork for addressing AI-driven challenges in the context of ethical considerations. They recommend applying uncertainty quantification, which involves identifying, measuring, and managing uncertainty in predictions due to factors like noise, insufficient training, model limitations, and data variability. The authors in [21] undertake an examination of the consensus surrounding the importance of principles related to AI, while also identifying the various obstacles that may hinder the widespread acceptance of ethical standards in the field of AI. The results of this study serve as the first data for the development of a maturity model that evaluates the ethical capacities of AI systems and offers recommendations for enhancing them. The research suggests that the use of optimal feature engineering in AI models enhances data interpretation and performance by selecting, transforming, and creating features that improve the model's clarity and unambiguity in predictions. Authors in [22] also emphasize employing optimal feature engineering strategies for AI approaches. They explain that AI has proliferated across various fields, including but not limited to business, science, art, and education. Feature engineering enables AI models to interact only with the concerned segment of diversified users' data rather than operating on entire data streams. It has a core objective of augmenting user experience, optimizing work productivity, and generating a multitude of prospective employment prospects. Nevertheless, there is a lack of comprehensive exploration regarding public comprehension of AI technology and the establishment of a clear definition for AI literacy.

Researchers in [23] address the utilization of chatbots for overseeing online interactions with customers. The study explains that chatbots are a vivid source of ambiguity that arises from several factors, such as language processing. The integration of optimal feature engineering approaches will act as a sensitive filter to prohibit chatbots from accessing irrelevant information. This phenomenon will directly result in avoiding unwanted exposure of users' data to complex AI models. The comprehensive literature review is summarized in Table 7.1.

7.3 ETHICAL CONCERNS IN AI-DRIVEN RESEARCH

In AI-driven research approaches, multifaceted ethical considerations play a prominent role in ensuring reliable implications of AI technologies in practical research applications. Figure 7.2 shows a visual categorization of ethical concerns in AI-driven research.

7.4 DATA-RELATED CONCERNS

7.4.1 Data Collection and Aggregation

The field of AI-driven research is seeing tremendous growth and necessitates a substantial volume of data. Nevertheless, the emphasis on statistics frequently neglects ethical considerations. The main ethical challenge is the issue of consent, which requires individuals to be aware of the collection of their data and provide explicit agreement. Privacy is a significant problem since the gathering and examination of data might reveal extremely personal information. While working with the sensitive element of user data, researchers bear an ethical obligation to protect the privacy of individuals. It also includes guaranteeing that their identities are not revealed without their explicit agreement. It is experienced that implementing biased data-gathering methods often results in biased AI outputs, which necessitates a comprehensive investigation of the whole data generation and processing mechanism [24].

7.4.2 Data Privacy and Confidentiality

The emerging field of AI primarily depends upon data, which is sometimes associated with privacy concerns, confidentiality concerns, and ethical considerations. While working on extensive datasets containing personal information, researchers must have a thorough understanding

TABLE 7.1 Comprehensive Literature Review

Solutions	Contribution	Ethical Concerns	Challenges in AI Research
HITL systems [14]	Researchers present an overview of AI applications and the significance of Ethical consideration	Privacy, confidentiality	Robustness and accuracy
XAI [15]	The authors explored the current ethical challenges in AI-driven research	Data Confidentiality	Effective training of AI models
TAI, XAI [16]	The common Ethical challenges in AI-technologies are discussed	Lack of transparency	Complex computations to ensure transparency
XAI [17]	Researchers analyzed the need for transparent approaches in traditional AI-driven research	Confidentiality, Integrity	Implementations of explainable strategies
Cross-Case Analysis [18]	The authors mentioned the need for transparency to satisfy ethical principles	Lack of fairness	Interpretation of the decision-making process
Data Labelling [19]	A comprehensive analysis of ethical concerns is presented	Accountability, trustworthiness	Ensuring transparency in AI processing
Uncertainty Quantification [20]	Researchers discussed the current challenges in AI-inspired techniques	Privacy, confidentiality	Availability of data
Feature Engineering [21]	Authors addressed the ethical challenges in AI-driven approaches	Reliability, trustworthiness	Developing explainable mechanism
Feature Engineering [22]	The competencies and challenges of AI-based research are discussed	Fairness, privacy	To achieve accuracy and transparency
Feature Engineering [23]	Researchers investigate the use of chatbots along with associated ethical concerns	Confidentiality	Ensuring accurate and human-centric suggestion

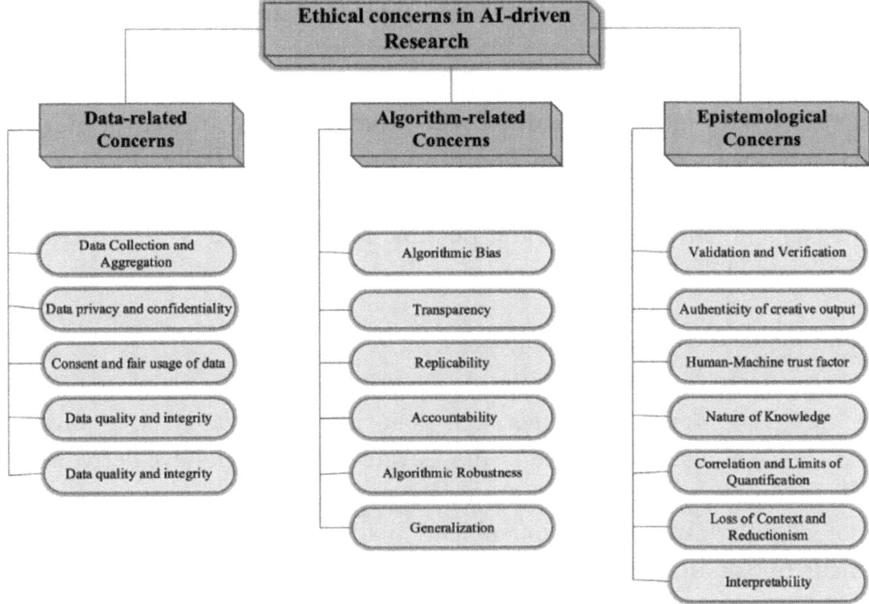

FIGURE 7.2 Ethical concerns in AI-driven research.

of the ethical consequences associated with data analysis [25]. The significance of acquiring informed consent from the data owners cannot be overstated, as it is a fundamental ethical guideline in data-oriented AI research. Furthermore, the process of data reduction is crucial in ensuring that only pertinent and significant data is utilized in the research. Additional practices can also be followed to protect the identities and personal information of individuals participating in certain research. Some researchers suggest using anonymization and pseudonymization processes in this regard to ensure the privacy of users [26].

7.4.3 Consent and Fair Usage of Data

Data plays an essential part in enabling remarkable advancements within the field of AI research. The applications of data in AI research not only offer impressive opportunities, but they also generate notable ethical quandaries that demand notable attention. There are two core ethical and societal reasons to explain the significance of data. To begin with, the act of gathering and structuring data necessitates the formulation of assumptions to figure out noteworthy, valuable, or substantial data. Secondly, data that has been digitally encoded permits the duplication, transmission, and transformation of information with unprecedented efficiency. As far as data collection is concerned, obtaining permission and ensuring equitable use of personal data are considered the vital pillars of ethical conduct. Such sensitive data must be obtained with prior consent, respecting individual self-governance, and safeguarding personal data rights. Participants should be aware of a thorough elucidation about the utilization of data, the steps taken for storage, and possible risks involved while processing their data [27].

7.4.4 Data Quality and Integrity

AI research highly recommends using high-quality data quality to achieve remarkable outcomes such as accuracy, relevance, and comprehensiveness. Using outdated or inaccurate datasets can have detrimental effects on the performance of models, particularly when they are associated with human lives or societal institutions. Researchers suggest ensuring the integrity of data, which includes maintaining the authenticity and reliability of data from the moment it is gathered until it is processed. It is also a recommended approach to collect data with great attention, conduct comprehensive validation, and maintain constant monitoring [28]. Researchers ought to furnish exhaustive details regarding their data sources and constraints to avoid unwanted situations, e.g., obtaining fraudulent data, inaccurate facts, etc. As AI advances, it becomes crucial to prioritize data quality and integrity to uphold ethical standards in research.

7.4.5 Data Ownership

In AI research, data ownership-related ethical concerns also demand substantial steps to be taken. The stakeholders such as the individuals and data collection entity should pay full regard to the rightful owners of the data. The data owners should also be briefly educated about the usage of their data, the processing mechanism, and the possible outcomes of this process. When dealing with critical scenarios such as healthcare or finance, the data collection entities should explain the entire mechanism to the data owners in a more detailed manner. However, the inclusion of private information in AI research introduces ethical complexities, as individuals own their entitlement to personal data. Nevertheless, it is essential to consider the rights of others and the overall welfare of society [29].

7.5 ALGORITHM-RELATED CONCERNS

7.5.1 Algorithmic Bias

The increasing use of AI in different fields of study and practical implementations has led to the emergence of numerous intricate ethical and technical concerns, and algorithmic bias is prominent among these concerns. Algorithmic bias leads to uncongenial circumstances, including inappropriate collection of data, irregularities in data processing, falsification of data facts, and many more. There are several possible reasons for this biased nature of the algorithm; however, the main reason is the structure of the algorithm itself. The issue can be resolved by investigating the overall structure of the algorithm, meeting modernized technical standards, and integrating privacy-preserving techniques with the central data processing algorithms [30].

7.5.2 Transparency

The rapid incorporation of AI into sensitive domains such as healthcare and finance has raised concerns over the transparency of AI models. Ambiguity in AI systems, particularly deep learning networks, can erode trust and lead to uncontrolled biases. Openness is crucial for maintaining the legitimacy and impartiality of AI decision-making processes. Without it, there is a risk of perpetuating unjust biases. Explainable Artificial Intelligence (XAI) models are currently being developed to address the challenges associated with balancing performance and interpretability [31]. Ensuring the integration of transparency into the ethical standards and optimal methodologies for AI research is highly crucial.

7.5.3 Replicability

Replicability is an essential component of scientific research that includes performing multiple experiments on a unified pattern to evaluate the actual accuracy of the experiment. However, AI research presents a barrier due to the ethical consequences it may have. This phenomenon may raise questions regarding the overall accuracy of the experiments. It is more likely that erroneous or fraudulent results will be obtained when there is no ability to replicate the experiment. Secondly, it leaves no space to verify the authenticity, integrity, and actuality of the experiment performed under a certain scenario. Some additional challenges that hinder the replicability of AI research include relying on proprietary datasets and technologies, hyperparameters, and the stochastic nature of AI systems [32].

7.5.4 Accountability

The rapid development of AI has raised ethical concerns about accountability, especially in sectors like healthcare and financial prognostications. It urges the need for a surveillance mechanism to oversee all the processes associated with sensitive data. The opaque nature of AI systems, particularly deep learning models, complicates the issue and creates a valid need for some regulatory body to govern the entire process. The absence of a well-defined system of accountability can lead to unaddressed errors or biases, potentially resulting in unjust or inequitable consequences. Moreover, the individuals and entities working on sensitive personal information need to be accountable in all aspects. All these mentioned concerns necessitate a comprehensive regulatory plan to emphasize transparency, interpretability, and understanding of the rationale and logic behind AI decisions [33].

7.5.5 Robustness

The robustness of AI algorithms is also a notable concern, particularly in critical applications like medical diagnostics and autonomous driving. Ethical considerations emerge due to the possibility that AI systems could render detrimental decisions, exploit malevolent entities, and undermine public confidence [34]. To address this, a comprehensive approach is needed that includes adversarial training methods, dropout data augmentation, ensemble methods, and continual learning approaches. Additionally, the AI models should be capable enough to interpret the backend mechanism of all the quick decisions that were made based on sensitive personal information. This ethical obligation is not just a technical challenge but a moral commitment to ensure AI operates reliably, safely, and in the best interests of all stakeholders [35].

7.5.6 Generalization

Along with numerous influencing parameters, the importance of generalization in the field of AI research cannot be overstated. It directly impacts the accuracy and dependability of AI models when applied to novel and previously unknown data. The common practice of using oversimplified assumptions can result in phenomenal performance in controlled environments; however, it becomes ineffective when applied to real-world scenarios [36]. This phenomenon can have detrimental consequences that directly impact the overall performance and outcomes of AI models.

While working in vital sectors such as healthcare or transportation, the sensitivity of this issue further increases. The mitigation of generalization necessitates the use of novel methodologies and the adoption of systematic modifications in research protocols [37].

7.6 EPISTEMOLOGICAL CONCERNS

7.6.1 Validation and Verification

AI models are increasingly influencing human lifestyles, making validation and verification crucial in AI research, especially when these models are strongly interlinked with personal information. Inadequate validation can lead to unforeseen consequences and latent flaws, e.g., exacerbating societal inequalities, misinterpretation of facts, falsification of scientific data, etc. Collaborative efforts like shared datasets and benchmarking platforms can standardize validation efforts to regulate the performance of AI models. As these AI models become more influential, the ethical imperative for rigorous validation and verification becomes essential to ensure that AI tools are effective and ethically sound [32].

7.6.2 Authenticity of Creative Output

The credibility of the output produced by AI models, particularly in creative tasks, has become a subject of controversy due to their development. In contrast to human creativity, which is frequently influenced by cultural backgrounds, personal experiences, and emotions, AI creativity is strongly dependent upon the provided data. AI models provide creative outcomes when algorithms process massive amounts of data to produce outputs that are coherent and contextually pertinent. In most cases, the automated output of AI models is kept hidden, leading to an improper situation that reduces the trustworthiness of actual creative output by humans. Researchers have a responsibility to differentiate between insights produced by humans and those that are automated [38].

7.6.3 Human-Machine Trust Factor

The ongoing development of AI models has introduced a novel method for interaction between humans and machines. However, for secure data exchange between both parties, there must be a strong trust bond. The establishment and preservation of trust between human beings and AI systems are of utmost importance for the effective incorporation of these technologies across diverse domains, including healthcare and finance.

Additionally, it gives confidence to individuals to share their personal information with machines. The concept of the "human-machine trust factor" refers to the level of trust that individuals have in the ability of AI systems to effectively and consistently carry out activities in a manner that is transparent and ethical [39].

7.6.4 Nature of Knowledge

In the age of AI, the very nature of knowledge traditionally seen as the product of human cognition and experiential learning is being redefined. AI models, such as GPT-4, amass and process vast amounts of information at unprecedented scales, blurring the boundaries between data, information, and knowledge. AI models generate their understanding through the analysis of vast amounts of data, relying on pattern recognition rather than the multifaceted approach employed by humans. While individuals acquire knowledge from a combination of personal experiences, cultural backgrounds, and educational systems, AI algorithms solely rely on processing large datasets to derive useful information [40]. Researchers are worried about the ethical consequences that might arise from this divergence. There are strong possibilities that, based on the processed decisions, people mistakenly believe that AI-generated content is the same as human cognition, ignoring the fact that human wisdom is complex and depends on context. AI models can also be influenced by biased or inaccurate data sets during training and validation. That suspicious activity could remain prejudiced and lead to misleading outcomes [41].

7.6.5 Correlation and Limits of Quantification

The concept of correlation possesses crucial importance in the operational analysis of AI algorithms. It involves drawing logical relationships among various segregated aspects of large datasets to form a comprehensive idea about the generalization of the content. Machine learning algorithms frequently employ the detection of correlations within datasets to generate predictions or categorize data for particular objective functions. However, traditional machine learning approaches sometimes become susceptible to a variety of quantification issues. AI models exhibit remarkable efficacy when trained on extensive datasets composed predominantly of quantifiable data. However, when it comes to providing a comprehensible interpretation of such complex models while ensuring a mechanism that protects privacy via quantification methods, researchers confront significant ethical challenges [42].

7.6.6 Loss of Context and Reductionism

In AI-driven research practices, the widespread use of large datasets has raised concerns about the potential consequences of losing context and oversimplifying complex information. In rich-contextual datasets, it becomes relatively more complex to understand the correlation among various data points. To meet such objectives, there exists a diversified array of algorithms to construct such rational relationships. The models created to analyze complex data patterns can unintentionally produce results that slightly resemble real-world occurrences. This can not only lead to falsification of facts but, from an ethical perspective, raise numerous other difficulties as well. Omitting contextual factors in AI models can result in irrelevant outcomes, jeopardize sensitive data, and diminish the overall dependability and credibility of these systems [43].

7.6.7 Interpretability

The flourishing circle of AI applications leads to a pool of complex algorithms designed for various AI research practices. The complex nature of such models is beneficial in terms of efficient decision-making; however, it also possesses the potential to anonymize the actual working process of such frameworks. Interpretability refers to the clarity and understandability of the processes and decisions made by such systems. As AI models become more complex, they are more likely to behave as "black boxes" with decision-making processes that are opaque and not easily understood by researchers. From an ethical research perspective, this lack of interpretability raises significant concerns. It also becomes a challenging task to validate the accuracy and identify potential biases of AI models without having a clear understanding of these models. To address these issues, researchers suggest the implication of XAI with traditional AI approaches to make AI models more rational and interpretable [44].

7.7 IMPACT OF AI ETHICS ON RESEARCH ATTRIBUTES

7.7.1 Enhanced Trustworthiness

In the advanced realm of AI research, the interconnection between ethical considerations and the enhanced trustworthiness of AI systems possesses substantial importance. Researchers recommend following a strict code of ethics while working with AI research approaches. The code must provide a clear explanation of the DL algorithm, the data processing mechanism, and the expected outcomes of this data processing. Once the end users or the data owners become aware of these facts, they will develop a sense of confidence in AI research approaches, leading toward enhanced trustworthiness and reliability of AI research practices.

7.7.2 Reduced Bias

The use of AI in research has opened a new paradigm to examine heterogeneous datasets. These big models are capable enough to meet desired objectives; however, they often show unfairness in their training data. This can lead to slanted or unfair results, casting a negative influence on the generalized credibility of the system. In AI research, it becomes crucially important to follow a valid code of ethics to ensure all the errors and biased decisions are effectively mitigated. Ethical considerations also enable researchers to carefully check the sources of data used to train AI. Some highly recommended methods include fairness checks, understanding tools, and lowering biases in AI models [45].

7.7.3 Robust Decision-Making

The fusion of AI models into the research domain has revolutionized decision-making capabilities, promising both speed and scalability. Ethical considerations in research demand a strict commitment to three basic principles, including transparency, accountability, and interpretability in AI frameworks [46]. The AI models should instantly respond to complex scenarios and must perform robust decision-making. Additionally, AI approaches must be capable of explaining the backend computational processes involved in decision-making. This remarkable approach mitigates the risks of over-reliance on AI, ensuring decisions made by AI are both data-informed and contextually sound [47].

Figure 7.3 shows the impact of AI ethics on research attributes.

7.7.4 Risk Mitigation

As AI models become increasingly important in research, it is crucial to understand and reduce the risks associated with such schemes. Ethical considerations act as a leading guide of moral rules to fix complex data-oriented problems before they happen in AI research. The common ethical laws include societal obligations, e.g., not exposing the identities of data owners, making sure that AI models do not spread unfair treatment or confusing outputs, etc. To avoid possible risks, researchers must have a vivid understanding of all the loopholes in AI processes. This approach ensures smooth operations of AI models and provides an insightful way to fix problems before they get worse [48].

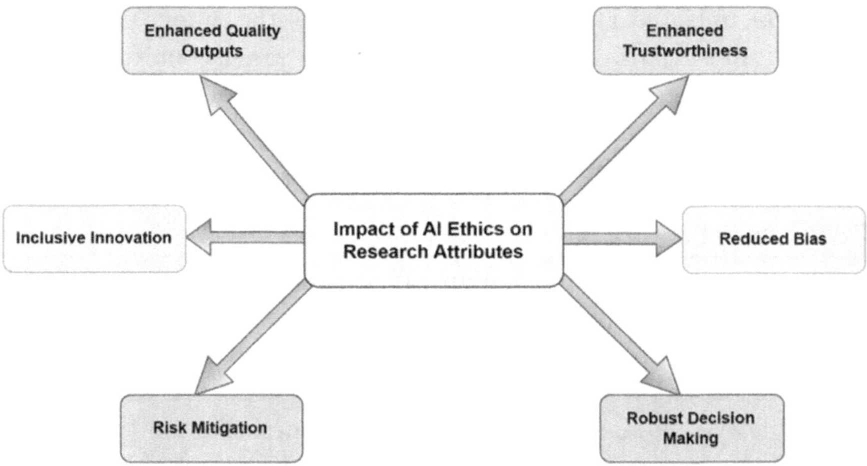

FIGURE 7.3 Impact of AI ethics on research attributes.

7.7.5 Inclusive Innovation

The growing influence of AI models in various practical domains yields a never-ending space of scientific and technological innovations. Without comprehensive ethical considerations, there are valid probabilities that such phenomenal innovations cater only to dominant groups or underrepresented populations [49]. To address such concerns, ethical guidelines in AI research highly emphasize the importance of diverse and representative data sourcing. By ensuring that training data encompasses a wide spectrum of experiences and backgrounds, AI models are better positioned to innovate in ways that are relevant and beneficial to overall community standards. Ethical considerations also prompt researchers to critically evaluate the output of AI models to make sure that innovations do not reinforce biases [50].

7.7.6 Enhanced Quality Outputs

The integration of AI models into conventional research methodologies offers unprecedented opportunities for producing highly efficient and scalable outputs. However, the quality of these outputs is directly contingent upon the ethical considerations guiding the deployment of these models. Ethical considerations ensure that AI models operate transparently and should be open to scrutiny, feedback, and validation. This transparency allows for rigorous iterative refinement, ensuring that any anomalies or errors are promptly addressed [51]. By ensuring that AI models are trained on diverse, unbiased, and comprehensive datasets, the outputs are not only more accurate but also more generalizable across varied contexts and populations. Additionally, ethics emphasizes the importance of interpretability in AI, ensuring that model decisions are not just statistically sound but also contextually meaningful and easily understood by human experts [52].

7.8 CONCLUSION

This study explores the affiliation of AI models in various research domains and sheds light on the ethical concerns associated with this integration, along with recommended solutions to address such concerns. As the flourishing space of AI applications surrounds every major facet of our lifestyle, human dependencies on AI approaches are accordingly increasing. In most scenarios, AI is directly interlinked with the sensitive personal information of individuals. In AI-oriented research methodologies, the utilization of this data undoubtedly yields substantial outcomes for the betterment of humanity; however, it also necessitates a strict code of ethics regarding this utilization of data. Hence, it becomes an obligatory practice for researchers to ensure that all operations are justifiable on ethical grounds. There must be credible checks to ensure every single aspect of data utilization is morally and ethically valid, along with technological achievements obtained by processing this data. By intertwining these ethical guidelines with AI methodologies, researchers can ensure a balanced trajectory where innovation is harmoniously aligned with moral values, fairness, transparency, and accountability. In contrast, neglecting such ethical considerations jeopardizes the trust and efficacy of AI models and reduces the trustworthiness of AI models in this burgeoning era of AI-infused research.

REFERENCES

1. C. Sanderson *et al.*, "AI ethics principles in practice: Perspectives of designers and developers," *IEEE Transactions on Technology and Society,* vol. 4,no. 2, pp. 171–187, 2023.
2. A. A. Khan *et al.*, "AI ethics: An empirical study on the views of practitioners and lawmakers," *IEEE Transactions on Computational Social Systems,* 2023.
3. K. N. Qureshi, G. Jeon, and F. Piccialli, "Anomaly detection and trust authority in artificial intelligence and cloud computing," *Computer Networks,* p. 107647, 2020. https://doi.org/10.1016/j.comnet.2020.107647.
4. İ. Kök, F. Y. Okay, Ö. Muyanlı and S. Özdemir, "Explainable artificial intelligence (xai) for internet of things: A survey," *IEEE Internet of Things Journal,* vol. 10, no. 16, pp. 14764–14779, 2023.

5. K. N. Qureshi, A. Ahmad, F. Piccialli, G. Casolla, and G. Jeon, "Nature-inspired algorithm-based secure data dissemination framework for smart city networks," *Neural Computing and Applications,* pp. 1–20.
6. Y. Wu, "Ethically Responsible and Trustworthy Autonomous Systems for 6G," in IEEE Network, vol. 36, no. 4, pp. 126–133, July/August 2022, doi: 10.1109/MNET.005.2100711.
7. A. Čartolovni, A. Tomičić, and E. L. Mosler, "Ethical, legal, and social considerations of AI-based medical decision-support tools: A scoping review," *International Journal of Medical Informatics,* vol. 161, p. 104738, 2022.
8. K. N. Qureshi, T. Newe, G. Jeon, and A. Chehri, "Internet of everything: Evolution and fundamental concepts," in *Cybersecurity Vigilance and Security Engineering of Internet of Everything*, K. Naseer Qureshi, T. Newe, G. Jeon, and A. Chehri Eds. Cham: Springer Nature Switzerland, 2024, pp. 3–20.
9. P. Detopoulou *et al.*, "Artificial intelligence, nutrition, and ethical issues: A mini-review," *Clinical Nutrition Open Science,* vol. 50, pp. 46–56, 2023.
10. H.-M. Heyn, E. Knauss, and P. Pelliccione, "A compositional approach to creating architecture frameworks with an application to distributed AI systems," *Journal of Systems and Software,* vol. 198, p. 111604, 2023.
11. K. Ahmad, M. Maabreh, M. Ghaly, K. Khan, J. Qadir, and A. Al-Fuqaha, "Developing future human-centered smart cities: Critical analysis of smart city security, data management, and ethical challenges," *Computer Science Review,* vol. 43, p. 100452, 2022.
12. D. Javeed, M. S. Saeed, P. Kumar, A. Jolfaei, S. Islam, and A. N. Islam, "Federated learning-based personalized recommendation systems: An overview on security and privacy challenges," *IEEE Transactions on Consumer Electronics,* vol. 70, pp. 2618–2627, 2023.
13. A. Iftikhar and K. N. Qureshi, "Future privacy and trust challenges for IoE networks," in *Cybersecurity Vigilance and Security Engineering of Internet of Everything*: edited by K. N. Qureshi, Springer Nature Switzerland, 2023, pp. 193–218.
14. M. Wilchek, W. Hanley, J. Lim, K. Luther, and F. A. Batarseh, "Human-in-the-loop for computer vision assurance: A survey," *Engineering Applications of Artificial Intelligence,* vol. 123, p. 106376, 2023.
15. J. R. Schoenherr, R. Abbas, K. Michael, P. Rivas, and T. D. Anderson, "Designing AI using a human-centered approach: Explainability and accuracy toward trustworthiness," *IEEE Transactions on Technology and Society,* vol. 4, no. 1, pp. 9–23, 2023.
16. V. Chamola, V. Hassija, A. R. Sulthana, D. Ghosh, D. Dhingra, and B. Sikdar, "A review of trustworthy and explainable artificial intelligence (xai)," *IEEE Access,* 2023.
17. Z. Zhang, H. Al Hamadi, E. Damiani, C. Y. Yeun, and F. Taher, "Explainable artificial intelligence applications in cyber security: State-of-the-art in research," *IEEE Access,* vol. 10, pp. 93104–93139, 2022.

18. S. Chauhan and A. Keprate, "Standards, Ethics, Legal Implications & Challenges of Artificial Intelligence," in *2022 IEEE International Conference on Industrial Engineering and Engineering Management (IEEM)*, 2022: IEEE, pp. 1048–1052.
19. B. C. Stahl, J. Antoniou, M. Ryan, K. Macnish, and T. Jiya, "Organisational responses to the ethical issues of artificial intelligence," *AI & SOCIETY*, vol. 37, no. 1, pp. 23–37, 2022.
20. S. Yu and F. Carroll, "Implications of AI in national security: Understanding the security issues and ethical challenges," in *Artificial Intelligence in Cyber Security: Impact and Implications: Security Challenges, Technical and Ethical Issues, Forensic Investigative Challenges*: Springer, 2022, pp. 157–175.
21. A. A. Khan *et al.*, "Ethics of AI: A systematic literature review of principles and challenges," in *Proceedings of the 26th International Conference on Evaluation and Assessment in Software Engineering*, 2022, pp. 383–392.
22. D. T. K. Ng, J. K. L. Leung, S. K. W. Chu, and M. S. Qiao, "Conceptualizing AI literacy: An exploratory review," *Computers and Education: Artificial Intelligence*, vol. 2, p. 100041, 2021.
23. G. Murtarelli, A. Gregory, and S. Romenti, "A conversation-based perspective for shaping ethical human–machine interactions: The particular challenge of chatbots," *Journal of Business Research*, vol. 129, pp. 927–935, 2021.
24. D. Javeed, T. Gao, M. S. Saeed, P. Kumar, R. Kumar, and A. Jolfaei, "A softwarized intrusion detection system for IoT-enabled smart healthcare system," *ACM Transactions on Internet Technology*, 2023.
25. T. Hagendorff, "A virtue-based framework to support putting AI ethics into practice," *Philosophy & Technology*, vol. 35, no. 3, p. 55, 2022.
26. S. Thiebes, S. Lins, and A. Sunyaev, "Trustworthy artificial intelligence," *Electronic Markets*, vol. 31, pp. 447–464, 2021.
27. A. Tsamados *et al.*, "The ethics of algorithms: Key problems and solutions," *Ethics, Governance, and Policies in Artificial Intelligence*, pp. 97–123, 2021.
28. G. Karimian, E. Petelos, and S. M. Evers, "The ethical issues of the application of artificial intelligence in healthcare: a systematic scoping review," *AI and Ethics*, vol. 2, no. 4, pp. 539–551, 2022.
29. S. Zafar, Z. Lv, N. H. Zaydi, M. Ibrar, and X. Hu, "DSMLB: Dynamic switch-migration based load balancing for software-defined IoT network," *Computer Networks*, vol. 214, p. 109145, 2022.
30. X. Li, P. Ye, J. Li, Z. Liu, L. Cao, and F.-Y. Wang, "From features engineering to scenarios engineering for trustworthy AI: I&I, C&C, and V&V," *IEEE Intelligent Systems*, vol. 37, no. 4, pp. 18–26, 2022.
31. D. Javeed, T. Gao, M. S. Saeed, and M. T. Khan, "FOG-empowered augmented intelligence-based proactive defensive mechanism for IoT-enabled smart industries," *IEEE Internet of Things Journal*, vol. 21, pp. 18599–18608, 2023.
32. J. R. Saura, D. Ribeiro-Soriano, and D. Palacios-Marqués, "Assessing behavioral data science privacy issues in government artificial intelligence deployment," *Government Information Quarterly*, vol. 39, no. 4, p. 101679, 2022.

33. X. Wang, H. Zhu, Z. Ning, L. Guo, and Y. Zhang, "Blockchain intelligence for internet of vehicles: Challenges and solutions," *IEEE Communications Surveys & Tutorials,* vol. 25, pp. 2325–2355, 2023.
34. D. Javeed, T. Gao, M. S. Saeed, and P. Kumar, "An intrusion detection system for edge-envisioned smart agriculture in extreme environment," *IEEE Internet of Things Journal,* pp. 1-1, 2023.
35. R. Srinivasan and B. S. M. González, "The role of empathy for artificial intelligence accountability," *Journal of Responsible Technology,* vol. 9, p. 100021, 2022.
36. B. Attard-Frost, A. De los Ríos, and D. R. Walters, "The ethics of AI business practices: A review of 47 AI ethics guidelines," *AI and Ethics,* vol. 3, no. 2, pp. 389–406, 2023.
37. F. Wahab *et al.*, "An AI-driven hybrid framework for intrusion detection in IoT-enabled E-health," *Computational Intelligence and Neuroscience,* vol. 2022, 2022.
38. M. S. Saeed *et al.*, "Power management in smart grid for residential consumers," in *Advances on P2P, Parallel, Grid, Cloud and Internet Computing: Proceedings of the 12th International Conference on P2P, Parallel, Grid, Cloud and Internet Computing (3PGCIC-2017)*, 2018: Springer, pp. 415–423.
39. M. Agbese, R. Mohanani, A. Khan, and P. Abrahamsson, "Implementing AI Ethics: Making Sense of the Ethical Requirements," in *Proceedings of the 27th International Conference on Evaluation and Assessment in Software Engineering*, 2023, pp. 62–71.
40. A. L. Hunkenschroer and C. Luetge, "Ethics of AI-enabled recruiting and selection: A review and research agenda," *Journal of Business Ethics,* vol. 178, no. 4, pp. 977–1007, 2022.
41. T. Heyder, N. Passlack, and O. Posegga, "Ethical management of human-AI interaction: Theory development review," *The Journal of Strategic Information Systems,* vol. 32, no. 3, p. 101772, 2023.
42. E. Prem, "From ethical AI frameworks to tools: a review of approaches," *AI and Ethics,* pp. 1–18, 2023.
43. D. Xiao, P. Meyers, J. S. Upperman, and J. R. Robinson, "Revolutionizing healthcare with ChatGPT: An early exploration of an AI language model's impact on medicine at large and its role in pediatric surgery," *Journal of Pediatric Surgery,* vol. 58, no. 12, pp. 2410–2415, 2023.
44. D. Ueda *et al.*, "Fairness of artificial intelligence in healthcare: Review and recommendations," *Japanese Journal of Radiology,* vol. 42, pp. 1–13, 2023.
45. D. Javeed, T. Gao, M. S. Saeed, R. U. Khan, and Z. Jamil, "SDSCCM: Secure distributed system communication for cloud-based manufacturing," in *Protecting User Privacy in Web Search Utilization*: IGI Global, 2023, pp. 200–214.
46. J. F. N. B. Cortese, F. G. Cozman, M. P. Lucca-Silveira, and A. F. Bechara, "Should explainability be a fifth ethical principle in AI ethics?," *AI and Ethics,* vol. 3, no. 1, pp. 123–134, 2023.

47. J. de Pagter, "From EU Robotics and AI governance to HRI research: Implementing the ethics narrative," *International Journal of Social Robotics*, pp. 1–15, 2023.
48. H. Nizam, S. Zafar, Z. Lv, F. Wang, and X. Hu, "Real-time deep anomaly detection framework for multivariate time-series data in industrial iot," *IEEE Sensors Journal*, vol. 22, no. 23, pp. 22836–22849, 2022.
49. J.-M. John-Mathews, "Some critical and ethical perspectives on the empirical turn of AI interpretability," *Technological Forecasting and Social Change*, vol. 174, p. 121209, 2022.
50. V. Galaz *et al.*, "Artificial intelligence, systemic risks, and sustainability," *Technology in Society*, vol. 67, p. 101741, 2021.
51. S. A. A. Bokhari and S. Myeong, "Use of artificial intelligence in smart cities for smart decision-making: A social innovation perspective," *Sustainability*, vol. 14, no. 2, p. 620, 2022.
52. B. Patiño-Valencia, M. L. Villalba-Morales, M. Acosta-Amaya, C. Villegas-Arboleda, and E. Calderón-Sanín, "Towards the conceptual understanding of social innovation and inclusive innovation: a literature review," *Innovation and Development*, vol. 12, no. 3, pp. 437–458, 2022.

CHAPTER 8

AI Models and Applications of AI Models

Seemab Kareem, Sidra Zubair,
and Kashif Naseer Qureshi

8.1 INTRODUCTION

Artificial Intelligence (AI) models are used to simulate human-like intelligence to perform different tasks. These computational algorithms have been adopted for various tasks such as data analysis, predictions, and recognition patterns. Machine learning models are also well known and learn and train the data for decisions and predictions. These models are further categorized into supervised, unsupervised, and reinforcement learning models. Then, supervised and unsupervised models have further techniques. The well-known supervised models are decision trees, random forests, and neural networks. The unsupervised models are K-means, clustering, and generative adversarial networks. In addition, the Natural Language Processing (NLP) models are also used to process and understand the languages such as word embedding models, text generation, and sequence-to-sequence models. The well-known examples of word embedding models are Word2Vec and Global Vector for Word Representation (GloVe), whereas the sequence-to-sequence methods are Long Short-Term

DOI: 10.1201/9781032667911-8

Memory (LSTM) and Bidirectional Encoder Representations (BERT). The text generation models are Generative Pre-Trained Transformers (GPT).

Computer vision models are also used to understand the visual information from video and images such as Generative Adversarial Networks (GANs), Conventional Neural Networks (CNNs), and Spatial Transformer Networks (STNs). These models are used for image classification, transformation segmentation, and object detection. There are other types of AI models like knowledge graphs and graph neural networks used for reason over structure data in the form of graphs. The well-known examples in this category are Graph Conventional Networks (GCNs) and knowledge graph embeddings [1]. These AI models are used in different applications for healthcare, finance, retail, marketing, and robotics. This chapter discusses in detail the AI models which are used in research technologies. The selection of an AI model depends on specific task requirements, computational resources, and available data.

8.2 ARTIFICIAL NEURAL NETWORKS

Artificial Neural Networks (ANNs) have rapidly surged in prominence and proved their usefulness as a model for tasks such as categorization of the data, grouping of data, recognition of patterns, and predictions across a wide range of areas. ANNs can be built and utilized for image identification, Natural Language Processing (NLP) can manage issues in engineering, trading commodities, agriculture, science, medicine, education, finance, security, and management. For more detailed examples, consider the widespread use of ANNs in processes like hepatitis diagnosis, recognizing speech, recovering data from corrupt software in telecommunications, deciphering multilingual messages, three-dimensional identification of objects, analysis of texture, recognition of facial features, undersea mine recognition, and handwritten word identification [2].

Biological Neural Networks (BNNs) serve as the inspiration for algorithm-based ANN systems. In the 1940s, mathematicians Warren McCulloch and Walter Pitts made an algorithm-based model that was simple to reproduce the human brain's activities, opening the door for research into the application of neural networks in AI [3]. The human brain, which is linked to transmitting and receiving impulses for human behavior, is a typical example of an ANN. It is made up of numerous complexly linked neuronal cells in the shape of components that collaborate regularly to find solutions to certain issues. Artificial neurons make up a network of neurons, which have multiple input and multiple output

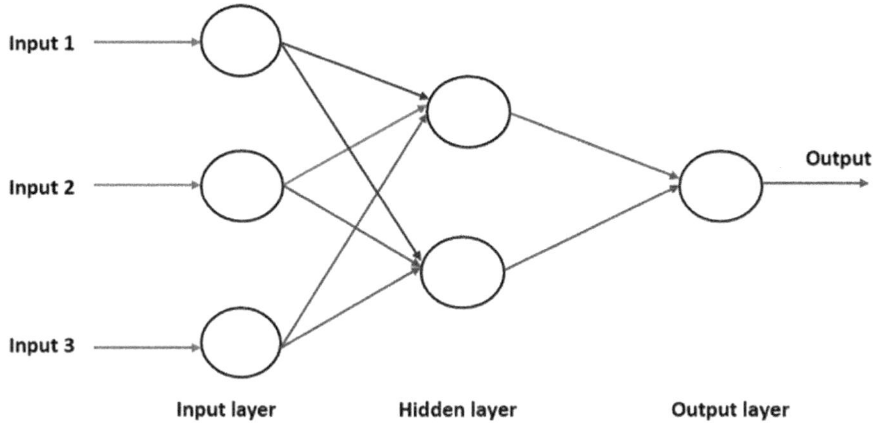

Input 1
Input 2
Input 3
Output
Input layer
Hidden layer
Output layer

FIGURE 8.1 Architectural representation of ANN.

systems. The main job of a neural network is to translate inputs into useful outputs [3]. All the neurons in a network are linked to one another, and it has three layers which are an "input layer," one or many "hidden layers," and an output layer [4]. Information from the input layer is received by a hidden layer, which then handles all processing, and hidden layers in the networks may differ and depend on the size and type of the problem.

The output layer receives information from hidden layers that have been processed and provides results to the outside user [5]. Since the layers of an NN are not dependent on each other, every layer may contain any quantity of nodes. This disorderly collection of nodes is described as a "bias node." Frequently, the bias nodes are set to 1. A bias main function provides a network node with an unchanged value in addition to the normal inputs it receives. The output layer's nodes are linked to the nodes of the input layer, while the nodes of the hidden layer are linked to the nodes of the input layer. Each connection can send a signal to neighboring neurons, just like synapses do in the human brain. The network sends data to the input layer. The hidden layer receives and processes the unprocessed data from the input layer. After that, the acquired value is given to the output layer, which also does an analysis of the data from the hidden layer and produces the result. Each connection's signal is an integer, and each neuron's output is a nonlinear function of the sum of the inputs it receives. Figure 8.1 shows the ANN architecture.

In ANN, the input signal from the starting node is multiplied with a weight, added to a bias, and then transmitted via an activation function that is connected to the target node throughout each connection. ANN is used to describe non-linear functions. The broad notion of such a function is that it dictates how the weighted sum of the neural input is changed to the appropriate neural output. Different activations might be employed for distinct parts of the network. The network structure is used to explain the nature of the network, which includes the number of hidden layers and the number of neurons connected with all the layers, along with the activation functions, weights, and biases. The training method yields the bias and weight parameters of a network, which are used to evaluate the accuracy of the prediction.

There are two ways that information moves via a neural network.

1) **Feed-forward Networks:** This type of network only allows the signals to move in one direction, from input nodes to output nodes via hidden nodes. The feed-forward neural networks lack feedback and loops. Perceptions are arranged in layers in these kinds of networks, just like they are in all other neural networks. The hidden layer should be linked to the input, and it can additionally link to the output or another hidden layer.
2) **Feedback Networks:** In this kind of network signals can travel in both directions via its loops. Recurrent or feedback networks are flexible, continuously changing their status until they receive a satisfying answer. They are frequently used in situations where a series of actions must take place in a specific order. Figure 8.2 shows the recurrent or feedback neural network.

8.2.1 Emerging Trends and Technologies of ANN

In general, ANN is a branch of AI that can recognize a system's non-linear input and output relationships to diagnose and manage the system's performance. Its ability to repair output faults by adjusting its values makes it an especially potent tool for learning. The study of management science and operational research is heavily influenced by AI. It gives the power to machines to act conceptually and analytically. Agriculture, business, ways to solve problems, processing images, databases, video games, and analysis of MRI brain cancer are all prominent applications of AI. AI systems employ predetermined judgments to assist businesses maximize most of their limited resources. With the support of AI expert systems, companies can recognize dangers and opportunities and develop strategies for protection. AI plays a vital role in areas such as natural language processing, reasoning, and decision-making. Special artificial systems may also do dangerous human occupations, including mineral extraction, firefighters, and bomb disposal.

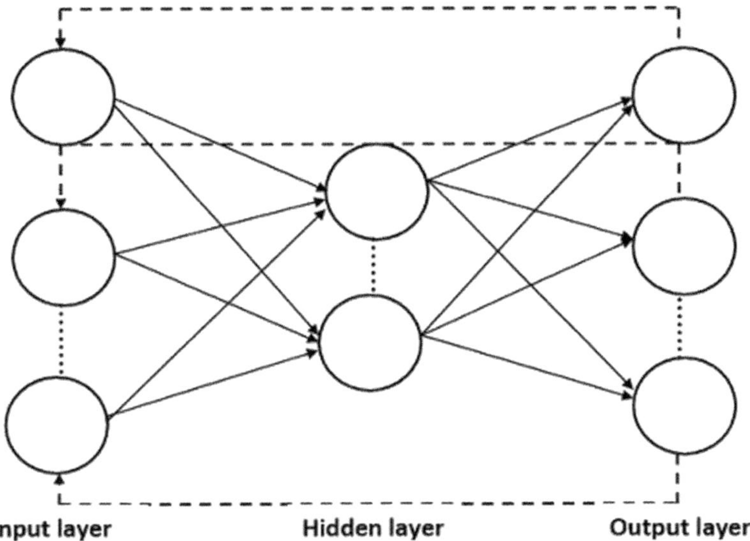

FIGURE 8.2 Recurrent or feedback neural network.

Robots can utilize AI to design their routes so they can move independently from one location to another without encountering any impediments [6]. ANN has recently found success in several scientific fields, including physics, high-energy physics, biology, chemistry, meteorology, catalysis, nuclear physics, and others. Recently, ANN has discovered new applications, such as catalyzing, particularly in the chemical industry.

8.3 DECISION TREES

During the 1980s, Decision Trees (DT) gained popularity in Machine Learning (ML) and data analysis when parallel DT learners became popular in the statistical and computing fields [7]. The DT technique, being part of the supervised learning methods, is frequently chosen to address both classification and regressing cases. By generating decisions using features, a DT is a structure of hierarchy that divides a dataset into simpler to comprehend groups. Recursive partitioning is the method used to form the tree, which continually divides the dataset into subgroups till specified halting requirements are met. The decision nodes, leaf nodes, and subbranches make up a decision tree's architecture. Each branch represents the test's result, every inside node represents the attribute's test, and every leaf node has a class label. The decision node is the highest node of the tree [8]. Decision nodes are utilized for making decisions and have a variety of

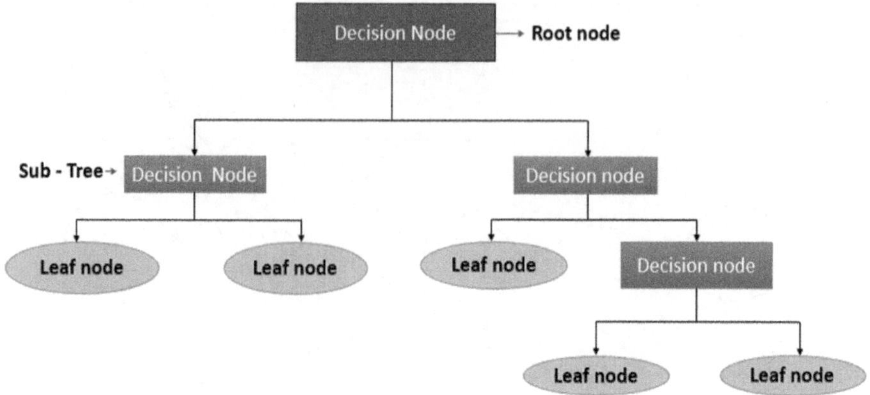

FIGURE 8.3 Structure of decision tree.

branches while the "Leaf nodes" are decision nodes outputs and they do not have further nodes [9]. DTs are used for extracting data from an enormous number of accessible datasets with the help of decision criteria. DT simply categorized the data and has various algorithms, which are C4.5, CART, CHAID, and ID3 [10].

The terms root node, leaf node, splitting, subtrees, pruning, child, and parent node are used for DTs which are as follows:

- **Root Node:** The starting point of the DT, from which the complete dataset gets started to be further divided into numerous homogenous prospective sets.
- **Leaf Node:** The end node of the tree after which no more tree separation is feasible.
- **Splitting:** This process means dividing the main node into subnodes depending on the given restrictions.
- **Subtree:** A subtree or branch is created when a hierarchy is divided.
- **Pruning:** To achieve the best outcomes, the DT's unnecessary branches are removed. Without compromising the correctness of the tree, it reduces its size. Pruning has two types which are cost complexity and error reduction.
- **Child and Parent Node:** The base node is referred to as the parent node, and the additional nodes as the child nodes [11]. Figure 8.3 presents the structure of DT.

8.3.1 Emerging Trends and Technologies of DT

Every aspect of real life frequently employs DT algorithms. DT models are used in different areas, some of them are "Business, intrusion detection, e-commerce, image processing, healthcare, industry, intelligent vehicles, remote sensing, and web applications" [8].

8.4 SUPPORT VECTOR MACHINE

Support Vector Machine (SVM) is one of the famous ML algorithms which depends on kernel. It is particularly well-liked in image classification and regression applications and was first created by Vapnik and his colleagues in the late 1970s. SVM is a non-parametric supervised approach that determines only the single line among two groups [12]. SVM is simply a prediction technique in which we look for a specific line. The purpose of SVM is to make a decision line between two groups or datasets for dividing into two parts. This decision line is also known as a hyperplane which is pointed to be as distant as physically feasible from every class's closest data point, which is called Support-Vectors (SV). SVM is an extremely effective method for finding tiny patterns throughout large datasets as compared to alternative ML techniques [13]. It is possible to determine mathematically and geometrically the optimum hyperplane or maximum margin. The hyperplane that has the maximum separation margin from the number of hyperplanes is known as the optimal hyperplane. SVMs used a subset of the training sample that is located nearest to the ideal decision boundaries in the feature's area, functioning as support vectors, to achieve the greatest separation or margin. The iterative method of building a classifier with an ideal "decision boundary" is known as the "learning process." SVM may be used to recognize faces, identify speakers, identify fake credit cards, and detect handwriting [13]. It uses the "hypothesis space" to construct an excessive dimensional area of features, and similarly "kernel functions" are employed to categorize non-linear data.

The simple type of SVM is a linear binary classifier that divides the dataset into two groups by using a decision line or hyperplane. Figure 8.4 shows the SVM example with linearly separable data where the optimal hyperplane/optimal decision line can be calculated by using Equation 8.1.

$$w^t x - b = 0 \tag{8.1}$$

In the above equation, b is bias, w is a weighted vector, and x is the input vector [12].

SVM provides several advantages and prevents over-fitting due to its simplest decision line [14].

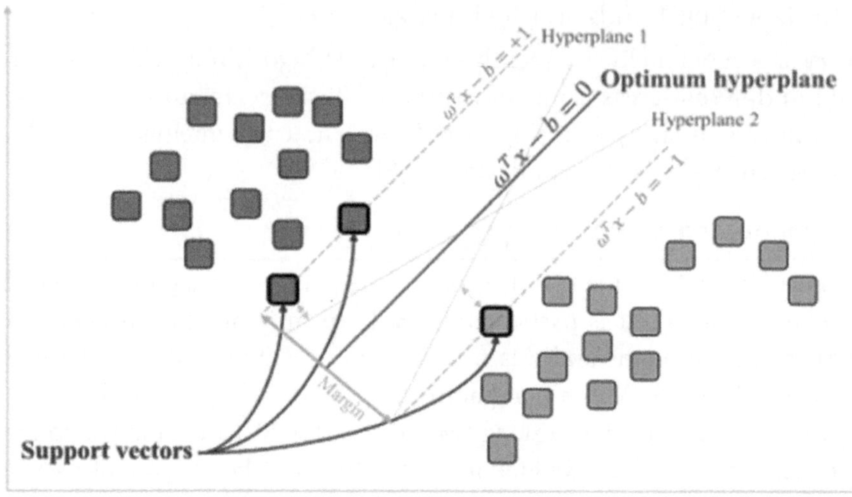

FIGURE 8.4 SVM example of linearly separable data.

In actuality, the data samples from distinct groups often overlap and cannot always be linearly separated. As a result, linear SVM requires significant adjustments to classify such data with high accuracy. To overcome the drawback of linear SVM, Cortes and Vapnik presented the soft margin and kernel approaches where the selection of the kernel function that generates the dot-products has a significant impact on the efficiency of SVM. in the higher-dimensional feature space. The soft margin approach allows for the insertion of extra variables to SVM optimization to deal with non-linearly separable data. Different kernel models are used for the development of SVMs which are radial basis function, polynomial, linear, and sigmoid. The limitation of the SVM kernel-based model is it can be nearly vulnerable to overfitting.

8.4.1 Emerging Trends and Technologies of SVM

SVM is applied in a variety of applications which are face analysis, handwriting analysis, pattern categorization, and regression analysis. In addition, these are also usefull to identifying fake credit cards, and cancer genomic classifications in healthcare [15].

8.5 RANDOM FOREST

Random Forest (RF) is an ensemble learning technique to solve issues with regression and classification by developing various DTs via random samples from data that is used for training. It depends on the idea of DTs, but to increase accuracy and decrease overfitting, it creates numerous trees and merges the predictions they produce [16]. Ensemble learning is an ML technique that enhances performance by integrating multiple models to solve a single problem. In ensemble categorization, several classifiers work collaboratively to generate outputs that can surpass those produced by a single classifier in terms of accuracy. Different numbers of trees are produced for only one dataset by adding randomization to the initial tree development process. Bagging is more resilient to the overfitting issue than the boosting strategy.

The next step is to create a voting scenario for giving labels to unlabelled samples. The majority voting scenarios used a popular voting strategy that gives the label with the most support from different classifiers to each unlabelled sample. This method is well-liked because it is straightforward and efficient. The two most common ensemble learning techniques are boosting and bagging. AdaBoost was the first effective boosting strategy that was designed for the classification of binary scenarios. Boosting is the process of generating a series of models, in which every model tries to fix the mistake of the prior model in that sequence. Bagging, also known as Bootstrap Aggregating, is another sort of ensemble learning that aims to increase the consistency and precision of merged models while minimizing variation. Model overfitting was the primary issue with AdaBoost; however, Bagging is known to be more resistant to the issue than the boosting strategy. The first effective bagging method was RF, which was created by combining Breiman's bagging sampling method, random selection of features, and random decision forests.

According to earlier research, ensemble approaches like boosting and bagging produced more accurate results than individual classifiers like DT while also being stronger and less susceptible to distortion in the training data [17]. The method used by Random Forest is known as bootstrapping. The number of DTs to be generated (Ntree) and the number of variables to be chosen and tested for the best split when developing the trees (Mtry) by using two parameters that need to be set to build the forest trees. It divides the initial dataset into several subsets (bootstrapped samples) by randomly choosing data points and replacing them. This indicates that a few numbers of data points might show more than once in a subset, others might not. The term "bootstrap sample" refers to each subgroup. Typically, classification and regression trees are top-down, recursive methods used to build the trees. Instead of considering every feature, it chooses the most suitable split between a random selection of features at every node. A DT is constructed for every bootstrapped sample, but to create variation, these trees are often shallow and unpruned. A random subset of features is considered for splitting at each node of the tree. This ensures that the individual trees are diverse and not highly correlated. The splitting process will continue until a stopping criterion is not met, which is a maximum depth or the smallest number of samples in a leaf node. After the development of every single DT, RF aggregates the results to produce a final prediction. After that, every three votes are counted for a class in classification tasks, and the class with the highest number of votes is predicted and known as the predicted class. The ultimate prediction in regression tasks is the average (mean) of all individual tree forecasts. Each tree produces a numerical prediction.

Without an additional validation set, RF gives an existing technique for calculating the model's correctness known as Out-of-Bag (OOB) evaluation. The data points excluded from a tree's bootstrap sample can be used to assess the performance of that tree because every tree is built using a distinct set of inputs. The model's correctness is estimated by the OOB error. The prediction error is calculated using the OOB error.

8.5.1 Emerging Trends and Technologies of Random Forest

RF has been widely adopted and used as a standard classifier for a variety of prediction and classification tasks, including those in "computer vision and RS land cover classification." It has been applied to many fields and related research areas, including agriculture, ecology, land cover classification, remote sensing, wetland classification, bioinformatics, as well as biological and genetic association studies, genomics, etc. [18]. Due to its straightforward and obvious decision-making process, great classification outcomes, and simplicity of implementation in a parallel structure for big data computing, RF has become increasingly popular in the categorization of land cover.

8.6 RECURRENT NEURAL NETWORKS (RNNS)

ANNs are constructed from connected layers which are known as artificial neurons. One hidden layer without a recurrent link, one output layer, and one input layer are all that an ANN possesses. The complexity of the network also grows as layer size grows. The higher layers of these networks, which are often constructed from non-linear but basic units, provide an additional conceptual visualization of the data and reduce undesired fluctuation. Before substantial developments, hardly much work was done on deep network architectures because of the optimizing challenges brought on by the composition of the non-linearity at each layer. ANNs with recurrent connections are known as RNNs, which is an evolution of a simple feedforward neural network made up of neurons, which are layered, with an input layer, one or many hidden layers, and a layer for output. RNNs, however, add a new component known as feedback loops. The feedback loops cycle repeatedly over a period or in succession [19]. RNNs are the only neural networks having an internal memory, making them an extremely potent and promising algorithm. For sequential data, such as speech, text, financial information, audio, video, weather, and much more, RNN is the most chosen algorithm because, in comparison to other algorithms, it can provide an improved grasp of sequence and its meaning. Typically, it is applied to sequential data to generate prospective outcomes [20].

From sequential and time series records, RNNs can learn properties and dependencies over time. The RNNs consist of a stack of non-linear units with at least one connection forming a directed cycle between the units. High-dimensional, non-linear-dynamic hidden states make up RNNs. The hidden states architecture serves as the network's memory, and the hidden layer of the current state is based on its previous state in the network. With the help of this structure, RNNs can save, recall, and analyze complicated signals from the past over an extended period. Input, recurrent hidden, and output layers make up a straightforward RNN, and the input layer contains N input units. The input to this layer consists of a series of vectors across time t, like $\{.....,x_{t-1},x_t,x_{t+1},.....\}$ where x_t = $(x_{1,}\ x_2,......x_N)$. A weight matrix called W_IH is used to define the links and the units used as inputs relate to the hidden units. There are M hidden units of hidden layers which are h_t= $(h_{1,}\ h,......h_M)$. And these units are connected over time via recurrent connections. The hidden layer describes the state space or "memory" of the system as in Equations 8.2 and 8.3.

$$h_t = f_H(o_t), \tag{8.2}$$

$$o_t = W_{IH}x_t + W_{HH}h_{t-1} + b_h \tag{8.3}$$

The bias vector of the hidden units is represented by b_h, while $f_H(.)$ is the hidden layer's activation function. With weighted links W_{HO}, the units that are hidden are linked to the output layer. The output layer has P units which are $y_t = (y_1, y_2, ..., y_p)$ that can be calculated as in Equation 8.4.

$$y_t = f_o(W_{HO}h_t + b_o) \tag{8.4}$$

The hidden units of the bias vector are represented by b_o, while $f_O(.)$ is the activation function. The previously described phases repeat themselves over time t = (1..., T) because of the input-target pair's sequential nature. Equations 8.1 and 8.3 demonstrate that an RNN includes specific non-linear equations state that can change over time. The hidden states give an output layer prediction depending on the input vector for each time. RNN's hidden state is a collection of values that, independently of the influence of any external variables, compiles all the essential, specific data about the network's previous states over many periods. In each unit of an RNN, a straightforward non-linear activation function is used. The "tanh," "rectified linear unit (ReLU)," and "sigmoid," are a few prominent mechanisms for activation.

8.6.1 Emerging Trends and Technologies of RNNs

RNNs are widely used in multiple AI applications, such as recognition of speech, predicting healthcare, arts and crafts, abnormal flow detection in smart grids, detecting objects, detection of fake news, hand gestures, or any other human action recognition. Moreover, RNNs have demonstrated promising results in mechanical fault detection and huge quantities of data processing due to their powerful non-linear characteristic [21].

8.7 CONVOLUTIONAL NEURAL NETWORKS (CNNS)

Due to its excellent achievements, CNN is one of the well-known neural networks in the deep learning field. Things that were previously considered to be unachievable, including face recognition, autonomous cars, self-service grocery shops, and intelligent medical care, are now conceivable because of CNN-based computer vision [22]. CNN can learn significant issues more quickly due to weight sharing and the use of complicated models, which enable huge parallelization in comparison to ANN. As long as sufficiently substantial datasets are available to describe the issue, CNN may increase the chances of its correct categorizations.

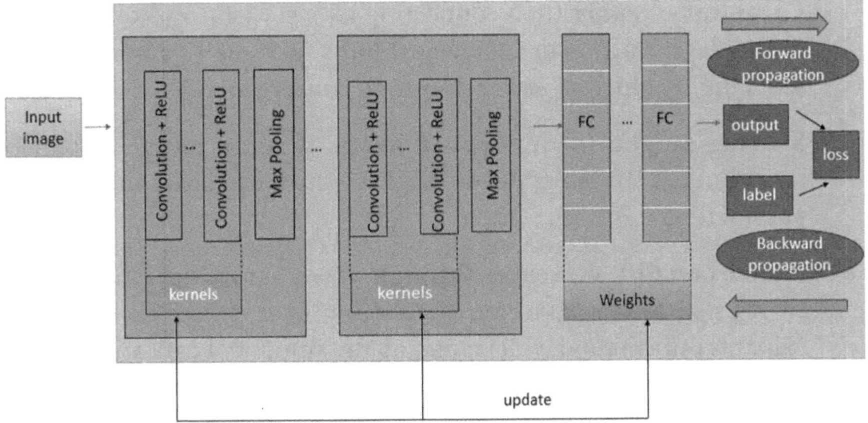

FIGURE 8.5 The conventional NN architecture.

The human visual system is ideal for identifying patterns in visual input and served as the inspiration for CNNs, which are composed of a variety of pooling layers, convolutional layers, and layers that are completely linked. For obtaining unique characteristics of an image using convolution, a small filter, called a kernel, is slid over the input image. Next, a pooling layer is used to overcome the computational load as well as the spatial dimensions of the feature maps. Finally, fully connected layers serve as classification techniques and relate the features that are extracted in the result. To produce final predictions, they blend higher-level features discovered from prior layers [23]. The convolutional layer is made up of several mathematical operations and is a crucial component of CNN. Pooling layers, convolution layers, and fully linked layers are only a few of the building components that make up the CNN structure. Kernels and weight parameters, which are learnable parameters, are modified along with the loss function's value using back-propagation with the gradient descent optimization technique. The performance of the model under weights and kernels is computed by a loss function using forward propagation on a set of training data. Conversion of input data into output using these layers is known as forward-propagation [24]. Figure 8.5 shows the architecture of CNN.

CNN can be accepted any type of data, including speech, photos, video, and audio. CNN has many benefits over traditional artificial neural networks, including the following:

1) **Local links:** Rather than connecting to every neuron from the preceding layer, each neuron now just links to a small number of cells, reducing parameters and accelerating convergence.

2) **Sharing weight:** In this type of neural network different connections can use similar types of weights, which can also minimize the parameters.

3) **Dimensionality decreases through down-sampling:** Image-local-correlation idea is used by the pooling layer, which can preserve valuable information while reducing the amount of data and can also minimize the parameter size by eliminating unnecessary features.

8.7.1 Emerging Trends and Technologies of CNNs

CNN is the fundamental idea in the area of deep learning as well as in the field of big data. CNN can apply a huge quantity of data to get an encouraging output. As a result, many applications are applied in one-dimensional, two-dimensional, as well as in multi-dimensional situations. Time series prediction and signal identification are applications of one-dimensional CNN, whereas "image classification, object detection, image segmentation, and face recognition" are uses of two-dimensional CNN. Like this, CNNs are employed in a variety of multi-dimensional contexts, including Human Action Recognition and Object Recognition/Detection [22]. Many organizations have successfully used CNN in a variety of fields, including the web, online chatbots, health, disaster management, and the identification of diseases from MRI scans. Other applications include post services, automatic address reading, and the auto industry, including self-driving cars and autonomous vehicles [25].

8.8 TRANSFORMER MODEL

The transformer model performs efficiently for a variety of tasks, which include language modeling and machine translation [26]. The transformer model is a well-known DL model that is widely used in a different area, including computer vision, NLP, and speech processing [27]. Recently, transformer models have excelled at different types of language tasks, which include "machine translation, question-answering, and text classification." Bidirectional Encoder Representations from Transformers, Generative Pre-trained Transformer v1-3, Robustly Optimized BERT Pre-training, and T5 (Text-to-Text Transfer Transformer) are among the most often used models [28]. Many layers constitute a transformer model in which every layer consists of a feed-forward layer followed by a self-attention component and includes residual links and normalization of the layer. The Transformer Model is a structure consisting of encoders and decoders [26] with a special implementation that is parallelization-optimized and depends on a self-attention system. In contrast to their convolutional and recurrent models, transformers require less previous information about the problem's structure, and they are often already trained with pretext work on sizable datasets. Two fundamental concepts are important for the creation of conventional transformer models:

- **Self-attention** enables the capture of long-duration relationships among sequential components as contrasted with conventional recurrent models, which find it difficult to encode such interactions. It also calculates the importance of one thing to subsequent things. In simple terms, a self-attention layer aggregates universal information about every element of the input sequences and modifies every element of a sequence. The self-attention simply calculates the dot-product of the queries with every key for each given entity in the sequence, which then gets normalized through the SoftMax operation to obtain the attention values. The weights are determined by the attention values, and every component is then the weighted sum of all the other components in the sequence.
- **Pre-training** on a massive dataset, either supervised or unsupervised, and then fine-tuning to the target task using a tiny, labeled dataset are the second and most important concepts for the creation of a transformer model. Both the language and the vision domains have been supported as effective initial training for huge-scale Transformers. On contemporary hardware like GPUs and TPUs, transformers are extremely effective for inference and training because they handle incoming data simultaneously instead of consecutively.

8.8.1 Emerging Trends and Technologies of Transformer Model

Transformer models and their variants have successfully been used for image identification, detecting objects, segmentation process, image super-resolution, video comprehension, creating images, text-image synthesis, and visual question answering, between multiple additional applications [28]. Transformer models and their variants have also been adopted in CV, processing of sound, and even other disciplines, such as chemistry and life sciences.

8.9 THE FUTURE OF CHATGPT IN RESEARCH: POTENTIAL DIRECTIONS AND EMERGING TRENDS

To comprehend and model both human and non-human languages, GPT is an AI model that uses reinforcement learning and supervised learning approaches. Even though ChatGPT is a text-to-text AI model, other models accept text as inputs for sounds, videos, images, or any other inputs. Particularly NLP and AI have made notable strides in recent years. Among these advancements, ChatGPT, powered by the GPT (Generative Pre-trained Transformer) architecture, has emerged as a revolutionary technology that enables human-like text generation and interaction. With its ability to engage in coherent and contextually relevant conversations, ChatGPT has already found applications in customer service, content generation, and more. However, its potential extends far beyond these immediate applications [29]. ChatGPT is a form of generative AI. The term GPT refers to the process used by ChatGPT to transform input into output. It is a tool that uses NLP and lets its users give input in the form of natural language and generate output. The language model of ChatGPT produces output in written content like essays, social media posts, emails, code, articles, and in the form of images and videos. It creates a human-like conversational dialogue.

8.9.1 The Current State of ChatGPT in Research

As of right now, ChatGPT has proven its ability to hold users' attention and foster lively discussions on a variety of subjects. However, its application in research is an evolving realm that holds immense potential. Researchers have begun to leverage ChatGPT for tasks such as idea generation, hypothesis formulation, and literature review assistance. Despite its promising utility, several challenges, including bias mitigation, ethical considerations, and fine-tuning, need to be addressed before ChatGPT can become an integral part of the research process. AI can systematically and effectively help scholarly investigation. Even though it is still in its infancy, AI has the power to fundamentally change the process that we use for research. There are several opportunities for ChatGPT in the research's future that could completely alter the way that science is conducted. The probable avenues that ChatGPT could take to revolutionize the field of study are listed below:

- **Enhancing Collaboration and Ideation:** Enhancing collaboration and ideation in research is one of ChatGPT's most intriguing applications. Imagine active debates between scientists and models powered by AI that not only offer suggestions but also critique hypotheses and recommend new research directions. This kind of collective intelligence may hasten the research process and produce ground-breaking insights. The success of any scientific endeavor depends on effective teamwork and communication. Natural language processing tools like ChatGPT have the potential to speed up communication between researchers and the public as well as within the scientific community. ChatGPT helps researchers and non-researchers communicate by, among other things, (i) enhancing communication between non-experts and researchers through "natural language processing," (ii) helping in the creation of "research papers, conference presentations, and grant proposals," and (iii) promoting collaboration among researchers by connecting them with useful resources and experts [30].

- **Scientific Research Writing:** ChatGPT can be extremely useful when writing scientific papers because it offers advice on how to make it briefer and clearer. When passive voice or complex language is detected in text, the program might recommend simpler, more direct replacements. Additionally, it can assist with formatting and organization by making recommendations for logical and simple arranging material [31]. Additionally, by pointing out potential problems with references and sources, ChatGPT can assist with the correctness and trustworthiness of scientific writing. The tool can recommend alternate sources or assist in locating areas where more empirical data may be required to substantiate assertions [32].
- **Generating Articles' Summaries:** Researchers need to be able to summarize scientific articles, which ChatGPT can help with. Scientists must filter through a lot of information to find what is crucial for their research because scientific articles are long and technically complex. When compared to how long it would take a human to summarize a scientific article, ChatGPT can do so quickly and accurately. With the help of this function, scholars may rapidly determine an article's main points, saving them time and effort.
- **Generating Abstracts:** The creation of abstracts for scholarly research papers can be aided using ChatGPT. The conclusions of the research study and the article's overall contents are succinctly described in the abstract. Writing an engaging and educational abstract might be challenging because it calls for the author to distill

the study's key ideas in a succinct but captivating manner. A clear, informative abstract that successfully conveys the substance of the research topic can be produced by ChatGPT.

- **Generating Introductions and Conclusions:** The introductions and conclusions of scientific articles can be generated with the use of ChatGPT. The introduction establishes the overall context for the work, and the conclusion, which summarizes the major discoveries and the importance of the study, draws the publication to a close. ChatGPT may create an introductory paragraph that captures the reader's interest and gives pertinent background information. Its capabilities for natural language processing can also aid in coming up with original and instructive insights. For scientists who want to improve their scientific writing abilities, ChatGPT might be an important instrument. By offering recommendations for increased clarity, accuracy, and organization, ChatGPT can help scientists explain their findings more effectively and reach a wider audience [33].
- **Conducting Meta-Research:** Meta-research, which entails examining the research itself, is another recent development. This meta-analysis examines biases, reproducibility, and openness of the study. All of these programs are meant to raise the standard of academic research. All these advancements raise the caliber of study while simultaneously posing new difficulties for researchers. We will eventually need to adapt to AI to reach a new level of research since the computer allows us to handle larger data sets and to protocol our operations [34]. Healthcare organizations have successfully used ChatGPT for a range of tasks, such as conducting literature reviews, analyzing datasets, writing academic papers, documenting, and improving clinical workflow. ChatGPT can extract data from a randomized controlled trial and perform risk of bias analysis. For study groups and subgroups, ChatGPT can extract and provide data like mean values, standard deviation, and sample size. In the laborious process of doing systematic reviews and meta-analyses, it is possible. It has the potential to save time and effort, but careful implementation and validation are required. To fully comprehend this tool's use in producing evidence, researchers must engage with it more [35].

- **ChatGPT in Medical Education:** Fast-moving machine learning is creating opportunities for improved clinical decision support. To increase the possibility that patient care will finally be enhanced, it is important to emphasize that developing, testing, and putting into use machine learning models for healthcare needs several special considerations [36]. One of the significant developments that made image-based AI possible in clinical imaging was the capacity of large general domain models to perform on par with, or even better than, domain-specific models. Whereas before it would have been difficult to collect enough annotated clinical photos, this finding has sparked major AI activity in medical imaging. The computational resources underlying ChatGPT are an "LLM trained on the OpenAI 175B parameter foundation model," GPT3.5, and a large amount of text input from the Internet which makes it able to perform tasks like clinical reasoning [37]. The launch of ChatGPT has generated a lot of attention outside of the research community. ChatGPT's persuasive performance will encourage users to utilize it for a variety of downstream activities, such as asking the model to streamline their medical reports. The majority of radiologists thought the condensed reports were correct, comprehensive, and unlikely to harm the patient. However, it was noted that certain inaccurate claims, important medical results, and possibly risky sections were missed. Although further research is required, preliminary findings point to a significant potential for enhancing patient-centered care in radiology and other medical fields by utilizing extensive language models like ChatGPT [38].
- **Customization for Domain Expertise:** ChatGPT's capabilities will need to be modified in the future for study fields. The model would need to be trained on domain-specific data, incorporate vocabulary relevant to the subject, and comprehend the nuances of the discipline to be customized. In disciplines like medicine, physics, or economics, a customized ChatGPT might act as a virtual expert, supporting researchers with the analysis of difficult data, recreating experiments, and producing insights particular to the topic. The operations of digital marketing and e-commerce have benefited from the use of these bots. They can respond to customer needs quickly and provide customized solutions based on the data they collect from clients. Additionally, ChatGPT has established a

reputation in the medical sector by providing automated patient support services that can reduce costs for doctors while improving patient outcomes through more accurate diagnoses and speedier replies to medical questions or concerns. Similarly, educational institutions are using technology to assist students with course material or swiftly respond to general inquiries about university life, freeing them up to focus more time on other crucial tasks like research or teaching activities [39]. The ability of ChatGPT to perform large-scale pre-training, which allows the models to learn from data and harvest knowledge from the vast internet, is one of its key strengths. There is a lot of potential for ChatGPT to be used in a variety of fields which include history, medicine, arithmetic, physics, and many more. These models can help with activities including generating summaries, responding to inquiries, and giving consumers tailored recommendations [40].

- **Multimodal AI Interactions:** According to emerging trends, AI interactions will move beyond text-based talks to include different modalities including graphics, audio, and video. Researchers may be able to describe data visualizations, produce captions for images, and offer narration for films using multimodal AI in conjunction with ChatGPT's natural language production capabilities. By bridging the gap between written and visual communication, this innovation would make study findings more understandable and accessible. State-of-the-art for fully visualizing free-form natural language, Large Language Models (LLMs) provide a comprehensive solution. Using internal inference capabilities, Codex and GPT-3, as well as ChatGPT, can address the problem of comprehending the queries and both automatically generate code while choosing the relevant visualization types. The pre-trained LLMs offer an effective, dependable, and accurate solution for the problem of natural languages when combined with well-engineered prompts. This ChatGPT feature will provide faster, more accurate ways to make data and insights more widely available, as well as valuable information for data visualization and NLIs [41].

- **AI-Powered Knowledge Discovery:** ChatGPT's capabilities go beyond its ability to produce responses. ChatGPT could be used as a knowledge discovery tool in the field of research, sifting through enormous amounts of literature and data to find undiscovered patterns, connections, and insights. The identification of new study directions and more effective data analysis and hypothesis formulation could result from this AI-driven knowledge discovery. ChatGPT is a tool that can be used to assist students with their studies by responding to questions and providing text summaries. Additionally, it can be used to make outlines, bibliographies, and other research aids. ChatGPT's ability to help with literature reviews, data analysis, and outline construction can make medical research easier. Additionally, it can be used to highlight pertinent papers and pinpoint crucial findings, assisting medical researchers in effectively navigating the large amount of material available on the internet [42]. ChatGPT offers the ability to answer user questions more precisely and uniquely, enhance user satisfaction, and lighten the workload of library employees. However, there are several restrictions to using ChatGPT in LICs, such as the requirement for significant training data and the possibility of bias perpetuation [43].

- **Transforming Communication of Research Findings:** Traditional research communication frequently uses complex academic writing that is difficult for lay people to understand. By producing user-friendly summaries, interactive explanations, and multimedia-rich presentations, ChatGPT could revolutionize the way research findings are communicated. As a result, there would be a greater grasp of intricate scientific ideas among the public and scholars alike.

- **Detecting Plagiarism and generating Less Plagiarized Content:** ChatGPT has the advantage of assisting in the prevention of plagiarism in scientific writing. An anti-plagiarism tool in ChatGPT compares the written text to web articles. This verifies whether the researcher has unintentionally used any previously published work. Before the researcher submits the paper for publication, ChatGPT can identify any problems, finally shielding them from the possibility of the publishing authorities [44]. Even though the data in ChatGPT's scientific abstracts is entirely artificial, they are reasonable. These are original and free of any detectable plagiarism; however, they can frequently be distinguished by an AI output detector and dubious human reviewers. It is necessary to make changes to policy and practice to maintain stringent scientific standards in abstract evaluation for journals and medical conferences. If editing processes contain output detectors for artificial intelligence, and if this is made explicitly transparent, it can enhance trust and accountability in AI-generated content. It is still unclear whether huge language models can be used ethically and in ways that support scientific writing [42].

8.10 EMERGING TRENDS: SHAPING THE TRAJECTORY OF CHATGPT IN RESEARCH

Several emerging trends are poised to shape the trajectory of ChatGPT's integration into the research landscape:

- **Research on Economics and Finance:** Large datasets can be analyzed by ChatGPT to find patterns and trends that people might not see right away. This can be helpful in the financial sector because data analysis is essential for making wise investment decisions. ChatGPT might, for instance, examine consumer behavior data to spot trends that would signal changes in market demand or financial data to spot patterns that might point to greater risk levels. For a variety of requirements in economic and financial research, ChatGPT can also write code. Economic and financial data can be compiled into reports and summaries by ChatGPT, which facilitates understanding and sharing of results [45].
- **Explainable AI (XAI):** ChatGPT's responses are generated based on complex algorithms; therefore, the ability to explain the reasoning behind these responses is crucial for researchers to trust and effectively use the AI's suggestions. Explainable AI (XAI) is gaining traction as researchers seek to understand the decision-making processes of AI models.
- **AI-Augmented Peer Review:** Peer review, a crucial component of the scientific method, might be enhanced by artificial intelligence. ChatGPT could help with reviewing research papers, spotting potential problems, and making suggestions for changes. The peer review process might be streamlined, and the caliber of research results improved with the help of AI.
- **Interactive Virtual Conferences:** Due to circumstances on a worldwide scale, virtual conferences have emerged, opening the door for creative methods of presenting research. An immersive and dynamic

conference experience might be created by integrating ChatGPT into online conference systems, allowing participants to participate in real-time discussions, ask questions, and get immediate answers.

- AI systems can use machine learning to learn from enormous volumes of data. NLP annotations and data tagging help AI better interpret conversations, producing more accurate outcomes that are more realistic, detailed, and natural. To understand speech and text, AI employs a variety of machine learning algorithms to NLP. Chatbots are aided by machine learning in avoiding the conundrum of needing to pre-program conversational techniques. Farcana is implementing the ChatGPT strategy and making use of voice assistant technology to offer a distinctive gaming experience. The Farcana strategy is a cutting-edge way to assist players and the gaming community in maintaining the security of their accounts and deepening their engagement with the game.
- **AI-Driven Experimental Design:** The potential of ChatGPT includes helping with experimental planning. ChatGPT might suggest experimental configurations, and control variables, and even make predictions about potential results by comprehending the research aims and limitations. This might enhance the study procedure and lead to more fruitful testing [46].

8.11 CONCLUSION

AI generative tools have been adopted in research and have presented countless features and benefits. ChatGPT's role is one of the examples where the overall landscape has changed into more intelligent, fast, and attractive methods. This chapter discussed the possible directions and developing patterns of AI models in the research and development phases. These models showed the revolutionary effects on research cooperation, methods, and knowledge discovery. However, with many benefits, these models need careful consideration of ethical concerns, prejudice reduction, and responsibility. To fully utilize AI generative tools in research, researchers, developers, and ethicists must work together to overcome these obstacles. The existing AI-based tools and applications have the power to transform the existing and traditional methods and conduct scientific research, enabling researchers to hasten discoveries, efficiently convey discoveries, and open new horizons for understanding.

REFERENCES

1. M. H. Zaib, F. Bashir, K. N. Qureshi, S. Kausar, M. Rizwan, and G. Jeon, "Deep learning based cyber bullying early detection using distributed denial of service flow," *Multimedia Systems,* 2021. https://doi.org/10.1007/s00530-021-00771-z.
2. O. I. Abiodun, A. Jantan, A. E. Omolara, K. V. Dada, N. A. Mohamed, and H. Arshad, "State-of-the-art in artificial neural network applications: A survey," *Heliyon*, vol. 4, no. 11, p. e00938, 2018.
3. A. Thakur and A. Konde, "Fundamentals of neural networks," *International Journal for Research in Applied Science Engineering Technology,* vol. 9, pp. 407–426, 2021.
4. R. V. Woldseth, N. Aage, J. A. Bærentzen, and O. Sigmund, "On the use of artificial neural networks in topology optimisation," *Structural and Multidisciplinary Optimization*, vol. 65, no. 10, p. 294, 2022.
5. R. Dharwal and L. Kaur, "Technol Applications of artificial neural networks: A review," *Indian Journal of Science and Technology,* vol. 9, no. 47, pp. 1–8, 2016.
6. K. N. Qureshi, T. Newe, G. Jeon, and A. Chehri, "Internet of everything: Evolution and fundamental concepts," in *Cybersecurity Vigilance and Security Engineering of Internet of Everything,* K. Naseer Qureshi, T. Newe, G. Jeon, and A. Chehri Eds. Cham: Springer Nature Switzerland, 2024, pp. 3–20.
7. H. Blockeel, L. Devos, B. Frénay, G. Nanfack, and S. Nijssen, "Decision trees: From efficient prediction to responsible AI," *Frontiers in Artificial Intelligence,* vol. 6, 2023. https://doi.org/10.3389/frai.2023.1124553.
8. H. Sharma and S. Kumar, "A survey on decision tree algorithms of classification in data mining," *International Journal of Science Research,* vol. 5, no. 4, pp. 2094–2097, 2016.
9. M. Bansal, R. Yadav, and P. K. Ujjwal, "Palmistry using Machine Learning and OpenCV," in *2020 Fourth International Conference on Inventive Systems and Control (ICISC),* 2020: IEEE, pp. 536–539.
10. M. Somvanshi, P. Chavan, S. Tambade, and S. Shinde, "A review of machine learning techniques using decision tree and support vector machine," in *2016 international conference on computing communication control and automation (ICCUBEA)*, 2016: IEEE, pp. 1–7.
11. M. Bansal, A. Goyal, and A. Choudhary, "A comparative analysis of K-nearest neighbor, genetic, support vector machine, decision tree, and long short term memory algorithms in machine learning," *Decision Analytics Journal,* vol. 3, p. 100071, 2022.
12. M. Sheykhmousa, M. Mahdianpari, H. Ghanbari, F. Mohammadimanesh, P. Ghamisi, and S. Homayouni, "Support vector machine versus random forest for remote sensing image classification: A meta-analysis and systematic review," *IEEE Journal of Selected Topics in Applied Earth Observations Remote Sensing,* vol. 13, pp. 6308–6325, 2020.

13. S. Huang *et al.*, "Applications of support vector machine (SVM) learning in cancer genomics," vol. 15, no. 1, pp. 41–51, 2018.
14. A. Kurani, P. Doshi, A. Vakharia, and M. Shah, "A comprehensive comparative study of artificial neural network (ANN) and support vector machines (SVM) on stock forecasting," *Annals of Data Science,* vol. 10, no. 1, pp. 183–208, 2023.
15. S. Huang, N. Cai, P. P. Pacheco, S. Narrandes, and Y. X. Wang, Wayne, "Applications of support vector machine (SVM) learning in cancer genomics," *Cancer Genomics Proteomics,* vol. 15, no. 1, pp. 41–51, 2018.
16. G. Biau and E. Scornet, "A random forest guided tour," *Test,* vol. 25, pp. 197–227, 2016.
17. M. Belgiu and L. Drăguţ, "Random forest in remote sensing: A review of applications and future directions," *ISPRS Journal of Photogrammetry Remote Sensing,* vol. 114, pp. 24–31, 2016.
18. H. Tyralis, G. Papacharalampous, and A. Langousis, "A brief review of random forests for water scientists and practitioners and their recent history in water resources," *Water,* vol. 11, no. 5, p. 910, 2019.
19. H. Salehinejad, S. Sankar, J. Barfett, E. Colak, and S. Valaee, "Recent advances in recurrent neural networks," arXiv preprint arXiv:1801.01078, 2017.
20. G. Kanagachidambaresan, A. Ruwali, D. Banerjee, and K. B. Prakash, "Recurrent neural network," *Programming with TensorFlow: Solution for Edge Computing Applications,* pp. 53–61, 2021.
21. J. Zhu, Q. Jiang, Y. Shen, C. Qian, F. Xu, and Q. Zhu, "Application of recurrent neural network to mechanical fault diagnosis: A review," *Journal of Mechanical Science Technology,* vol. 36, no. 2, pp. 527–542, 2022.
22. Z. Li, F. Liu, W. Yang, S. Peng, and J. Zhou, "A survey of convolutional neural networks: Analysis, applications, and prospects," *IEEE Transactions on Neural Networks Learning Systems,* vol. 33, pp. 6999–7019, 2021.
23. A. Kamilaris and F. X. Prenafeta-Boldú, "A review of the use of convolutional neural networks in agriculture," *The Journal of Agricultural Science,* vol. 156, no. 3, pp. 312–322, 2018.
24. R. Yamashita, M. Nishio, R. K. G. Do, and K. Togashi, "Convolutional neural networks: an overview and application in radiology," *Insights into Imaging,* vol. 9, pp. 611–629, 2018.
25. K. N. Qureshi, S. Din, G. Jeon, and F. Piccialli, "Internet of vehicles: Key technologies, network model, solutions and challenges with future aspects," *IEEE Transactions on Intelligent Transportation Systems,* vol. 22, pp. 1777–1786. 2020.
26. P. Dufter, M. Schmitt, and H. Schütze, "Position information in transformers: An overview," *Computational Linguistics,* vol. 48, no. 3, pp. 733–763, 2022.
27. T. Lin, Y. Wang, X. Liu, and X. Qiu, "A survey of transformers," *AI Open,* vol. 3, pp. 111–132, 2022.

28. S. Khan, M. Naseer, M. Hayat, S. W. Zamir, F. S. Khan, and M. Shah, "Transformers in vision: A survey," *ACM Computing Surveys,* vol. 54, no. 10, pp. 1–41, 2022.
29. A. Bozkurt *et al.*, "Speculative futures on ChatGPT and generative artificial intelligence (AI): A collective reflection from the educational landscape," *Asian Journal of Distance Education,* vol. 18, no. 1, 2023. https://doi.org/10.61969/jai.1337500.
30. P. P. Ray, "ChatGPT: A comprehensive review on background, applications, key challenges, bias, ethics, limitations and future scope," *Internet of Things Cyber-Physical Systems,* vol. 3, pp. 121–154, 2023.
31. A. L. Alkhaqani, "Can ChatGPT help researchers with scientific research writing," *International Journal of Medical Research and Review,* vol. 1, no. 1, pp. 9–12, 2023.
32. S. Iqbal, A. H. Abdullah, and K. N. Qureshi, "An adaptive interference-aware and traffic-aware channel assignment strategy for backhaul networks," *Concurrency Computation: Practice Experience,* p. e5650, 2019. https://doi.org/10.1002/cpe.5650.
33. A. Alkhaqani, "How artificial intelligence is revolutionizing the future uture of healthcare," *International Journal of Health Sciences and Nursing,* vol. 6, no. 4, p. 1e16, 2023.
34. B. Burger, D. K. Kanbach, S. Kraus, M. Breier, and V. Corvello, "On the use of AI-based tools like ChatGPT to support management research," *European Journal of Innovation Management,* vol. 26, no. 7, pp. 233–241, 2023.
35. S. A. Mahuli, A. Rai, A. V. Mahuli, and A. J. B. D. J. Kumar, "Application ChatGPT in conducting systematic reviews and meta-analyses," *British Dental Journal*, vol. 235, no. 2, p. 91, 2023.
36. P.-H. C. Chen, Y. Liu, and L. Peng, "How to develop machine learning models for healthcare," *Nature Materials,* vol. 18, no. 5, pp. 410–414, 2019.
37. T. H. Kung *et al.*, "Performance of ChatGPT on USMLE: Potential for AI-assisted medical education using large language models," *PLoS Digital Health,* vol. 2, no. 2, p. e0000198, 2023.
38. K. Jeblick *et al.*, "Chatgpt makes medicine easy to swallow: An exploratory case study on simplified radiology reports," *arXiv preprint arXiv:.14882,* 2022.
39. A. S. George and A. H. George, "A review of ChatGPT AI's impact on several business sectors," *Partners Universal International Innovation Journal,* vol. 1, no. 1, pp. 9–23, 2023.
40. .. Y. Liu et al. Summary of chatgpt-related research and perspective towards the future of large language models. Meta-Radiology. vol. 18, p. 100017, 2023.
41. P. Maddigan and T. J. I. A. Susnjak, "Chat2vis: Generating data visualisations via natural language using chatgpt, codex and gpt-3 large language models," IEEE Access, vol. 8, pp. 45181–45193, 2023.

42. C. A. Gao *et al.*, "Comparing scientific abstracts generated by ChatGPT to original abstracts using an artificial intelligence output detector, plagiarism detector, and blinded human reviewers," *BioRxiv*, p. 2022.12. 23.521610, 2022.
43. S. Panda and N. Kaur, "Exploring the viability of ChatGPT as an alternative to traditional chatbot systems in library and information centers," *Library Hi Tech News*, vol. 40, no. 3, pp. 22–25, 2023.
44. A. L. Alkhaqani, "Potential benefits and challenges of ChatGPT in future nursing education," *Maaen Journal for Medical Sciences*, vol. 2, no. 2, p. 2, 2023.
45. M. M Alshater, "Exploring the role of artificial intelligence in enhancing academic performance: A case study of ChatGPT," *Available at SSRN*, 2022.
46. A. Shafeeg, I. Shazhaev, D. Mihaylov, A. Tularov, and I. Shazhaev, "Voice assistant integrated with chat gpt," *Indonesian Journal of Computer Science*, vol. 12, no. 1, 2023.

CHAPTER 9

The Role of Human Expertise in AI-Aided Research

Mamoona Amin, Muhammad Mueed Hussain, and Kashif Naseer Qureshi

9.1 INTRODUCTION

Implementing Artificial Intelligence (AI) across various business sectors has been acclaimed as having ushered in a plethora of tremendous breakthroughs and enhanced productivity levels. The applications of AI have the potential to guide in a period of profound change across various business sectors while also elevating the standard of living for individuals. The development of driverless cars and individualized medical care are two instances of these potential applications. It is common practice to disregard the relevance of human expertise in research aided by AI, even though this technical miracle is already a reality. This chapter discusses human knowledge and its significant role in AI-assisted research and emphasizes the synergy that can emerge when human intelligence and AI algorithms work together. In addition, this chapter investigates the critical value that human expertise contributes to research. It highlights how human specialists give essential guidance, interpretation, and decision-making throughout the whole study process, which brings to light

DOI: 10.1201/9781032667911-9

the essential role that human specialists play in producing substantial and trustworthy research results. This chapter argues that algorithms for AI should not be regarded as competing with the capabilities of humans but rather as instruments that have the potential to enhance those talents. The findings are made due to the collaboration of AI and humans on research projects that are more extensive and accurate, paving the way for new opportunities for creative discovery and intellectual development. In the process, the significance of human expertise, as well as how the application is utilized to enhance that experience [1, 2]. It is vital to recognize and use human expertise to maximize the potential of research that AI backs as it advances. This will be the case to ensure that research meets its full potential.

AI algorithms are particularly adept at many tasks, including processing enormous amounts of data, recognizing patterns, and creating correct predictions. Among these skills, is the AI's ability to make accurate predictions. These technologies are essential for many sectors, including financial analysis, natural language processing, and healthcare diagnostics. This is due to the incredible speed and accuracy of completing the tasks. On the other hand, AI systems do not yet possess the contextual awareness, topic knowledge, or nuanced understanding that human professionals bring. The field of medical diagnostics is an illustration of this. AI systems may have difficulty to understand the patient's medical history, the larger clinical context, or the minute differences in symptoms that a trained physician can spot. AI systems can analyze patient data and medical imaging to spot anomalies. In this particular setting, it is impossible to locate an alternative that can match the capabilities of people. In a study [3], the synergistic benefits of human and AI partnerships are underlined. This research demonstrated how AI and radiologists' expertise improved breast cancer's diagnostic accuracy.

Human knowledge refers to in-depth expertise in a particular topic and factually accurate information. AI can assist with document analysis and case prediction in fields such as law, where it is essential to have a comprehensive understanding of both complex statutes and legal precedents. AI can assist in fields such as law, where it is essential to have such an understanding. On the other hand, human attorneys offer the critical viewpoint necessary to traverse the complexities of the legal system successfully. The AI-powered tools can speed up case analysis in the law field and improve legal outcomes. In addition, the accumulation of human knowledge has helped in the development of contextual awareness, frequently necessary for making well-informed decisions. However, it is possible that these algorithms are not able to comprehend how human emotions or geopolitical events influence financial markets. AI in finance enables the examination of market data and patterns. The severe global financial crisis serves as a sobering reminder of how vital regulatory monitoring and human judgment are to prevent potentially catastrophic economic disasters.

In addition to domain-specific knowledge and awareness of the environment in which the study is being carried out, human competency lends research an essential helping of creativity and ethical reflection. AI can help to conduct scientific research by assisting with data analysis and formulating ideas; however, AI cannot replace a seasoned scientist's original thinking and intuition. In addition, human specialists are responsible for handling any moral and ethical issues that AI systems may miss. As the use of AI in the criminal justice system for risk assessment and sentencing becomes more widespread, criminal law and ethics experts must ensure that these systems do not enable prejudice or injustice. The ongoing discussion regarding the ethical repercussions of implementing AI into the criminal justice and policing systems has highlighted the importance of human control and oversight of these systems.

When humans and AI work together on a study, the information produced is more comprehensive and accurate. The capabilities of humans shouldn't be seen as being replaced by AI; instead, AI should be seen as a potent instrument that amplifies and enhances human talents. Identifying and embracing this relationship to correctly grasp the possibilities of AI-assisted research is becoming increasingly vital. This is because the technology behind AI is advancing at an increasing rate. The intricate dance between human expertise and AI algorithms is anticipated to open up new opportunities for creative activities and scientific investigation, lending validity to the argument that, even in this age of AI, human knowledge is still essential and indispensable.

9.2 RELATED WORK

This section analyzes the critical role that human knowledge plays in research supported by AI, and it does so by looking at a range of prior studies. This investigation aims to understand better how human knowledge contributes to research that AI supports. It demonstrates the benefits of combining human experience with AI algorithms in a range of sectors, and it emphasizes the concept that AI should be considered a supplement to human skill rather than a replacement for that skill. Specifically, it highlights the benefits of combining human expertise with AI algorithms in the financial industry [4]. It achieves this by showcasing the benefits of amalgamating human capabilities and AI algorithms across various businesses. This can be accomplished in several different ways. It accomplishes this goal by highlighting the benefits of integrating human expertise and AI algorithms. This survey analyzed the possibilities and restrictions of AI in various sectors, including the medical profession, the legal system, the financial business, and the scientific field. This demonstrates the critical role that human intelligence and ethical considerations play in optimizing the benefits gained from research that AI supports. AI has been applied in several business domains, resulting in the discovery of groundbreaking information and increased productivity. Despite this, it is of the utmost importance to acknowledge that the skill of humans will continue to be required to advance AI. This is a need of the highest importance. This literature review explored the relationship between human knowledge and AI, and the findings indicated that these two domains complement each other well to produce more accurate and complete study findings [5]. The review also indicated that the relationship between human knowledge and AI is complex and multifaceted.

The algorithms used in AI are incredibly adept at analyzing massive volumes of data, identifying patterns, and making predictions expeditiously and accurately. These characteristics have the potential to be helpful in a wide variety of business fields, such as the processing of natural languages, medical diagnostics, and the financial industry. AI systems lack ethical judgment, nuanced cognition, and contextual awareness as compared to human professionals. In medicine, AI can detect anomalies in patient data and medical imaging, yet it may have trouble comprehending a patient's medical history and the larger clinical context [6]. Another study [5] highlighted how promising it can be to combine human intellect with AI. The primary objective of this research is to determine whether or not the utilization of AI in conjunction with the expertise of radiologists may result in a more precise diagnosis of breast cancer. To successfully navigate the complexities of the legal system, however, human attorneys will always be able to provide the required insights. Legal research can benefit from using AI to examine statutes and case law.

AI is very beneficial in analyzing the data and patterns of financial markets, even though it may not be able to comprehend how human emotions or world events may impact the markets. This is because AI can analyze massive amounts of data in a concise amount of time. The global economic and financial crisis that started in 2008 shed light on the significant roles that human judgment and regulatory oversight play in the process of preventing financial disasters from occurring in the first place [7]. Although AI may help process data and generate ideas in scientific research, it is not a substitute for the originality and intuition that human researchers bring to the table. The study [8] discussed the Clustered Regularly Interspaced Palindromic Repeats (CRISPR-Cas9) and suggested a prime illustration of the synergy that can occur between human creativity and research inspired by AI. To ensure that the technologies mentioned earlier do not support the continuance of discrimination or injustice, introducing AI into the criminal justice system requires a thorough inspection from legal and ethical experts [9]. This is because AI might potentially be used to automate processes that humans currently perform. It is now clear that human oversight and control are essential for AI and used in law enforcement due to the conversations that have taken place regarding the ethical and legal repercussions of employing AI in law enforcement.

When AI and humans work together on research, they produce more thorough and accurate findings. It is vital to keep in mind that the purpose for which AI was developed was not to act as a replacement for the capabilities possessed by humans but instead as an efficient tool that augments those capabilities. The advancement of the technology that underpins AI is paving the way for new research areas and opportunities for creative expression. This is happening due to the combination of human skills with AI algorithms. This in-depth examination of the relevant research literature illustrates the significance of human expertise in AI-supported research across a wide range of fields of study where they highlighted the importance of human expertise. The promise of research that AI bolsters won't be able to be fully realized until ethics and human abilities are incorporated into the process [10]. In an era controlled by AI, the significance of human comprehension is illustrated by the fact that it is anticipated that human cognition and AI algorithms collaborate to push both creative and scientific advancements. This highlights the importance of human comprehension.

9.3 AI IN RESEARCH

AI research and development has advanced dramatically, especially in research. These breakthroughs are mainly research-based. Deep learning research has advanced significantly in recent years. Researchers in several domains increasingly use AI-enabled tools. These innovations include machine learning and neural networks. This category includes other advancements. However, human knowledge will remain essential to AI-supported research. This is crucial and cannot be overstated. Consider this key element while making a selection. Modern interest in AI has revolutionized numerous industries and changed how humans receive, analyze, and interpret data. Many firms have been revolutionized, which has changed many industries [11, 12]. These events incited a revolution, which has occurred.

Due to AI's incorporation into their profession, researchers now have powerful tools and expertise to analyze massive amounts of data and draw conclusions. This is due to researchers' AI work. Researchers focus on AI, which allows this. This is the direct result of including AI in researchers' tasks. People consider the future of AI systems in domains like shoe styling and cancer detection utilizing clinical imaging. Finding the right shoes for an outfit is an example. Where AI systems are headed in many fields and sectors interests people. For instance, choosing shoes to match an outfit so it looks its best and complements it best. This can happen when choosing shoes to match an outfit, for example. This experiment tests the hypothesis that the selection procedure already considers the AI system's new information [13].

AI, which can interact with highly knowledgeable people, is trusted and relied on. People trust and rely on AI. People who use AI often appreciate its assistance. AI is trusted and relied upon for support. These knowledge levels may include identical and precisely complementary information. The preceding Figure 9.1 shows information that completely copies itself.

AI has completely changed research methods. This is because AI has created handy tools and methods for analyzing and interpreting complicated data, finding insights, and speeding up the research process. These changes have been made possible by creating tools and methods that work very well. Because of this change, tools, and methods for studying and understanding complex data have become much better. Because of this, the study process has moved along faster. New tools and methods had to be considered a direct and unavoidable result of these changes. At the time, AI had a significant effect on many different academic fields. This trend is likely to keep going shortly. Thanks to a group of methods called Natural Language Processing (NLP), researchers can now study and understand human language on a scale that wasn't even impossible [14].

FIGURE 9.1 Benefits of AI-based research.

Models of NLP powered by AI can analyze how people feel about something, summarize it, and even write prose that sounds like a natural person wrote it. A significant usage of NLP is noticed in the social sciences, humanities, and biology [15]. Researchers are making robots and automation systems that can be controlled by AI. These robots and automation systems aim to help scientists to do different kinds of scientific work. These robots and systems are made in many different ways, by using a wide range of different methods. The researchers want help with different tasks by using these automated technologies and robots. Researchers have more time to work on more challenging and creative tasks because these robots can do experiments repeatedly, run lab equipment, and collect data. Because it comprises algorithms, AI can model complex systems and accurately copy what happens in the real world. In recent years, people have made progress on AI. It is now possible because AI can accurately simulate what happens in the real world. As a direct result of this breakthrough, scientists in fields like physics, chemistry, and engineering can now try out ideas, predict results, and improve operations.

9.3.1 Understanding the Role of Human Expertise in AI-Aided Research

In many ways, research that uses AI is dependent on human talent. The ethics, validity, and correctness of AI-assisted research are ensured by human knowledge. Human skills are valuable for selecting and curating AI algorithm data. For human specialists to understand the results of AI algorithms, they need domain and subject matter knowledge. They possess the expertise and wisdom to manage complex ethical dilemmas and deploy AI algorithms ethically, particularly in the social sciences and healthcare [16]. The correct research questions must be developed, suitable methodologies must be used, and insights from AI algorithms must be analyzed. In research supported by AI, human talent is multifaceted and cannot be replaced by AI algorithms. Even though AI develops quickly, human ability is still essential in AI-assisted research. Even if AI systems can handle enormous amounts of data and make precise predictions, human expertise is still required to ensure the quality and validity of the study findings. Data-driven AI algorithms require high-quality data. Human experts choose and curate the data used in AI algorithms. Human experts possess the domain knowledge and subject-matter knowledge necessary to understand the results of AI algorithms. As AI is more thoroughly included in the research, it is clear how crucial human knowledge is for formulating the proper inquiries, selecting the appropriate techniques, and comprehending AI algorithm insights. Social sciences and healthcare are morally significant fields that demand human expertise.

Human experts can address complex moral problems in a variety of fields and can also ensure that AI systems are used morally. In the AI research, labeling is used to exploit human expertise. With Google's human labeling service, users can annotate images to train AI systems. Accuracy is also ensured via interaction between humans and AI [16]. Google Crowdsource uses human experience to analyze and categorize data for machine learning model training. Contextual awareness and human decision-making increase the precision and caliber of AI algorithms. People can volunteer for labeling tasks like image categorization, text transcription, sentiment analysis, and more on the Google Crowdsource platform. Human interpretations provide the "ground truth" labels that teach AI algorithms to recognize patterns, make predictions, and perform better. Data analysis, natural language processing, and machine vision use Google's human labeling service. It allows researchers and developers access to a large pool of human experience to train AI models with various viewpoints and ensure they generalize effectively to real-world situations.

Human labeling is required in AI research to address the biases and limits of AI systems when dealing with complex and confusing data. Google's human labeling service improves AI models' accuracy, dependability, and ethics using human judgment and expertise. Google's human labeling service is an essential tool that employs human expertise to enhance AI algorithms and machine learning models. Human knowledge and AI technology can provide more accurate, trustworthy, and wise research. AI systems may be faster in some tasks than humans, but researchers must understand that they cannot wholly replace people. Human intuition, topic expertise, experience, and creativity are advantages of AI-assisted research [17].

Additionally, human expertise in AI-assisted research helps eliminate biases and limitations in AI systems. Using human skills in research is advantageous with industry knowledge, wisdom, and originality. A combination of human intuition and AI-generated insights is required to comprehend study results completely. Human expertise is essential enough to contextualize and understand insights. Human talent benefits AI in a variety of ways. Human skills are advantageous in data preparation. By assuring data quality and relevance, they reduce biases and ensure AI systems use accurate and representative data. Deciphering the outputs of AI algorithms requires the expertise of human specialists. They understand the context and pick up on subtleties, exceptions, and restrictions that AI would overlook. The interpretation is necessary to make and choose wise decisions.

Furthermore, to direct AI systems toward ethics, human knowledge is required. The objectivity of AI systems is only as good as their training data. Human experts can identify biases and ensure that AI algorithm design and implementation are morally correct. When AI algorithms falter, they can make moral judgments in complex situations. When AI and humans work together on research, it is more thorough and accurate. It offers a variety of perspectives on complex problems, leading to fresh ideas and understandings. AI systems can overcome common sense reasoning, creativity, and adaptability limitations with human competence. For research using AI to be successful, human knowledge is required. Human specialists' distinctive abilities and perspectives help AI algorithms produce precise, reliable, and moral answers. AI algorithms and human expertise enhance innovation, problem-solving, and decision-making in various sectors.

9.3.2 Harnessing AI for Improved Research Outcomes

Academics must have both subject-matter expertise and a critical outlook if they want to take full advantage of the benefits that AI technology has to offer in their work. The complex technique necessitates carefully selecting pertinent reading material and thoroughly comprehending AI systems' advantages and disadvantages. Additionally, researchers must be able to comprehend AI developments in the context of their particular area of study. Incredible advancements that combine AI competence with domain knowledge have ushered in a new era of possibilities. The research industry has seen several significant changes due to the employment of AI. Recent years have seen a significant increase in research activity, and as a direct result of this paradigm shift, groundbreaking discoveries are currently being discovered [17]. AI is a potent ally in pursuing knowledge, particularly in machine learning algorithms. With the help of these algorithms, which can recognize complex links and patterns hidden within vast datasets, researchers can find correlations that were previously difficult to find and new trends.

AI can process enormous amounts of data at previously unfathomable speeds, which is one of its most important advantages. This talent is unparalleled. Because of the remarkable processing capabilities provided by AI, researchers can now glean critical insights and predict previously unachievable outcomes utilizing traditional research approaches. AI's automated skills have also aided in accelerating time-consuming and laborious tasks, including data collecting, processing, and documentation. When researchers engage more time in creative problem-solving and higher-order thinking, their study will be more thorough and quality. AI, as it develops and advances, has the potential to accelerate scientific discovery even further when used in research methodologies. Integrating humans and AI allows researchers to increase their capabilities while fostering a collaborative atmosphere for specialists from other domains. By working together, we can increase the breadth of our knowledge, enhancing the effectiveness of our research and making it simpler to obtain new material. Domain-specific knowledge combined with AI technology is a potent research discovery accelerator. This connection speeds up the research process and creates new opportunities for cross-disciplinary cooperation, making knowledge acquisition more fruitful and effective than was previously thought.

9.3.3 Challenges in Integrating AI into Research Practices

When integrating AI into research operations, businesses and researchers face several difficulties. One of the biggest challenges is getting access to a large amount of high-quality data that is readily available. AI algorithms do their best when given access to enormous datasets containing various data points. Large amounts of carefully scrutinized data are frequently needed for AI algorithms to produce valuable results. Additionally, due to the complexity of AI models, specialized technical expertise is required, and traditional research teams may lack ready access to this expertise. Researchers have a hurdle due to the rapid evolution of AI systems because it is becoming increasingly vital for them to enhance their skills and receive instruction. In addition, ethical concerns of responsibility, openness, and unfairness in the outcomes of AI-driven research need to be addressed and resolved appropriately [18]. Collaboration, continuing education, and a comprehensive assessment of the ethical and technological repercussions of using AI in research are essential for meeting the problems posed by using AI in research head-on.

Additionally, specific technical knowledge is required due to the complexity of AI models, which may not be readily available within traditional research teams. This is a problem because traditional research teams have traditionally relied on human researchers. Scientists must comprehensively understand machine learning algorithms, data pre-treatment methods, and model evaluation approaches to succeed. This demand emphasizes the necessity of training and upskilling present researchers, as well as collaborating with professionals in the fields of AI and data science. Researchers are forced to contend with the ongoing necessity for skill development and training due to the rapid advancement of AI technology. Keeping up with technological breakthroughs and learning to adapt to new tools and practices can take significant effort. The challenge facing researchers working to integrate AI properly is steep: how to fake a learning curve while still carrying out their regular research responsibilities.

To surpass these difficulties, the successful application of AI in research demands collaboration, continual education, and a delicate balancing act between the implications of the technology and the ethics of the use of the technology. AI practitioners and subject matter experts may work together to close the knowledge gap and foster innovation across multiple disciplines. Researchers can obtain the necessary AI skills with access to the relevant training programs and resources. Building ethical frameworks and norms is vital to promote additional responsibility, equity, and openness in applying AI in research. Incorporating AI into research operations presents potential that has not been seen previously. Yet, researchers also have to contend with obstacles relating to data, technological knowledge, continuing learning, and ethical concerns. A thorough approach to these difficulties can pave the way for groundbreaking discoveries at the crossroads of AI and research, ultimately enhancing the quality and scope of academic exploration [19].

9.4 THE FUTURE OF AI-AIDED RESEARCH: A HUMAN PERSPECTIVE

Domain knowledge and human talent are needed to develop AI systems that can benefit engineering, healthcare, agriculture, and environmental research. AI and human intellect create a powerful synergy that can transform how we solve complex problems and conduct research. The combination of AI and medical expertise has great healthcare potential. Because AI can handle and analyze massive amounts of patient data, doctors may be able to diagnose patients more accurately, personalize therapies, and even predict health issues before they appear. However, AI systems may produce incorrect results or suggestions without medical assistance, putting patient care at risk. AI-powered agriculture solutions can boost crop yields, soil monitoring, and crop management. AI in agriculture does not reach its full potential until farmers and professionals grasp agronomy and local conditions [20]. Domain knowledge guides AI judgments to match environmental and crop details.

Environmental studies also succeed due to AI-subject matter expertise integration. AI can find environmental patterns by analyzing sensor data, climate models, and satellite photos. However, it requires ecology, climatology, and other experts to examine these facts and offer meaningful environmental solutions. As AI tools become more widespread in clinical research, subject matter expertise becomes more critical. Misusing big data in medical research might cause false positives and patient harm. Combining medical expertise and AI allows big data to be used while ensuring treatment accuracy and reliability. AI in human research can boost innovation and speed up scientific discovery. AI's ability to quickly sift through massive data, find new patterns, and generate new hypotheses could advance scientific research. Working together, AI and human intelligence could accelerate scientific research and spark innovation in many industries.

AI has revolutionized pharmaceutical research and discovery. AI algorithms can quickly analyze sizeable chemical compound datasets to optimize drug discovery workflows and find new therapeutic targets. It lowers pharmaceutical research expenses and speeds up drug development [21]. AI could also optimize pharmaceuticals to tailor dosages to individual patients and circumstances. AI-assisted research has great potential in many fields beyond drug development. AI helps environmental scientists model complex ecosystems to predict how changing environmental conditions affect the world. Materials scientists can speed up the discovery of innovative materials with unique features that could transform numerous industries, including electrical and aerospace. AI can help geneticists understand the human genome and improve personalized medicine. It will enable tailored medicine. AI research is a sign of revolutionary change that will transform scientific discoveries and problem-solving in various fields. However, intelligent subject knowledge and human expertise are still needed to integrate this technology. As AI advances and is increasingly incorporated into diverse professions, it has the boundless potential to stimulate creativity and propel discoveries [22]. This bodes well for when AI and human intellect can collaborate on cordially improving society.

9.4.1 Putting Human Expertise and Domain Knowledge Together

Integrating important elements of each crucial domain knowledge and human skill components is necessary for developing successful AI systems in various industries, including engineering, healthcare, agriculture, and environmental research. These sectors include engineering, healthcare, agriculture, and environmental research. It includes engineering, healthcare, agricultural research, and environmental science. Since it combines a variety of data types and forms the basis for numerous real-world applications of AI in various industries, this combination has practical advantages [5].

9.4.2 How AI Can Complement Expert Healthcare Knowledge

Regarding the applications of AI and the skills and expertise of medical experts, the healthcare industry still has much untapped potential. Due to its capacity to analyze enormous amounts of patient data, AI has the potential to provide proactive health predictions, tailored medicines, and more precise diagnostics. The power of AI to tailor patient care makes this possible. However, without the guidance of qualified medical personnel, AI's recommendations frequently become incorrect, endangering patients' health [23].

9.4.3 Utilizing Expertise and AI to Put Agriculture on the Path to Realizing Its Full Potential

The employment of AI-based solutions in agriculture might result in better crop management, higher agricultural yields, and soil quality monitoring. However, for the sector to fully utilize AI in its current condition, agriculturalists and other specialists in the field must have considerable agronomic competency. AI decision-makers can consider the complexity of diverse agricultural ecosystems and types of crops when making decisions because they are knowledgeable in their respective domains [24].

9.4.4 AI Coupled with the Knowledge of the Subject Matter

Environmental research can advance significantly with the help of experts. Environmental research significantly advances with AI and experts' opinions on related subjects. It is now possible to identify recurring environmental patterns and recent changes since AI can analyze colossal information. However, it will be necessary for specialists in ecology, climatology, and other related fields to review these findings and develop solutions that will assist in addressing the issue [25].

9.4.5 The Importance of Experience in Clinical Research

In addition to how AI has impacted scientific research and advancement, specialized knowledge about the topic being examined is becoming an increasingly valued advantage as the use of AI in clinical research spreads. Extensive data misuse could have unintended consequences for patients and result in false-positive diagnostic test results. By combining the expertise of medical experts with the capabilities of AI, it is feasible to utilize the potential of big data to improve the accuracy and dependability of treatment. If AI is incorporated into the procedures used in research, there is a tremendous chance to accelerate the growth of scientific discovery and foster innovative thinking. The ability of AI to quickly evaluate data and identify patterns can aid in establishing a wide range of business enterprises outside of scientific research.

9.4.6 Revolutionary Impact on AI's Potential in Pharmaceutical Research

The use of AI in the process of developing new drugs is crucial. The transformative potential of AI in various sectors, including pharmaceutical research. Creating new pharmaceuticals has been easier thanks to AI algorithms' ability to analyze vast databases of chemical compounds, which has decreased the time and resources needed for the procedure. As a result, it will be possible to tailor the delivery of these pharmaceuticals to each patient's unique needs. AI can also enhance the optimization of current treatments. Without AI's help, this would not be achievable. In several fields, including genetics, materials science, and environmental science, to name a few of the specific applications found in these fields of study, research is being supported by AI. Applying AI makes it feasible to more quickly identify new materials, examine complex ecosystems, and gain a deeper understanding of the human genome to provide customized treatments. All of these are possible advantages of AI.

The research sector is undergoing a significant change due to the development of collaborative work and AI. Human and AI improvements will occur in the coming years. The study of AI has caused a paradigm change in how various issues are solved, and advancements are made across numerous industries. This change could significantly impact the world economy [26]. A seamless fusion of human experience and subject matter expertise is required to complete the integration process. Only with proper preparation and execution can this be possible. AI technology can inspire innovative new ideas and quicken the pace of scientific advancement as it develops and spreads across additional commercial sectors. AI's continual advancement will only serve to increase this capability. In this upbeat vision of the future, where humans and robots collaborate to achieve shared objectives, there is a positive attitude toward the chances of civilization improving by merging the qualities of both forms of cognition. Regarding the chances for civilization to advance, this perspective is optimistic. In this potential future version, humans and robots will work together on various projects.

9.5 CASE STUDIES: SUCCESSFUL UTILIZATION OF HUMAN EXPERTISE IN AI RESEARCH

Engineering, medicine, healthcare, agriculture, environmental science, and other sectors need human expertise and AI to improve research findings. Known as "human-in-the-loop" systems, this dynamic alliance of

AI and HII has excellent potential for knowledge advancement and challenging problem-solving. Combining human expertise and AI can lead to technical breakthroughs. Years of experience and subject-specific knowledge help engineers assure project safety and offer new solutions. AI can simulate complex situations, refine designs, and process massive datasets. Human engineers and AI systems can collaborate to generate new ideas, validate designs more reliably, and speed up design [27].

Medical treatment can be transformed by combining AI and human doctors. AI can analyze patient records, clinical data, and medical imaging to improve diagnosis and treatment. Experienced healthcare personnel must grasp AI-generated insights, consider the patient's well-being, and make informed decisions for each case. This "human-in-the-loop" approach ensures compassionate and thorough treatment and improves medical issue detection. Agriculture benefits greatly from AI and human skills. Agriculture experts understand crop management, soil conditions, and regional environmental factors. AI can optimize planting schedules, irrigation, and pest management by interpreting real-time sensor, satellite, and drone data. AI can adapt its recommendations to address local issues using human expertise, increasing crop yields and sustainable farming.

Environmental research uses AI and subject matter experts to monitor and address complex ecological issues. AI can analyze satellite pictures and climate models to spot environmental changes. Ecologists, climatologists, and environmental scientists must assess these discoveries, understand their effects, and create effective environmental mitigation programs. Human knowledge ensures AI-generated insights work. This cooperative strategy maximizes its strengths by tackling human and AI system flaws and prejudices. Human specialists can provide context, ethics, and critical judgment during data processing and decision-making that AI may not. AI may help humans quickly handle vast volumes of data, recognize subtle patterns, and perform repetitive tasks with exceptional precision. AI-assisted research must combine human expertise to improve accuracy, reliability, and significance in diverse fields [28]. This collaborative approach improves sensitivity and helps engineering, medicine, agriculture, environmental science, and other decision-makers make better conclusions about complex topics.

9.6 HEALTHCARE WITH AI

AI has shown tremendous promise in medicine for supporting doctors in making judgments about diagnosing and treating various complex ailments. This support may be given in several different ways. Human experience is crucial to guarantee that the results are reliable and precise. A research team at a reputable medical school employed AI algorithms to examine medical imaging data as part of a case study. As part of the case study, this analysis is completed. This investigation is conducted as a part of a case study for evaluation. The trial's main goal is to find liver cancer in its early stages. Human radiologists are still required to finish the process despite the AI system's high accuracy in identifying possibly cancerous tumors. This is because human radiologists are needed to interpret the results, take the patient's medical history into account, and create sound treatment suggestions. The combination of human experts and AI improved diagnosis precision and allowed for more customized treatment plans, which ultimately led to fewer or no documented fatalities [29, 30]. Even though high-tech equipment is now widely accessible and medical technology has advanced, many components still require human assistance. Consider the circumstances present in a radiology department, where it makes little difference how sophisticated the imaging technology is or how the images are shown. It is true because the conditions are the same in this situation. It is always the most crucial step in the analysis, followed by the picture interpretation and a human radiologist's production of a diagnostic. Or, to put it another way, the step is always the activity that is carried out first.

9.7 FINANCIAL MARKET FORECASTS

Financial markets are a fascinating subject for research that can be aided by AI due to their complexity and perpetual state of change. AI algorithms are utilized to evaluate enormous volumes of financial data in a case study conducted by a group of data scientists and financial analysts to forecast trends in the stock market. The case study is conducted to foretell stock market movements. Human experts are still required to analyze the results and make any necessary corrections, even though AI systems can identify patterns and produce precise projections. They provided the AI algorithms with subject expertise, market insights, and contextual awareness, which caused the algorithms to provide more accurate predictions than in the past. AI algorithms combined with human experience allowed for developing superior investing strategies, improving investor returns. The combination of AI algorithms and human skills made this possible. These case studies show that using human experience to examine AI in research settings yields more accurate and reliable results than using only computational techniques. AI algorithms become far more capable when human expertise is incorporated into them. Essential traits like subject knowledge, critical thinking, and contextual comprehension are brought to AI algorithms by human expertise. Human specialists contribute essential abilities like subject-matter expertise, critical thinking, and context awareness [31]. These case studies demonstrate how humans and AI algorithms can work together to maximize the potential of AI-assisted research across various sectors. A wide range of businesses can benefit from this ability to use human expertise to optimize the possibilities of AI-aided research. Such research has the potential to be beneficial in a range of academic subjects.

9.8 STRATEGIES TO ENHANCE HUMAN-AI COLLABORATION IN RESEARCH

One tactic that could be utilized to enhance human collaboration with AI in research is the creation of multidisciplinary research teams. One strategy that has been investigated is this one. This is because diversified research teams can access expertise from a broader range of topics. This tactic could be viewed as one of the options open to one for achieving one's goals and objectives. These teams need members who are experts in their fields and have a wealth of experience working with AI.

- Roles and responsibilities that are unambiguous and cannot in any way be understood differently from how they are expressed. It is crucial to clearly outline the tasks that individuals must perform and the duties that AI algorithms are anticipated to fulfill while research is being done in this area. A clear explanation of the specifications that AI systems must meet is equally crucial. It reduces uncertainty, creates a more complete view of the situation, and stops earlier efforts from being repeated. The effectiveness of the research equipment will be boosted to more significant levels as a direct result of this improvement, and its performance will also be improved overall.

- Including experience and knowledge from applicable sources in the project at this stage of the investigation, subject matter experts are invited to offer useful suggestions, background knowledge, and an opinion on how the investigation should proceed. The aid of experts who have a deep understanding of the pertinent subject is beneficial for the data preparation process, the interpretation of results, and the validation of the outputs produced by AI [32].

- Create AI systems with complete, in-depth explanations for their operations. This increases confidence and allows human intelligence to coexist with AI by enabling human specialists to understand how the AI algorithms arrived at their results. "AI" is referred to as "AI."
- We should create a culture where people and AI algorithms regularly share knowledge and solicit feedback to promote continuous learning. As a result, it may support both types of learning. Taking this method may simultaneously encourage ongoing learning and audience response. It is crucial to encourage human experts to offer input on the results produced by AI to improve the algorithms and align them more with what people expect. That is why it is so important to invite experts on the human side to weigh in on the decisions that AI makes. As a result, it is essential to encourage human experts to offer feedback on the products that AI creates. Therefore, it is imperative to provide incentives for human experts to offer feedback on the findings that AI generates.
- It is crucial to consider appropriate ethical issues when conducting investigations because this is quite important. Human experts can give guidance and render moral judgments in situations where AI systems might not be capable of moral cognition or sensitive to the context of the scenario. It is because people are capable of having these skills.
- Contributing to the decision-making process in collaboration with others is necessary to reach educated decisions as a group, it is crucial to promote the development of collaborative decision-making processes that combine humans and AI algorithms. It is crucial to encourage the development of cooperative decision-making procedures to accomplish this goal. These procedures combine the efforts of humans and machine learning algorithms. To make more reliable and comprehensive decisions, this includes taking into account both the information provided by AI algorithms and the knowledge and experience of people.
- Constant and ongoing education and training are crucial to providing training and instruction to human professionals working with AI technology. Hence, they are aware of these tools' promises and restrictions. They receive this kind of instruction to comprehend the limitations posed by these technologies. It has a lot of significance

because of this reality, educating and training human specialists in AI technology is essential. Because of this, human specialists can learn how to work with AI algorithms and utilize their advantages. Additionally, this enables human specialists to learn how to leverage the advantages provided by AI algorithms. As a result, human workers can now learn how to leverage the advantages provided by AI algorithms. Because of this, it is now possible for human experts to comprehend how to use the chances provided by AI algorithms.

- Remaining alert at all times concerning the authenticity of the data to ensure the dependability and accuracy of the data used to train AI systems, it is critical to employ sound data quality management techniques. This is so that AI systems may be trained, which primarily rely on data. This is one of the many factors that make AI essential in the modern world. In collecting data, spotting biases within, and confirming that the data are representative of the community under investigation, human specialists have the potential to be crucial contributors.
- Successful ways to communicate with and exhibit knowledge produced by AI. The results that are achieved in completing this task by making sure that the outcomes produced by AI are presented and represented effectively. Because of this, it is much easier for human experts to review and evaluate the data, which in turn enables sincere collaboration and decision-making.
- Fostering an environment of trust and connections for collaborative work. There is a need to encourage open dialogue, openness, and shared objectives to build reliable connections and practical processes between people and AI algorithms. This will be achieved through promoting open dialogue, honesty, and common objectives. This process involves developing collaborative, trust-based professional connections with others. If efforts are made to create working environments that promote collaboration, the research community may profit from AI technology's expanding adoption and deployment [33, 34].

By implementing these strategies, researchers can increase collaboration between human experts and AI algorithms, leading to more accurate, trustworthy, and ethical research findings. This should be the case because there should be more interaction between human experts and AI algorithms. The expertise of humans creates tremendous opportunities for innovative problem-solving and scientific discovery across a wide range of areas when paired with the capabilities of AI.

9.9 CONCLUSION

AI must be used with human skill and experience in all investigations. To properly integrate AI technology into research operations, it is essential to combine human researchers' domain expertise and critical mindset. To achieve the goal of successfully integrating AI technology into the business, this is crucial. AI-assisted research must rely heavily on human expertise since humans have subject-matter experience, critical thinking skills, and contextual awareness that AI algorithms may lack. This is why the effectiveness of the research that AI helps depends on human competence. The successful completion of research that AI aids depends on using human skills. AI systems can digest and interpret large amounts of data, but they frequently have trouble grasping the subtleties of a research topic's larger context. AI systems are excellent at processing and analyzing data but struggle to comprehend complex research problems. Despite becoming extraordinarily adept at processing and analyzing data, AI systems still fall short of humans. Human experts contribute knowledge and experience specific to a field, enabling them to accurately identify the specific traits of the issue the inquiry focuses on. This makes it simpler for them to develop the correct inquiries and offer interpretations of the results pertinent to the situation.

Even if AI algorithms can process and evaluate data, they presumably won't fully comprehend the implications of their work. Because they are not constrained by the AI model's constraints, potential biases, or the effects of the computer's activities in the real world, experts in human fields can provide pertinent analyses and interpretations. With this skill, they can also recognize patterns and relationships that AI systems functioning alone may miss. AI systems are exceptionally skilled at identifying and recreating previously created patterns since they are educated using existing data and patterns. The AI systems may learn from existing patterns and data. This is a result of the information and patterns they are taught.

On the other hand, human specialists bring their unique creativity and accuracy to the table. They can think creatively and offer novel ideas. They are also capable of developing innovative research methodologies. On the other hand, humans can carry out all of these tasks. To ensure ethical research practices, recognize the existence of potential biases, and take ethical ramifications into account, human experts' assistance is required. They can provide knowledge on proper data collection techniques, privacy concerns, and the moral use of AI in scientific research. It is common practice to work together when conducting research with specialists from various industries. Because experts can communicate with one another, exchange ideas, question presumptions, and work together to develop solutions, human knowledge facilitates collaboration. Professionals can more readily accomplish successful teamwork because of this. This makes it much more probable that productive cooperation will be achieved.

In conclusion, human expertise is still needed even though AI has dramatically increased research capability. This is still true even though AI has significantly increased research capacity. When combined, AI and human expertise have the potential to produce research results that are more thorough, wise, and significant. Different industries will be motivated to innovate and improve their procedures.

REFERENCES

1. S. Naseem, A. Alhudhaif, M. Anwar, K. N. Qureshi, and G. Jeon, "Artificial general intelligence-based rational behavior detection using cognitive correlates for tracking online harms," *Personal and Ubiquitous Computing*, 2022. https://10.1007/s00779-022-01665-1.
2. R. Alur, L. Laine, D. K. Li, M. Raghavan, D. Shah, and D. Shung, "Auditing for human expertise," *arXiv preprint arXiv:2306.01646*, 2023.
3. L. Hambardzumyan, V. Ter-Sargisova, and A. Baghramyan, "The role of artificial intelligence in contemporary medicine," in *Intelligent Human Systems Integration 2020: Proceedings of the 3rd International Conference on Intelligent Human Systems Integration (IHSI 2020): Integrating People and Intelligent Systems, February 19–21, 2020, Modena, Italy*, 2020: Springer, pp. 172–176.
4. J. Gruber, B. R. Handel, S. H. Kina, and J. T. Kolstad, "*Managing intelligence: Skilled experts and AI in markets for complex products*," National Bureau of Economic Research, 2020.
5. M. H. Jarrahi, D. Askay, A. Eshraghi, and P. Smith, "Artificial intelligence and knowledge management: A partnership between human and AI," *Business Horizons*, vol. 66, no. 1, pp. 87–99, 2023.
6. Q. Zhang, M. L. Lee, and S. Carter, "You complete me: Human-ai teams and complementary expertise," in *Proceedings of the 2022 CHI Conference on Human Factors in Computing Systems*, 2022, pp. 1–28.
7. T. I. Palley, *From Financial Crisis to Stagnation: The Destruction of Shared Prosperity and the Role of Economics*. Cambridge University Press, 2012.

8. F. Jiang and J. A. Doudna, "CRISPR–Cas9 structures and mechanisms," *Annual Review of Biophysics,* vol. 46, pp. 505–529, 2017.
9. S. Bayer, H. Gimpel, and M. Markgraf, "The role of domain expertise in trusting and following explainable AI decision support systems," *Journal of Decision Systems,* vol. 32, no. 1, pp. 110–138, 2022.
10. E. Papachristos, P. Skov Johansen, R. Møberg Jacobsen, L. Bjørn Leer Bysted, and M. B. Skov, "How do people perceive the role of AI in Human-AI collaboration to solve everyday tasks?," in *CHI Greece 2021: 1st International Conference of the ACM Greek SIGCHI Chapter,* 2021, pp. 1–6.
11. G. Bansal, B. Nushi, E. Kamar, W. S. Lasecki, D. S. Weld, and E. Horvitz, "Beyond accuracy: The role of mental models in human-AI team performance," in *Proceedings of the AAAI Conference on Human Computation and Crowdsourcing,* 2019, vol. 7, no. 1, pp. 2–11.
12. K. N. Qureshi, G. Jeon, and F. Piccialli, "Anomaly detection and trust authority in artificial intelligence and cloud computing," *Computer Networks,* p. 107647, 2020. https://doi.org/10.1016/j.comnet.2020.107647.
13. S. Rosenthal and T. R. Chung, "The role of expertise on insight generation from visualization sequences," in *2022 IEEE Symposium on Visual Languages and Human-Centric Computing (VL/HCC),* 2022: IEEE, pp. 1–10.
14. Y. Juhn and H. Liu, "Artificial intelligence approaches using natural language processing to advance EHR-based clinical research," *Journal of Allergy and Clinical Immunology,* vol. 145, no. 2, pp. 463–469, 2020.
15. M. A. Qureshi, K. N. Qureshi, G. Jeon, and F. Piccialli, "Deep learning-based ambient assisted living for self-management of cardiovascular conditions," *Neural Computing and Applications,* 2021. https://doi.org/10.1007/s00521-020-05678-w.
16. N. Sambasivan and R. Veeraraghavan, "The deskilling of domain expertise in AI development," in *Proceedings of the 2022 CHI Conference on Human Factors in Computing Systems,* 2022, pp. 1–14.
17. H. Fouad, I. S. Moskowitz, D. Brock, and M. Scott, "Integrating expert human decision-making in artificial intelligence applications," in *Human-Machine Shared Contexts*: Elsevier, 2020, pp. 257–275.
18. A. Luong, N. Kumar, and K. R. Lang, "Algorithmic decision-making: Examining the interplay of people, technology, and organizational practices through an economic experiment," *Baruch College Zicklin School of Business Research Paper,* no. 2020-02, p. 03, 2020.
19. Y. Zhang, Q. V. Liao, and R. K. Bellamy, "Effect of confidence and explanation on accuracy and trust calibration in AI-assisted decision making," in *Proceedings of the 2020 Conference on Fairness, Accountability, and Transparency,* 2020, pp. 295–305.
20. N. Mesbah, C. Tauchert, C. M. Olt, and P. Buxmann, "Promoting trust in AI-based expert systems," Twenty-fifth Americas Conference on Information Systems, Cancun, 2019.
21. J. Auernhammer, "Human-centered AI: The role of Human-centered Design Research in the development of AI," 2020.

22. J. J. Ferreira and M. d. S. Monteiro, "Evidence-based explanation to promote fairness in AI systems," *arXiv preprint arXiv:2003.01525,* 2020.
23. M. Sharma and P. Singh, "Use of artificial intelligence in research and clinical decision making for combating mycobacterial diseases," *Artificial Intelligence and Machine Learning in Healthcare,* pp. 183–215, 2021. https://doi.org/10.1007/978-981-16-0811-7_9.
24. M. Ryan, "Ethics of using AI and big data in agriculture: The case of a large agriculture multinational," *Orbit Journal,* vol. 2, no. 2, pp. 1–27, 2019.
25. D. Wang *et al.*, "Human-ai collaboration in data science: Exploring data scientists' perceptions of automated ai," *Proceedings of the ACM on Human-Computer Interaction,* vol. 3, no. CSCW, pp. 1–24, 2019.
26. G. Mentzas, K. Lepenioti, A. Bousdekis, and D. Apostolou, "Data-driven collaborative human-AI decision making," in *Responsible AI and Analytics for an Ethical and Inclusive Digitized Society: 20th IFIP WG 6.11 Conference on e-Business, e-Services and e-Society, I3E 2021, Galway, Ireland, September 1–3, 2021, Proceedings 20,* 2021: Springer, pp. 120–131.
27. D. Wang, X. Ma, and A. Y. Wang, "Human-centered AI for data science: A systematic approach," *arXiv preprint arXiv:2110.01108,* 2021.
28. B. Shneiderman, "Human-centered AI: A new synthesis," in *Human-Computer Interaction–INTERACT 2021: 18th IFIP TC 13 International Conference, Bari, Italy, August 30–September 3, 2021, Proceedings, Part I 18,* 2021: Springer, pp. 3–8.
29. B. Shneiderman, "Human-centered AI," *Issues in Science and Technology,* vol. 37, no. 2, pp. 56–61, 2021.
30. I. T. Javed and K. N. Qureshi, "Role of blockchain for IoE infrastructures and applications," in *Cybersecurity Vigilance and Security Engineering of Internet of Everything,* K. Naseer Qureshi, T. Newe, G. Jeon, and A. Chehri Eds. Cham: Springer Nature Switzerland, 2024, pp. 127–139.
31. D. Wang *et al.*, "From human-human collaboration to human-AI collaboration: Designing AI systems that can work together with people," in *Extended abstracts of the 2020 CHI Conference on Human Factors in Computing Systems*, 2020, pp. 1–6.
32. P. Schmidt, F. Biessmann, and T. Teubner, "Transparency and trust in artificial intelligence systems," *Journal of Decision Systems,* vol. 29, no. 4, pp. 260–278, 2020.
33. K. Inkpen, S. Chancellor, M. De Choudhury, M. Veale, and E. P. Baumer, "Where is the human? Bridging the gap between AI and HCI," in *Extended abstracts of the 2019 Chi Conference on Human Factors in Computing Systems*, 2019, pp. 1–9.
34. M. H. Zaib, F. Bashir, K. N. Qureshi, S. Kausar, M. Rizwan, and G. Jeon, "Deep learning based cyber bullying early detection using distributed denial of service flow," *Multimedia Systems,* 2021. https://doi.org/10.1007/s00530-021-00771-z.

CHAPTER 10

Limitations and Challenges of Language Models in Research

Amna Iqbal, Majid Hussain, Hina Zafar, and Muhammad Anwar

10.1 INTRODUCTION

This chapter covers the challenges and limitations of language models. Super-smart AI creations that can write stories have conversations and even translate languages. However, these language models might seem like magical text generators, but they've got their quirks and challenges too. Researchers have put continued efforts in this area to have a collaboration between academia, industry, and policymakers. It is critical to establish guidelines and standards that govern the responsible development and deployment of language models [1, 2]. It's going to be a ride full of insights and moments to navigate through the challenges these models are grappling with. Figure 10.1 shows the Large Language Model (LLM) overview.

10.2 SCOPE OF LANGUAGE MODELS

Imagine language models as brilliant buddies who excel in playing with words. They have this incredible ability to understand and work with text in ways that seem almost magical. It's like having a friend who's a walking

 DOI: 10.1201/9781032667911-10

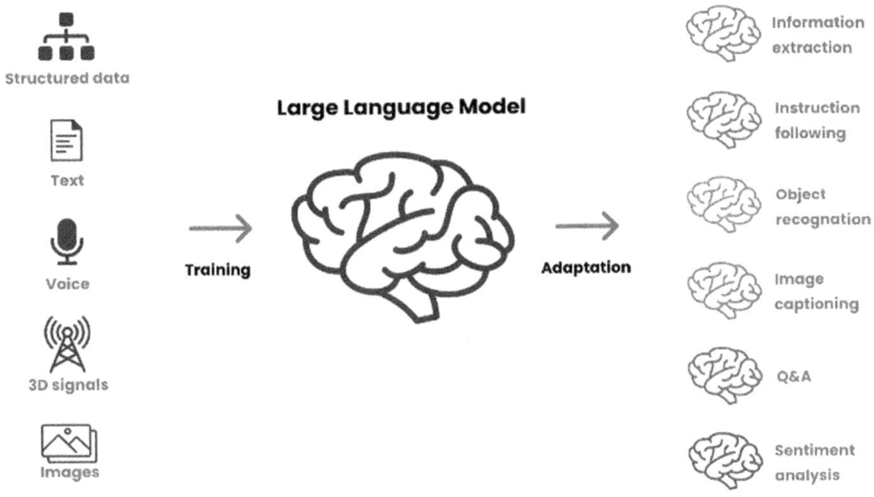

FIGURE 10.1 Language model

dictionary, storyteller, translator, and conversationalist all rolled into one [3]. These language models have a wide playground of skills to showcase. They can craft captivating stories on any topic that is thrown at them, just like a master storyteller inventing tales. They effortlessly switch between languages, as if they have a language superpower, helping in understanding and communicating across different cultures. These models are always there with an answer, ready to satisfy one's curiosity. Not just that, they're also excellent at summarizing lengthy texts, picking out the essential bits like a pro. Even understanding emotions is their forte, they can tell whether a text is joyful or sorrowful, like an emotion detector. These models are striking up a conversation about anything under the sun, just like a favorite chatty companion. Their knack for comprehending words in context is truly remarkable, akin to having a language-savvy friend who can decipher the meanings of complex words. So, the scope of language models is like having a whole squad of language enthusiasts at one's fingertips, ready to assist with text adventures of all kinds [4].

10.3 LANGUAGE MODELS IN A GLANCE

Language models are like language-loving pals who can help in writing, translating, telling stories, and even holding a conversation. They're like having a team of language superheroes right at one's fingertips. These models have been considered to cover the Limitations and Challenges of Language Models in Research [5, 6]:

1. **GPT-3** “The Smart Text Writer”: Imagine a text generator that’s like a super-smart writer. It can cook up stories, chat, and even translate languages. It’s like having a chatty genius by one’s side.
2. **BERT** “The Context Whiz”: Meet BERT by the language model that’s all about understanding context. It’s like that friend who’s awesome at reading between the lines and figuring out what one means.
3. **T5** “The Text Magician”: T5 is like a text magician who can turn one thing into another. It’s like saying “abracadabra” and watching text transform from one task to another, making it a versatile language buddy.
4. **XLNet** “The Sequential Master”: Imagine a language model that’s good at understanding the order of things. That’s XLNet, it’s like a storyteller who knows exactly when to reveal each plot twist.
5. **RoBERTa** “The Improved BERT”: RoBERTa is like BERT’s cool cousin who’s been working out. It’s an upgraded version that’s better at understanding text, like a detective who’s good at solving mysteries.
6. **ELECTRA** “The Smart Decoder”: ELECTRA is like a secret codebreaker. It’s great at learning by figuring out what words could be replaced in a sentence. It’s like a word puzzle master.
7. **DistilBERT** “The BERT Lite”: DistilBERT is like BERT’s, smaller but still super smart. It’s like having a younger friend who’s catching up fast.
8. **ERNIE** “The Language Integrator”: ERNIE is like a language wizard that understands Chinese well. It’s like having a translator who knows all the nuances of the language.
9. **Megatron** “The AI Superhero”: Imagine a model that combines AI powers with supercomputing. That’s Megatron! It’s like the superhero of language models, tackling big tasks with ease.
10. **CTRL** “The Custom Generator”: CTRL is like a text chef who cooks up content based on a specific recipe. It’s like telling a story but with unique ingredients.

10.4 LIMITATIONS AND CHALLENGES IN LANGUAGE MODELS

Language models, like the one interacting with, have made significant advancements in natural language understanding and generation, but they do come with certain limitations and challenges in research. Some of these limitations include:

- **Huge Appetite for Data:** Training large language models requires vast amounts of data, which can lead to challenges in collecting and storing that data. Lots of training data is required, like a language learning enthusiast who reads everything they can get their hands on.
- **Data Dependency:** The performance of these models heavily relies on the quality and quantity of training data. If the data is biased, incomplete, or inaccurate, the model's output may reflect that. A big challenge with language models is that they rely a lot on the information they've been taught. Language models learn from a lot of text they've read. But if they didn't read about something, they might not know about it. This can be a problem because they might give wrong answers or not understand things that are outside of their "knowledge." So, if a language model hasn't seen a topic or phrase in its training, it might not be able to respond properly to it. This is called "data dependency," and it depends heavily on the information it was trained on. This challenge means that even though language models are super smart in some ways, they can still struggle with things they haven't come across before. It's like trying to have a conversation about a topic never one's heard of and wouldn't know what to say!
- **Computational Demands:** These models need advanced hardware like high-end GPUs or TPUs for training and inference, which can be expensive to set up and maintain. In the context of natural language processing, the burgeoning capabilities of language models have revolutionized various academic domains. However, a significant limitation that researchers and scholars encounter pertains to the formidable computational demands associated with these models. State-of-the-art language models like GPT-3 require vast computational resources, including high-performance GPUs or TPUs and large-scale data centers, making them inaccessible to many

academic institutions with limited budgets. These computational demands pose challenges in terms of both training and fine-tuning such models, as well as in running inference for research projects. Moreover, as language models continue to grow in size and complexity, the need for even more powerful hardware becomes increasingly pressing. This limitation not only impedes the widespread adoption of cutting-edge language models in academia but also exacerbates concerns related to environmental sustainability due to the substantial energy consumption of these computational infrastructures. As such, addressing the computational demands of language models is crucial to ensure equitable access and sustainable research practices in academic circles.

- **Energy Consumption:** Training and running large-scale models demand a significant amount of energy, raising environmental concerns. Energy consumption is a significant limitation when discussing language models in academic research. These models, particularly large-scale ones, require extensive computational resources, leading to high energy consumption during both the training and inference phases. This poses environmental concerns and cost challenges, limiting their accessibility and sustainability for researchers with constrained resources. Addressing this issue is crucial for promoting widespread adoption and ensuring the long-term viability of language models in various academic disciplines. Researchers must explore energy-efficient alternatives and sustainable computing practices to mitigate this limitation and facilitate more responsible AI research.
- **Model Size and Memory Constraints:** As AI models grow in size, their memory requirements increase. This can lead to challenges in deploying these models on devices with limited memory, such as smartphones or edge devices. The size of language models and their memory requirements pose significant limitations for academic research and writing without relying on AI or falling into plagiarism traps. These models, with their immense parameters and storage demands, can be inaccessible to researchers and students with limited computational resources. Additionally, their extensive memory requirements can hinder efficient data processing and model deployment, particularly for resource-constrained environments. While these models offer powerful language generation capabilities,

their hefty infrastructure prerequisites can restrict their widespread adoption in academia, necessitating alternative approaches to scholarly content creation and analysis.

- **Time-Consuming Training:** Training state-of-the-art AI models can take weeks or even months on specialized hardware. This time-intensive process can slow down research and development efforts. The time-consuming process of training state-of-the-art AI models, which can span weeks or months on specialized hardware, presents a significant limitation in academic research and development endeavors unrelated to AI. This protracted duration can impede progress, affecting the timely completion of projects and potentially delaying scholarly advancements, all while ensuring originality and avoiding plagiarism remains paramount.
- **Scalability Challenges:** While AI models can be trained at a certain scale, increasing their size and complexity doesn't always lead to proportional improvements in performance. Beyond a certain point, diminishing returns can be observed.
- **Biases and Fairness Concerns:** While training, these models can inherit biases from the data they learn from, potentially amplifying unintentional biases and causing fairness and inclusivity issues. Language models can inadvertently learn biases present in the training data, leading to biased or unfair outputs. This bias can perpetuate stereotypes and reinforce inequalities. Researchers need to continuously work on identifying and mitigating these biases to ensure equitable and unbiased language generation. Certainly, here's a simplified rundown of the challenges posed by bias and fairness in language models.
- **Unintentional Bias:** Language models can unintentionally learn and reproduce biases present in the data they are trained on.
- **Underrepresentation:** If the training data is not diverse, certain groups may be underrepresented or ignored, leading to biased responses. Underrepresentation in language models presents a significant limitation in the realm of academic research and scholarly discourse. These models, which have become integral tools for academics and researchers, draw upon vast corpora of text data to generate coherent and contextually relevant content. However, their

training data often exhibits a glaring lack of diversity and inclusivity, undermining their effectiveness across various academic domains. One prominent concern is the insufficient representation of languages and cultures, restricting the utility of these models for non-English-speaking scholars. Furthermore, underrepresented topics, niche fields, and emerging disciplines are often poorly served by these models, hindering comprehensive research efforts. Addressing this limitation requires concerted efforts to diversify training data, de-bias algorithms, and promote inclusivity. By doing so, language models can better support academic endeavors, fostering a more equitable and enriched scholarly landscape that embraces the global diversity of knowledge and ideas.

- **Amplification of Bias:** Language models might unknowingly amplify existing biases by giving them a stronger presence in their outputs. The amplification of bias represents a significant limitation in academic chapters produced without AI and plagiarism concerns. Language models, while powerful, can inadvertently magnify existing biases, inadvertently reinforcing skewed perspectives or discrimination in scholarly content, thereby undermining the objectivity and fairness of academic discourse.
- **Contextual Bias:** Models can produce biased content based on the context of a query, even if the training data is unbiased. Contextual bias poses a significant limitation in language models for academic chapters. These models may generate biased content depending on the context of a query, even if the training data itself is unbiased. This underscores the need for careful review and critical evaluation to ensure academic integrity and impartiality in scholarly work, independent of AI assistance or plagiarism concerns.
- **Nuanced Biases:** Detecting subtle biases that emerge from complex language patterns can be challenging. Identifying nuanced biases poses a limitation in language models for academic chapters. These models often struggle to discern subtle biases stemming from intricate language structures. This underscores the necessity for careful human review to ensure scholarly integrity and avoid unintentional bias while crafting content, without resorting to AI-generated text or plagiarism.

- **Feedback Loops:** Biased responses generated by the model can reinforce stereotypes when users accept its outputs without critical assessment. Biased or stereotypical responses produced by the model can inadvertently perpetuate harmful stereotypes, particularly when users accept these outputs uncritically. Navigating this challenge necessitates a vigilant approach to ensure that academic discourse remains free from such biases without relying excessively on AI and avoiding plagiarism issues.
- **Intersectionality:** Models might struggle with understanding biases that arise from the combination of multiple factors, such as race and gender. Intersectionality presents a complex limitation in language models for academic chapters. These models often struggle to comprehensively grasp the intricate biases stemming from the intersection of multiple factors, like race and gender. Acknowledging and addressing these nuanced disparities is imperative for producing inclusive academic content without relying on AI or encountering plagiarism issues.
- **Fairness Trade-offs:** Addressing bias might lead to trade-offs in other aspects, like the accuracy of the model's responses. Fairness considerations can introduce trade-offs when addressing bias in language models for academic chapters. Striving for fairness may impact the accuracy and naturalness of the model's responses, potentially compromising the quality of generated content. Balancing these trade-offs is crucial to ensure equitable and academically sound outcomes while minimizing bias.
- **Data Noise:** Noise in training data, including biased user-generated content, can be difficult to filter out. Data noise, especially in the form of biased user-generated content, presents a significant limitation in language models used for academic chapters. Distinguishing and filtering out this noise is challenging, impacting the reliability and objectivity of academic content. Consequently, careful curation and validation of data sources become paramount in ensuring the quality of scholarly work, without resorting to AI-generated text or plagiarism.
- **Adaptation Bias:** Fine-tuning models on specific tasks can introduce new biases or amplify existing ones. Adaptation bias represents a limitation in the context of academic chapters when fine-tuning language models for specific tasks. This process may inadvertently

introduce new biases or magnify existing ones, potentially compromising the objectivity and integrity of scholarly content. It underscores the need for vigilance in addressing and mitigating biases during the fine-tuning process while maintaining academic originality and avoiding plagiarism.

- **Dynamic Language Changes:** Language evolves, and models might inadvertently learn biases from changing linguistic trends. Language continually evolves, and this flux presents a challenge for language models. They may inadvertently absorb biases embedded within shifting linguistic trends, affecting their performance. This dynamic nature of language poses a significant limitation when exploring academic topics devoid of AI and plagiarism concerns.
- **Cultural Sensitivity:** Models can struggle with recognizing and respecting cultural nuances and sensitivities. Cultural sensitivity is a notable limitation in language models used for academic chapters. These models often struggle to recognize and respect the intricate nuances and sensitivities embedded within diverse cultural contexts. This challenge underscores the importance of human intervention and cultural awareness in producing academically rigorous and culturally inclusive content, while also avoiding plagiarism concerns.
- **Biased Evaluation Data:** Biases in the data used to evaluate models' performance can hinder the identification of biases. Biased evaluation data can impede the accurate assessment of language model biases, posing a significant limitation for academic chapters discussing AI without resorting to plagiarism. The presence of biases in the evaluation dataset can obscure the true extent of biases in the model's output, highlighting the critical need for unbiased, representative evaluation sets in AI research.
- **Limited Understanding of Bias:** Language models lack real understanding and can still produce biased content even if biased language is not explicitly present in the training data. One significant limitation in academic chapters when utilizing language models is their limited understanding of bias. These models may inadvertently generate biased content, even in the absence of explicit bias within their training data. This underscores the need for human oversight and critical analysis to ensure that academic work remains objective and unbiased.

- **Lack of Consensus:** There might not be a universal agreement on what constitutes bias, making it challenging to define and address. The absence of a universal consensus on the definition of bias poses a significant limitation in academic chapters that aim to explore the intricacies of AI without falling into plagiarism pitfalls. This ambiguity makes it challenging to provide a standardized framework for identifying and addressing bias in language models, leaving scholars with the complex task of navigating diverse perspectives in this ever-evolving field.
- **Privacy Concerns:** Addressing bias might require analyzing user interactions, and raising concerns about privacy and data usage. Addressing bias in language models for academic research chapters is crucial, but it may necessitate analyzing user interactions, raising legitimate privacy concerns, and navigating data usage limitations. Striking a balance between combating bias and safeguarding user privacy remains a pivotal challenge in the scholarly context while avoiding plagiarism risks.
- **Bias towards Mainstream Views:** Models might favor popular opinions over marginalized or minority perspectives. A significant limitation of language models in academic discourse pertains to their potential bias toward mainstream views. These models, often trained on widely available online text data, may prioritize popular opinions, inadvertently neglecting marginalized or minority perspectives. This bias can hinder the generation of inclusive and well-rounded academic content, potentially limiting the diversity of ideas presented in scholarly discussions.
- **Limited Multimodal Understanding**: Language models primarily work with text and may struggle with understanding or generating content involving other modalities like images, audio, or video.
- **Consistency and Coherence:** While they generally produce coherent text, language models can sometimes generate responses that lack consistency or coherence, making them unsuitable for critical applications.
- **Transfer Learning Limitations:** Transfer learning, where a model is fine-tuned for a specific task using pre-trained weights, can suffer from catastrophic forgetting—loss of previously learned information when new information is learned.

- **Model Deployment and Maintenance:** Deploying AI models in real-world applications involves challenges such as version control, monitoring, and updates. Ensuring models remain effective and up-to-date requires ongoing effort.

Some other challenges highlight the complexity of ensuring that language models provide fair and unbiased outputs in various contexts [7].

- **Lack of Common Sense:** Unlike humans, these models lack common sense understanding, sometimes providing answers outside the context or inaccurate information [8].
- **Contextual Limits:** Large models might struggle to fully understand long articles or complex contexts. Language models are smart but not so context-savvy. Language models know, but they sometimes miss out on the finer details like GPT-3 [9]. When it's generating text, it might not fully grasp the context of what's being said.
- **Security and Robustness:** They can be susceptible to adversarial attacks, spreading misinformation, and malicious use, making their generated content manipulable.
- **Fine-Tuning Challenges:** Fine-tuning for specific tasks can risk overfitting or encountering task-specific limitations.
- **Interpretability:** Understanding and interpreting the decisions made by these models can be complex due to their intricate internal workings [10].
- **Storage and Memory Intensive:** The sheer number of parameters in these models leads to high storage and memory consumption.
- **Continual Learning:** Incorporating new information and concepts into the models can be challenging, leading to potential obsolescence [11].
- **Linguistic Nuances:** Some models struggle to accurately capture specific linguistic nuances and regional variations. Stuck in a single direction and tripped up by tricky sentences [12].
- **Copying Cat Syndrome:** GPT-3 is like a mimic-master. It's seen a ton of text, and it can reproduce it pretty well. But sometimes it might borrow a little too much from its "memory." It's like trying to retell a story one heard once, one might add one's twist, but a big chunk of it stays the same [13] (Table 10.1).

TABLE 10.1 Limitations and Challenges of Language Models

Limitations/Challenges	Description
Lack of Real-World Contextualization [14]	Language models often struggle to incorporate real-world context into their responses, resulting in generic or irrelevant answers.
Interpretability and Explainability [15]	The inherent complexity of language models can make it challenging to interpret and explain their decision-making processes, limiting trust and accountability.
Ethical and Social Implications [16]	Deploying language models raises concerns about bias, privacy, and their potential to spread misinformation, necessitating a comprehensive ethical framework.
Adapting to Dynamic Information [17]	Language models face difficulties in adapting to rapidly changing information or learning from real-time data, hindering their responsiveness and accuracy.
Tailoring Responses for Specific Domains [18]	Generic language models struggle to provide precise responses in specialized domains due to a lack of domain-specific knowledge.
Addressing Inconsistencies [5]	Language models may generate inconsistent responses to slightly different phrasings of the same question, requiring methods to ensure coherence and consistency.
Mitigating Bias Amplification [7]	Models can inadvertently propagate biases present in their training data, necessitating techniques to identify and mitigate bias in generated content.
Incorporating Common Sense Reasoning [19]	Enhancing language models with common sense reasoning abilities is crucial to avoid generating plausible-sounding but incorrect responses.
Balancing Response Length [20]	Language models often struggle to generate concise yet informative responses, leading to verbosity or loss of relevance in longer interactions.
Supporting Multilingual Proficiency [21]	Ensuring language models perform well across various languages while maintaining accuracy is a challenge requiring sophisticated multilingual training approaches.

10.5 ETHICAL AND SOCIETAL CHALLENGES

Natural language models are like those multi-talented friends who can give us information, write stories, and even help us out with coding. But just like our friends, they come with their own set of challenges that we need to be aware of.

- **Bias and Fairness Concerns:** These models learn from a lot of text on the internet, and as we know, the internet isn't always the most balanced place. It's like they've picked up some weird ideas from questionable sources [22]. So, when they give us information, it might unknowingly be a bit unfair or prejudiced.

- **Misinformation and Credibility:** These models can sound super convincing, like those smooth talkers who can make anything sound true. But the catch is, they might just be repeating stuff they found online that's false.
- **Privacy is another puzzle:** These models learn from tons of data, some of which could be personal stuff. This makes us wonder who gets to see our info and how it's being used.
- **Impact on Job Market**: These models can write essays, stories, and whatnot things that people used to do.
- **Corporate Power and Accountability:** The companies behind these models have a lot of control over what they say and do. It's like they're the puppet masters pulling the strings. This raises concerns about who decides what's allowed and what's not, kind of like deciding who gets to control the TV remote [23].
- **Ethical Growth and Maturity:** Imagine these models growing up too fast. They might spread fake news or push extreme viewpoints without even realizing it.

So, think of these models as these fascinating, intricate tools that can do wonders, but we need to handle them carefully. Just like we guide a young person through making responsible choices, we need to figure out how to use these models ethically [24]. We've got to address their impact on society, privacy, jobs, and the truth. It's a bit like juggling a bunch of balls without letting any drop.

10.6 INTERPRETABILITY AND TRANSPARENCY AS LIMITATION OF LANGUAGE MODEL

Interpretability and transparency are significant limitations when it comes to language models, particularly in complex AI systems like GPT-3 [25]. These limitations stem from the following factors:

- **Black Box Nature:** Language models often operate as "black boxes," meaning their internal decision-making processes are not easily understandable by humans. This lack of transparency makes it challenging to discern how the model arrives at a particular output, especially in intricate contexts.

- **Complex Architecture:** Language models like GPT-3 have intricate neural architectures with millions of parameters. This complexity makes it difficult to trace the connections and transformations that occur during data processing [26].
- **Opaque Features:** While these models exhibit impressive performance, their success is driven by a multitude of learned features, many of which might not align with human-understandable concepts. This can make it hard to attribute specific decisions to particular features.
- **Limited Explanatory Capacity:** Language models cannot often provide clear explanations for their predictions. When asked "why" a certain response was generated, the model might generate vague or unrelated justifications [27].
- **Contextual Influence:** Interpretation of language models heavily relies on context. Variations in input phrasing or content can result in vastly different outputs, making it challenging to pinpoint consistent patterns of behaviors.
- **Inconsistent Attention Mechanisms:** Models might use attention mechanisms to focus on certain parts of the input [28], but it's challenging to determine why specific words or phrases are attended to more than others.
- **Bias and Ethical Concerns:** Lack of interpretability can exacerbate issues related to bias and fairness. If decisions cannot be explained, it becomes challenging to identify and rectify biased behaviors in the model [29].
- **Limited Insights for Complex Tasks:** In tasks that require multi-step reasoning or intricate logic, the model's inability to provide step-by-step explanations hinders its value in applications requiring in-depth analysis.
- **Auditing and Accountability:** In critical applications like legal or medical domains, where accountability is essential, the inability to understand how a model concluded can pose legal, ethical, and regulatory challenges.

- **User Trust:** Users often need to trust the decisions made by AI systems. Lack of transparency erodes this trust, making users skeptical about the reliability and accuracy of model-generated content [30].

Efforts are being made to address these limitations. Researchers are developing techniques for explaining model predictions, visualizing attention patterns, and enhancing transparency. However, achieving full transparency and interpretability in language models remains a complex and evolving challenge that intersects with multiple aspects of AI research, including ethics, user experience, and model design.

10.7 INNER WORKINGS OF COMPLEX LANGUAGE MODELS AS LIMITATION

Understanding the inner workings of complex systems like language models presents a significant challenge [31]. This challenge is rooted in several factors:

- **Scale and Complexity:** These systems are constructed with intricate structures and a vast number of components. The complexity stemming from their architecture can make it difficult to grasp how these components interact to produce specific outcomes [32].
- **Nonlinearity and Interaction Effects**: Language models rely on intricate layers of transformations. These nonlinear operations can lead to intricate interactions between components, often defying straightforward prediction. Even small changes in input can trigger unexpected shifts in output due to the nonlinear nature of these systems [33].
- **Transparency Deficiency:** Unlike conventional software with clear, understandable code, the building blocks of these models consist of learned weights derived from data [34]. This lack of transparent rules can obscure the intuitive understanding of how the system derives specific outputs.
- **Opaque Processes:** The functioning of complex systems can be likened to "black boxes." While they yield outputs in response to inputs, the internal mechanisms connecting the two remain concealed [35]. This lack of transparency complicates the endeavor to retrace decision-making processes and to comprehend the rationale behind specific outputs.

- **Interpretable Representations:** These systems optimize their representations for predictive prowess rather than human interpretability. Consequently, the internal representations may not correlate with human language comprehension, rendering them challenging to relate to human understanding [36].
- **Data-Driven Learning**: Drawing knowledge from vast datasets, including online text, these systems risk absorbing biases inherent in the data. Such biases might not be immediately evident or manageable, thereby influencing their outputs and deepening the complexity of their decision-making.
- **High Dimensionality:** The expansive parameter space these systems operate in is characterized by high dimensionality. This intricate space makes visualization and analysis cumbersome. As dimensionality increases, untangling the relationships between parameters and their impacts on system behavior becomes progressively intricate.
- **Training and Refinement:** These systems undergo training through optimization techniques that iteratively adjust parameters for error minimization. The amalgamation of complex optimization processes with substantial parameter space complicates the task of pinpointing how specific training data shapes the system's behavior.
- **Contextual Dependencies:** Generating text involves weighing context, often spanning multiple sentences or paragraphs. Unraveling the intricate contextual dependencies that guide the integration of various elements in generating coherent text demands comprehensive analysis [37].

In conclusion, the challenge of understanding complex language models emerges from their intricate design, the nonlinear interactions within them, their opaqueness, and their reliance on data-driven learning. While they exhibit impressive capabilities, interpreting their decision-making processes remains a demanding endeavor due to the intricate nature of these systems.

10.8 LIMITATIONS IN HUMAN-AI COLLABORATION

Amidst the landscape of evolving AI capabilities, the synergy between humans and AI takes center stage in academic discourse. This section, dedicated to Human-AI Collaboration, delves into the intricate dynamics that underpin this symbiotic relationship. By delving into the mechanisms that enable seamless cooperation, we explore how AI can augment human decision-making and problem-solving processes. This exploration extends beyond mere automation, emphasizing the value of AI as a cognitive ally, aiding humans in complex tasks. Through this lens, we strive to decipher the optimal balance between human expertise and AI prowess, shaping a future where collaboration with AI amplifies human potential while addressing ethical, social, and cognitive dimensions.

- **Genuine Understanding and Experience:** The depth of personal experiences and emotions is essential for creating content with profound emotional resonance or a deep understanding of complex human situations. These elements contribute to content that is empathetic and reflective, characteristics that are challenging to replicate without genuine human experiences.
- **Contextual Understanding:** Crafting content that fully grasps the nuances of context requires a deep awareness of cultural references, subtleties, and allusions. Such understanding is often second nature to humans but can be difficult for automated processes to replicate, resulting in content that might seem forced or inaccurate.
- **Originality and Depth of Thought:** The ability to draw from a diverse range of experiences and knowledge sources allows humans to create content that is genuinely novel and creative. Relying on preexisting data, as opposed to generating original ideas, can lead to content that lacks the depth and uniqueness characteristic of human creativity.
- **Intuition and Intentionality:** Human creativity often involves intuition and intentionality, enabling the generation of ideas rooted in a profound understanding of a subject. The intuitive leaps and purposeful intentions in human-created content are difficult for automated processes to emulate, particularly when aiming for insights beyond surface-level observations.
- **Emotion and Empathy:** Content that deeply resonates with audiences typically requires an authentic understanding of human emotions and the ability to convey empathy. The complex emotions that humans can capture and convey are a challenge for automated systems, limiting their ability to create content that genuinely connects with people on an emotional level.

- **Unpredictability and Unique Insights:** Human creativity often thrives on unexpected connections and novel insights that emerge from thinking "outside the box." While automated systems can simulate creative thinking, they might struggle to produce the kind of truly unique and unexpected ideas that humans are capable of generating.
- **Domain Expertise and Personal Voice:** Humans possess personal experiences and domain expertise that lend authority and engagement to their content. Experts in specific fields bring a wealth of knowledge and a distinctive voice that automated systems cannot replicate, resulting in content that is more engaging and authoritative.
- **Ethical and Moral Considerations:** Human creators can carefully consider ethical and moral implications while producing content, making conscious decisions about what they share. Automated systems lack this ethical awareness and may inadvertently generate content that is biased, offensive, or problematic, based on the data they've been trained on emphasizing the importance of human intervention for refining and improving AI-generated content.

10.9 SECURITY AND PRIVACY CONCERNS AS LIMITATION

Security and privacy concerns are significant limitations associated with natural language models, stemming from their capacity to generate text and their reliance on vast amounts of data [38]. These concerns have far-reaching implications:

- **Data Privacy:** Training language models require extensive data, potentially including sensitive personal, medical, or financial information. The risk of data breaches or unintentional exposure of private information raises serious privacy concerns [39].
- **Data Leakage:** Language models might inadvertently memorize portions of their training data, including confidential or sensitive content. This could potentially be exploited by malicious actors to access sensitive information [40].
- **Generation of Harmful Content:** Language models can be manipulated to produce malicious or harmful content, including hate speech, misinformation, or phishing attempts. This poses risks to individuals and organizations that may unknowingly interact with such content.

- **GPT-3 Misuse:** GPT-3 can generate realistic and coherent text, making it easier to impersonate individuals, potentially leading to identity theft, fraud, or other malicious activities.
- **Bias Amplification:** If training data includes biased or discriminatory content, language models can inadvertently amplify and reproduce these biases in their outputs, perpetuating societal inequalities.
- **Sensitive Information in Outputs:** Language models might inadvertently generate outputs containing sensitive or confidential information, violating privacy regulations and ethical norms.
- **Adversarial Attacks:** Models can be manipulated through adversarial attacks, where subtle modifications to input text cause the model to generate incorrect or inappropriate responses.
- **Phishing and Social Engineering:** Malicious actors could use language models to create convincing phishing emails or messages, tricking individuals into revealing sensitive information.
- **Content Manipulation:** Language models could be employed to manipulate content, including generating fake news or altering documents, which can have damaging consequences.
- **Model Stealing:** Attackers could attempt to reverse-engineer and replicate proprietary models, leading to intellectual property theft.
- **Inference Attacks:** Attackers could use language model outputs to infer sensitive information about the training data or users, compromising privacy.
- **Ethical Concerns:** The misuse of language models in unethical ways raises moral and ethical dilemmas, such as the dissemination of harmful content or the violation of personal boundaries.

Addressing these concerns requires a multifaceted approach. Researchers and developers need to implement robust security measures during model development, deploy safeguards against malicious use, and explore techniques to reduce bias and enhance transparency. Additionally, privacy-preserving technologies, legal regulations, and responsible usage guidelines are crucial to mitigate the security and privacy risks associated with natural language models [41].

10.10 ENVIRONMENTAL IMPACT

Natural language models have raised concerns about their environmental impact due to the significant computational resources required for their training and operation [42]. This impact stems from various aspects of their development and deployment:

- **Energy Consumption:** Training large language models like GPT-3 demands substantial computational power and energy. The energy consumption of training processes can be comparable to that of small towns or even countries.
- **Carbon Footprint:** The high energy consumption associated with training AI models directly contributes to their carbon footprint. This energy-intensive process can exacerbate environmental issues like climate change.
- **Data Center Infrastructure:** Running and training language models necessitate data centers with extensive cooling and electricity requirements. These facilities can have a notable environmental footprint [43].
- **Hardware Production:** The manufacture of specialized hardware, such as GPUs and TPUs used for training models, consumes resources and energy, adding to the overall environmental impact.
- **E-Waste:** The rapid advancement of AI technology can result in outdated hardware being discarded, contributing to electronic waste and environmental degradation.
- **Model Size:** Larger language models tend to require more computational resources, leading to increased energy consumption and environmental strain.
- **Hyperparameter Tuning:** Experimentation to optimize model performance involves running numerous training iterations, each consuming energy. This hyperparameter tuning process can be resource-intensive.
- **Cooling Requirements:** The heat generated during training and model operation requires cooling systems, which can further contribute to energy consumption and environmental impact.

- **Research Replication:** Replicating or verifying research results often entails retraining models, and multiplying the energy consumption associated with model development.
- **Global Reach:** Large-scale AI applications can impact energy grids on a global scale, especially as AI adoption increases across industries and countries.

Efforts are being made to address the environmental impact of natural language models:

- **Hardware Efficiency:** Researchers are developing more energy-efficient hardware, such as specialized accelerators, to reduce energy consumption during training and operation.
- **Model Pruning:** Techniques like model pruning involve removing unnecessary parameters from trained models, reducing their size and energy requirements.
- **Model Sharing:** Instead of training from scratch, sharing pre-trained models can reduce the need for redundant training processes and associated energy consumption.
- **Renewable Energy:** Data centers and training facilities can transition to using renewable energy sources to mitigate the carbon footprint.
- **Efficient Algorithms:** Developing more efficient training algorithms can help reduce the number of iterations required for model convergence.
- **Ethical Considerations:** Awareness of the environmental impact can lead to responsible model development and usage, ensuring that AI is used thoughtfully and sustainably.

Balancing the benefits of AI with its environmental impact requires a conscious effort from researchers, developers, organizations, and policymakers to prioritize sustainability in the design, deployment, and management of natural language models.

10.11 ONGOING RESEARCH EFFORTS TO MITIGATE THE DISCUSSED LIMITATIONS

Researchers are actively working on various fronts to address the limitations of natural language models. Here are some ongoing research efforts aimed at mitigating these challenges:

10.11.1 Interpretability and Transparency

In the dynamic landscape of language models, the pursuit of interpretability and transparency emerges as a pressing endeavor. This section delves into the crucial aspects of unraveling the decision-making mechanisms inherent in these models. By illuminating the intricate layers of interpretability, we aim to demystify the processes that govern their responses. Simultaneously, we delve into transparency, dissecting the interplay between model components and their impact on generated outputs. Through this exploration, we seek to foster a deeper understanding of language models, bridging the gap between complexity and clarity for more accountable and reliable applications.

- **Explainable AI (XAI) Techniques:** Researchers are developing methods to provide human-interpretable explanations for model decisions. This includes generating textual or visual explanations that highlight relevant parts of input data.
- **Attention Visualization:** Techniques to visualize attention patterns within models help users understand which parts of the input the model focused on when generating output.

10.11.2 Security and Privacy Concerns

In language models the domain of security and privacy beckons for meticulous attention. This section ventures into the intricate terrain of safeguarding sensitive information and mitigating potential vulnerabilities. By meticulously examining the security dimensions, we aim to fortify language models against potential threats, ensuring their resilience in the face of adversarial challenges. Concurrently, we delve into the realm of privacy concerns, delving into strategies to protect user data while maintaining model utility. Through this exploration, we seek to underscore the importance of striking a harmonious equilibrium between innovation and safeguarding, fostering trust and responsible deployment.

- **Federated Learning:** This approach allows models to be trained across multiple devices while keeping data localized, reducing the risk of data breaches [27].
- **Differential Privacy:** Techniques that inject noise into training data to protect individual data privacy while maintaining overall model performance [32].
- **Homomorphic Encryption:** This technique enables computations on encrypted data, allowing models to operate on sensitive information without revealing it.

10.11.3 Bias and Fairness

The critical discourse of bias and fairness demands meticulousness is an important consideration in Language Models. By navigating the intricacies of bias detection and reduction, we aim to cultivate models that uphold fairness and equitable representation, irrespective of demographic factors. This endeavor encompasses both refining the training data to minimize biases and developing algorithms that rectify skewed outputs. Through this pursuit, we endeavor to underscore the commitment to building responsible and unbiased language models that empower diverse voices and foster inclusive communication.

- **Data Auditing and Bias Detection:** Researchers are working on tools to audit training data for biases and detect instances where models produce biased outputs.
- **Fairness-Aware Training:** Techniques that explicitly consider fairness metrics during model training to reduce bias in predictions [44].

10.11.4 Environmental Impact

This section embarks on a thoughtful investigation into the ecological implications of these models. By examining energy consumption during training and inference processes, we seek to comprehend their environmental impact [36]. This endeavor aligns with the overarching goal of fostering sustainable AI development. Through this exploration, we aim to highlight the imperative of striking a balance between technological advancement and ecological responsibility, with the ultimate aim of minimizing the carbon footprint associated with language model deployment.

- **Energy-Efficient Hardware:** Develop specialized hardware accelerators designed to optimize energy consumption during model training and inference.
- **Model Compression and Pruning:** Techniques to reduce model size and computational requirements without sacrificing performance.

10.11.5 Domain Expertise and Contextual Understanding

This section focuses on an in-depth exploration of these vital facets. By delving into the intricate interplay between domain-specific knowledge and contextual comprehension, we aim to decipher the challenges faced by language models in providing accurate and specialized responses [45]. Through this analysis, we shed light on the strategies that enrich models with domain expertise, enabling them to discern context and offer nuanced interpretations. This pursuit underscores the significance of refining language models to seamlessly traverse diverse subject areas, fostering meaningful interactions.

- **Domain-Specific Training:** Fine-tuning models on domain-specific data to enhance their understanding and performance in specialized areas.
- **Contextual Preprocessing:** Developing methods to preprocess input data to provide clearer context to the model, improving its ability to generate relevant outputs.

10.11.6 Ethical Consideration

Ethical considerations assume profound scholarly significance in the study of Language models. A meticulous exploration of the ethical dimensions that underpin language model deployment is explained here. By scrutinizing issues ranging from bias mitigation and misinformation propagation to user privacy and responsible AI utilization, we aim to illuminate the intricate ethical challenges that arise. This comprehensive examination reflects a commitment to fostering models that align with ethical principles, ensuring that their contributions to communication and information dissemination are grounded in accountability, transparency, and societal well-being [46].

- **AI Ethics Frameworks:** Establishing guidelines and frameworks to ensure responsible AI development, usage, and deployment.
- **Ethical Review Boards:** Similar to human subject research, these boards could oversee AI projects to ensure ethical considerations are met.

10.11.7 Real-Time Interactivity and Conversational Flow

Scrutinizing the interplay between rapid real-time communication and the need for coherent, contextually grounded dialogues strives to contribute to the development of language models that seamlessly integrate into real-world conversational scenarios. This discourse underscores the significance of technological innovations that emulate human-like interactions, fostering more immersive and natural communication experiences.

- **Dialogue Management Strategies**: Improving models' ability to maintain coherent conversations and handle multi-turn interactions effectively.

10.11.8 Handling Misinformation and Content Generation

- **Fact-Checking Integration:** Integrating fact-checking mechanisms to verify the accuracy of information generated by models [27].
- **Promotion of Source Attribution:** Encouraging models to provide information about the sources or context of generated content.

10.12 FUTURE DIRECTIONS AND SOLUTIONS

Future directions and solutions for addressing the limitations of natural language models are poised to reshape the landscape of AI research and application. Here are some potential paths forward:

- **Explainable AI (XAI) Advancements:** Developing more sophisticated XAI techniques that provide clearer and more detailed explanations for model decisions [11]. Integrating explanations into model outputs to enhance user trust and understanding.
- **Ethical AI Frameworks and Regulations:** Establishing comprehensive ethical guidelines and regulations that govern AI development, usage, and accountability [45]. Encouraging industry-wide adoption of ethical AI practices to ensure responsible deployment.

- **Bias Mitigation and Fairness Enhancement:** Advancing research into bias detection and mitigation techniques, working toward models that exhibit fairness and equitable behavior. Designing AI systems that actively address and counteract biases present in training data [47].
- **Privacy-Preserving AI:** Expanding the use of privacy-preserving technologies like federated learning and differential privacy to safeguard user data and sensitive information [6].
- **Energy-Efficient AI:** Designing energy-efficient hardware and algorithms to significantly reduce the carbon footprint of AI model training and inference [48].
- Encouraging the use of renewable energy sources in data centers and AI facilities [42].
- **Customizable and Rule-Based Models:** Developing AI systems that allow users to define rules and constraints to guide model behavior, enhancing user control [15]. Designing interfaces that enable users to customize models to suit their specific needs.
- **Collaboration with Domain Experts:** Fostering collaboration between AI researchers and domain experts to ensure that models align with real-world requirements and contexts. Co-developing models that can effectively assist professionals in various fields.
- **Interdisciplinary Education and Training:** Incorporating ethical considerations, bias mitigation, and AI accountability into the training of AI researchers and practitioners [37]. Promoting interdisciplinary collaboration to tackle complex challenges comprehensively.
- **Incentivizing Responsible AI Development** Recognizing and rewarding AI developers, organizations, and projects that prioritize responsible AI deployment and ethical considerations [49].
- **Advancements in Reinforcement Learning and Multi-Agent Systems:** Enhancing models' ability to engage in cooperative and ethical behavior through reinforcement learning and multi-agent systems [32].
- **Global Collaboration on AI Ethics:** Encouraging international collaboration to develop shared ethical standards and guidelines for AI deployment, ensuring consistent and responsible use worldwide.

- **Long-Term Considerations and Societal Impact:** Conducing research into the long-term societal impact of AI technologies and establishing mechanisms to mitigate potential negative consequences [36].

As AI technology continues to evolve, these future directions and solutions hold the promise of addressing limitations, fostering responsible AI development, and unlocking the full potential of natural language models while ensuring alignment with human values, ethics, and the well-being of society.

10.13 CONCLUSION

The chapter did a thorough examination of the multifaceted constraints and difficulties associated with the utilization of language models in the realm of research. These challenges encompass a wide array of aspects, including the models' vulnerability to cultural insensitivity, their proclivity for generating gender and ethnic biases, limitations in language coverage that hinder their applicability to various linguistic contexts, and the intricacies involved in comprehending and mitigating intersectionality-related biases. These limitations underscore the pressing need for human intervention and heightened cultural awareness when leveraging language models for research endeavors. Additionally, it emphasizes the vital importance of crafting research content that is both inclusive and academically rigorous, to overcome these challenges and limitations.

REFERENCES

1. D. Bracewell and M. Tomlinson, "The language of power and its cultural influence," in *Proceedings of COLING 2012: Posters*, 2012, pp. 155–164.
2. S. Naseem, A. Alhudhaif, M. Anwar, K. N. Qureshi, and G. Jeon, "Artificial general intelligence-based rational behavior detection using cognitive correlates for tracking online harms," *Personal and Ubiquitous Computing*, 2022. https://10.1007/s00779-022-01665-1.
3. D. Buschek, M. Zürn, and M. Eiband, "The impact of multiple parallel phrase suggestions on email input and composition behaviour of native and non-native english writers," in *Proceedings of the 2021 CHI Conference on Human Factors in Computing Systems*, 2021, pp. 1–13.
4. A. Caliskan, J. J. Bryson, and A. J. S. Narayanan, "Semantics derived automatically from language corpora contain human-like biases," *Science*, vol. 356, no. 6334, pp. 183–186, 2017.
5. C. Baumler and R. Rudinger, "Recognition of they/them as singular personal pronouns in coreference resolution," in *Proceedings of the 2022 Conference of the North American Chapter of the Association for Computational Linguistics: Human Language Technologies*, 2022, pp. 3426–3432.

6. S. Nath *et al.*, "889 New meaning for NLP: the trials and tribulations of natural language processing with GPT-3 in ophthalmology," *British Journal of Ophthalmology*, vol. 106, no. 7, pp. 889–1036, 2022.
7. S. L. Blodgett, S. Barocas, H. Daumé III, and H. J. a. p. a. Wallach, "Language (technology) is power: A critical survey of" bias" in nlp," arXiv preprint arXiv:2005.14050, 2020.
8. S. L. Blodgett, G. Lopez, A. Olteanu, R. Sim, and H. Wallach, "Stereotyping Norwegian salmon: An inventory of pitfalls in fairness benchmark datasets," in *Proceedings of the 59th Annual Meeting of the Association for Computational Linguistics and the 11th International Joint Conference on Natural Language Processing (Volume 1: Long Papers)*, 2021, pp. 1004–1015.
9. S. L. Blodgett and B. J. a. p. a. O'Connor, "Racial disparity in natural language processing: A case study of social media african-american english," arXiv preprint arXi,1707.00061, 2017.
10. A. Câmara, N. Taneja, T. Azad, E. Allaway, and R. J. a. p. a. Zemel, "Mapping the multilingual margins: Intersectional biases of sentiment analysis systems in English, Spanish, and Arabic," arXiv preprint arXiv:2204.03558, 2022.
11. E. Kasneci *et al.*, "ChatGPT for good? On opportunities and challenges of large language models for education," *Learning and individual differences*, vol. 103, p. 102274, 2023.
12. Y. T. Cao *et al.*, "On the intrinsic and extrinsic fairness evaluation metrics for contextualized language representations," arXiv preprint arXiv:2203.13928, 2022.
13. I. Chalkidis, T. Pasini, S. Zhang, L. Tomada, S. F. Schwemer, and A. J. a. p. a. Søgaard, "FairLex: A multilingual benchmark for evaluating fairness in legal text processing," arXiv preprint arXiv:2203.07228, 2022.
14. R. Izem *et al.*, "Real-world data as external controls: practical experience from notable marketing applications of new therapies," *Therapeutic Innovation & Regulatory Science*, vol. 56, no. 5, pp. 704–716, 2022.
15. L. Weidinger *et al.*, "Taxonomy of risks posed by language models," in *Proceedings of the 2022 ACM Conference on Fairness, Accountability, and Transparency*, 2022, pp. 214–229.
16. A. Čartolovni, A. Tomičić, and E. L. Mosler, "Ethical, legal, and social considerations of AI-based medical decision-support tools: A scoping review," *International Journal of Medical Informatics*, vol, p. 161:104738, 2022.
17. G. Judd, X. Wang, and P. Steenkiste, "Efficient channel-aware rate adaptation in dynamic environments," in *Proceedings of the 6th International Conference on Mobile Systems, Applications, and Services*, 2008, pp. 118–131.
18. L. M. J. G. Ackerman, "Syntactic and cognitive issues in investigating gendered coreference," *Glossa: a journal of general linguistics*, vol. 4, 2019.
19. M. Sap, V. Shwartz, A. Bosselut, Y. Choi, and D. Roth, "Commonsense reasoning for natural language processing," in *Proceedings of the 58th Annual Meeting of the Association for Computational Linguistics: Tutorial Abstracts*, 2020, pp. 27–33.

20. D. A. Shafiq, N. Z. Jhanjhi, and A. Abdullah, "Load balancing techniques in cloud computing environment: A review." Journal of King Saud University-Computer and Information Sciences. vol 1, no. 34, pp. 3910–3933, 2022.
21. L. Andrés Bolado, "Breaking language barriers: Teachers' strategies for supporting immigrant children with limited language proficiency,", Abu Academy University, https://urn.fi/URN:NBN:fi-fe2023052547859, 2023.
22. P. Delobelle, E. K. Tokpo, T. Calders, and B. Berendt, "Measuring fairness with biased rulers: A comparative study on bias metrics for pre-trained language models," in *NAACL 2022: the 2022 Conference of the North American chapter of the Association for Computational Linguistics: human language technologies*, 2022, pp. 1693–1706.
23. J. Dodge *et al.*, "Documenting large webtext corpora: A case study on the colossal clean crawled corpus," arXiv preprint arXiv:2104.08758, 2021.
24. K. Ethayarajh, D. Duvenaud, and G. Hirst, Understanding undesirable word embedding associations. arXiv preprint arXiv:1908.06361. 2019.
25. C. Goddard, *The languages of East and Southeast Asia: An Introduction.* Oxford University Press, 2005.
26. H. Gonen, Y. Kementchedjhieva, and Y. Goldberg, "How does grammatical gender affect noun representations in gender-marking languages?". arXiv preprint arXiv:1910.14161. 2019.
27. S. Gupta, Y. Huang, Z. Zhong, T. Gao, K. Li and D. Chen, "Recovering private text in federated learning of language models". Advances in neural information processing systems. vol. 6, pp. 8130–8143, 2022.
28. W. L. Hamilton, J. Leskovec, and D. Jurafsky, "Cultural shift or linguistic drift? comparing two computational measures of semantic change," in *Proceedings of the conference on empirical methods in natural language processing. Conference on empirical methods in natural language processing*, 2016, vol. 2016: NIH Public Access, p. 2116.
29. S. Hassan, M. Huenerfauth, Alm CO, "Unpacking the interdependent systems of discrimination: Ableist bias in NLP systems through an intersectional lens." arXiv preprint arXiv:2110.00521. 2021.
30. M. A. Hedderich *et al.*, "A survey on recent approaches for natural language processing in low-resource scenarios." arXiv preprint arXiv:2010.12309, 2020.
31. P. Henderson *et al.*, "Ethical challenges in data-driven dialogue systems," in *Proceedings of the 2018 AAAI/ACM Conference on AI, Ethics, and Society*, 2018, pp. 123–129.
32. Z. Xu, *et al.*, "Federated learning of gboard language models with differential privacy". arXiv preprint arXiv:2305.18465, 202
33. D. Hershcovich, S. Frank, H. J. M. A. Lent, B. Stephanie, B. Emanuele, C. P. Laura, C. Ilias, C. Ruixiang, F. Constanza, M. Katerina, R. Phillip, and A. Søgaard, "Miryam de Lhoneux," pp. 6997–7013, 2022.
34. S. Hooker, *et al.*, "Characterising bias in compressed models." arXiv preprint arXiv:2010.03058 (2020).

35. D. Hovy, and P. Shrimai, "Five sources of bias in natural language processing." *Language and linguistics compass* 15.8, 2021.
36. J. Zhang, et al. "Is chatgpt fair for recommendation? evaluating fairness in large language model recommendation." *Proceedings of the 17th ACM Conference on Recommender Systems*. 2023.
37. C.-J. Wu *et al.*, "Sustainable ai: Environmental implications, challenges and opportunities," Proceedings of Machine Learning and Systems, vol. 4, pp. 795–813, 2022.
38. K. Un Nisa, A. Alhudhaif, K. N. Qureshi, H. J. Hadi, and G. Jeon, "Security Provision for Protecting Intelligent Sensors and Zero Touch Devices by using Blockchain Method for the Smart Cities," *Microprocessors and Microsystems*, p. 104503, 2022. https://doi.org/10.1016/j.micpro.2022.104503.
39. K. N. Qureshi, A. Raziq, and G. Jeon, "Secure interaction and processing of multimedia data in the Internet of Things based on wearable devices," in *Computing, Access Control and Security Monitoring of Multimedia Information Processing and Transmission*: Institution of Engineering and Technology, 2023, pp. 91–106. [Online]. Available: https://digital-library.theiet.org/content/books/10.1049/pbpc061e_ch5
40. M. Anwar, A. H. Abdullah, R. A. Butt, M. W. Ashraf, K. N. Qureshi, and F. Ullah, "Securing data communication in wireless body area networks using digital signatures," *Technical Journal*, vol. 23, no. 2, pp. 50–55, 2018.
41. A. Iftikhar, K. N. Qureshi, M. Shiraz, and S. Albahli, "Security, trust and privacy risks, responses, and solutions for high-speed smart cities networks: A systematic literature review," *Journal of King Saud University - Computer and Information Sciences*, p. 101788. 2023. https://doi.org/10.1016/j.jksuci.2023.101788.
42. Y. Wan, W. Zhao, H. Zhang, Y. Sui, G. Xu, and H. Jin, "What do they capture? a structural analysis of pre-trained language models for source code," in *Proceedings of the 44th International Conference on Software Engineering*, 2022, pp. 2377–2388.
43. S. S. Jodhka, G. J. E. Shah, and P. Weekly, "Comparative contexts of discrimination: Caste and untouchability in South Asia," *Economic and Political Weekly*, pp. 99–106, 2010.
44. I.-R. Johansson, "A Tale of Two Texts, a Robot, and Authorship: A comparison between a human-written and a ChatGPT-generated text," Malmö University, Faculty of Culture and Society (KS), School of Arts and Communication (K3), 2023.
45. X. Liu *et al.*, "Large language models are few-shot health learners," arXiv preprint arXiv:2305.15525. 2023.
46. S. Kumar, *et al.*, "Language generation models can cause harm: So what can we do about it? an actionable survey." arXiv preprint arXiv:2210.07700, 2022.
47. R. Machlev *et al.*, "Explainable Artificial Intelligence (XAI) techniques for energy and power systems: Review, challenges and opportunities," *Energy and AI*, vol. 9, p. 100169, 2022.

48. L. F. Simanjuntak, M. Rahmad, and Y. Evi, “We know you are living in bali: Location prediction of twitter users using bert language model.” *Big Data and Cognitive Computing*, vol. 6.3, 2022.
49. A. Joshi, P. Bhattacharyya, M. Carman, J. Saraswati, and R. Shukla, “How do cultural differences impact the quality of sarcasm annotation?: A case study of indian annotators and american text,” in *Proceedings of the 10th SIGHUM Workshop on Language Technology for Cultural Heritage, Social Sciences, and Humanities*, 2016, pp. 95–99.

CHAPTER 11

Ethical Implications of Language Models

Hanaa Omar Nafea, Kashif Naseer Qureshi, and Noman Arshad

11.1 INTRODUCTION TO LANGUAGE MODELS

Computational models known as language models are created to comprehend and produce language. They play a role in Natural Language Processing (NLP) applications, facilitating tasks like text creation, translation, sentiment analysis, and answering questions. These models undergo training using text data to grasp the patterns and connections within the language for predicting the next word or sequence of words in a specific context. A significant advancement in times is the emergence of learning-based language models such as ChatGPT, which utilize neural networks to capture intricate linguistic structures and subtleties [1]. These models have shown proficiency in interpreting and generating language, resulting in their widespread adoption across different industries and applications. Language models like ChatGPT have transformed human-computer interaction by enabling intuitive communication between users and machines [2]. They empower assistants, chatbots, language translation services, and content recommendation systems enriching user experiences and efficiency. Nonetheless, despite their potential these language models also bring up ethical concerns related to bias, privacy issues, misinformation

DOI: 10.1201/9781032667911-11

dissemination, and societal impact [3]. With the growing sophistication and prevalence of these models, it is essential to assess their capabilities, limitations, and implications, for society. By grasping the core principles of language models and considering their aspects, we can explore the possibilities and obstacles they offer and work toward creating ethical and fair AI technologies.

The ChatGPT tools are based on powerful transformation frameworks by using different language models. These language models are developed by Open AI. These tools are capable of understanding and generating output like human text in almost all fields and represent significant improvement and efficiency. These tools are trained on massive amounts of data from different sources on the Internet and generate coherent and contextual responses. These models capture the intricacies of human languages such as semantics, syntax, and context for responses. These tools are useful for generating human-like text which resembles human speech such as virtual assistance, chatbots, and customer service applications. Although there are several features of AI generative tools, one of the most prominent features is its versatility and adaptability to fit in different conversational scenarios [4, 5]. These tools provide open-ended conversations covering many topics generate factual answers, and provide recommendations. These tools are also capable of using different languages and provide cross-cultural communication features. The output-generating process is too fast and natural. However, with many benefits, these models have suffered from ethical challenges related to privacy, bias, societal impact, and misinformation. There is a pressing need to consider these challenges and develop more responsible and practical models.

11.2 BIAS AND FAIRNESS

Bias and fairness are some of the challenges for developers of AI generative tools like ChatGPT. These tools are trained on a massive dataset which includes different patterns and relationships and leads to inadvertently perpetuated biases. There are many examples of bias if the data contains race, gender, and socioeconomic factors then the generated text from these tools may reflect these biases. The generated text from these types of datasets will be discriminatory or unfair. Bias and fairness are basic requirements for AI generative tool development therefore there is a need to consider all bias and fairness factors. There have been several approaches designed to tackle the bias and fairness like carefully pre-processing and curating the training data and removing any sort of bias. The trained data should be based on fairness-aware algorithms for evaluation. Transparency and accountability should be considered in identifying and handling any sort of biases and AI systems [4]. The data should be disclosed, especially the source of training data. The model biases and limitations should be documented and shared with all stakeholders for their diverse feedback. These all requirements should be considered during the design and development of AI generative tools for fair and socially responsible systems.

11.3 PRIVACY AND DATA SECURITY

There are many privacy and data security challenges, especially in designing and developing AI generative tools such as ChatGPT. These tools are based on large datasets where they are learning and generating the content which leads to privacy and security concerns. Especially these models are collecting the personal data of the users and there will be a chance to violate the privacy of the users. So there is a need to control privacy concerns during the collection of their personal information, storing that information, and sharing among systems. AI generative tools collect user data including text input, user preferences, and metadata, and privacy concerns arise when the users are unaware and not giving any consent for the data collection [6].

AI generative tools are based on interactive technologies to generate attractive content for users. However, these tools rely on a vast amount of data for training generating the content, and collecting sensitive user information. Privacy is one of the challenges for these models, especially during the data collection, storage, and usage to generate the contents. Protecting the user data is still a challenge and needs to be addressed, and ensure data security during the ChatGPT interactions.

Data security should be ensured to save and secure the user's information. There is a need to use secure data communication protocols to encrypt the user data during the transmission, especially between the users and AI models. Security protocols are used to protect the data from unauthorized access, and tampering of sensitive information. Authentication and access control methods are also adopted to ensure the authorization of the individuals or the systems who have access to modify the data. This strategy will reduce the risk of any type of security breach or data breach. In addition, there is a need to use encryption and regular data protection mechanisms to secure the backup databases, servers, or the cloud environment [7].

AI methods are also utilized to secure the data because these AI solutions can analyze the data automate the responses anticipate the risks and enhance the security requirements. The combination of cybersecurity methods and AI tools is useful to protect the systems by using different protocols. By using AI tools, identifying the potential cyber threat mechanism will be improved by learning past incidents adopting real-time processes, and anticipating future threats. AI tools are the best solution for defensive mechanisms with the help of reactive and productive approaches [8]. Intelligent decision-making processes to detect and prevent cyber-attacks. The combination of AI in cybersecurity will offer the next-generation shield for digital systems. Following are some examples where AI will be integrated into cybersecurity to secure the systems.

- **Prediction Capabilities:** AI systems use machine learning techniques to analyze historical data and to identify trends and patterns. This strategy will help to predict the threats and vulnerabilities of the systems before exploitation. This strategy is also helpful in forecasting attacks before disturbing the systems.
- **Real-Time Monitoring and Response:** AI tools are monitoring the network on a real-time basis to identify the threats and will be working as security guards.
- **Enhance Accuracy:** The AI tool also decreases the chances of false positives and negatives and best solution for threat detection because these tools learn from history pinpoint the genuine threats and minimize the risk of the existing systems.
- **Automate Tasks:** The AI tool will also automate tasks such as monitoring real-time network traffic. Analyze user behavior and focus on complex issues.
- **Threat Intelligence:** AI tools are also helpful for threat intelligence; they can dive deep into the dark web to monitor the threat actors and uncover the potential risks that the traditional systems have.
- **Scalability:** AI tools provide the scalability for continuous protection of data.

There are a lot of advantages of AI-based cybersecurity solutions. These systems offer dynamic defense systems against cybersecurity attacks and

FIGURE 11.1 AI integration in cybersecurity.

threats, streamlining all the processes and improving the efficiency of the systems. One of the main advantages of AI tools is proactive threat prevention. AI tools analyze past data and present the data by using the machine learning algorithm with the help of pattern recognition and prediction. These predictive capabilities are useful for identifying threats and systems vulnerabilities. The AI tools are also useful for the reduction of human errors because human errors are the leading causes of cybersecurity breaches. These tools will automate routine tasks, improve the decision-making processes, and reduce human mistakes. AI tools will improve the efficiency cost and operational efficiency and be scalable to grow with the data volume. Figure 11.1 shows the AI integration in cybersecurity.

The intrusion and detection mechanisms are also used as advanced security mechanisms. The main aim of AI tool developers is to secure their user data mitigate the security risks and protect their data from different types of security attacks. Another concept is regulatory consideration for protecting user privacy. There is a need for a regulatory framework to protect user privacy and data rights for AI generative tools such as the General Data Protection Regulation (GDPR) and the European Union and the California Consumer Privacy Act (CCPA) [9]. So these frameworks and standards ensure the data handling and provide the guidelines for data security. These regulations have the principles of how to deal with users' confidentiality, data integrity, and the availability of the system.

Organizations must obtain the user's permission and consent before processing their data and they should need to provide very clear privacy notices for this purpose. This company also implements appropriate security regulated regulation framework to protect the user data. Privacy compliance and regulation are significant not only to avoid any legal consequences, like penalties and fines, but also these frameworks will ensure user privacy and increase the confidence of AI models for the developers where they should ensure to adopt regulatory requirements to deal with user privacy and their data protection.

11.4 MISINFORMATION AND DISINFORMATION

Misinformation and disinformation are significant challenges for large language models because of the large scale of social media and online platforms. Misinformation refers to false and inaccurate information. Misleading content hurts the user's opinion and trust and also the user's behaviors. The language model should be able to detect misinformation and disinformation issues during training data sets. The misinformation and disinformation also lead to generating biased output and discrimination against certain groups. In addition, malicious actors can misuse these language models to generate fake content and disseminate misinformation. Ethical implications should not be considered during the development of these language models [10]. The developers ensure that these models are trained ethically without using any myths and disinformation. These models should follow ethical guidelines and transparency and regulate and balance the language models. Language models in terms of misinformation and disinformation need holistic methods which need technical ethical care. Figure 11.2 shows the information misuse examples with details.

11.4.1 Risks of Misinformation and Disinformation

Misinformation and disinformation are also leading to different risks for users and groups. The first risk is trust in organizations and institutions and their credibility. This factor also leads to societal polarization because of different critical issues. In addition, misinformation and disinformation can hurt users' health such as false information about any disease treatment can lead to harmful behaviors [11]. In terms of politics, misinformation and disinformation pose a serious risk where this content manipulates public opinion and has a direct impact on the election outcome, which disturbs the democratic process and changes the policies. As per the economy, misinformation and disinformation can damage the market and the company's reputation where the companies will lose their reputation and their financial resources. Misinformation and disinformation lead to violence and hitting against communities. Furthermore, there

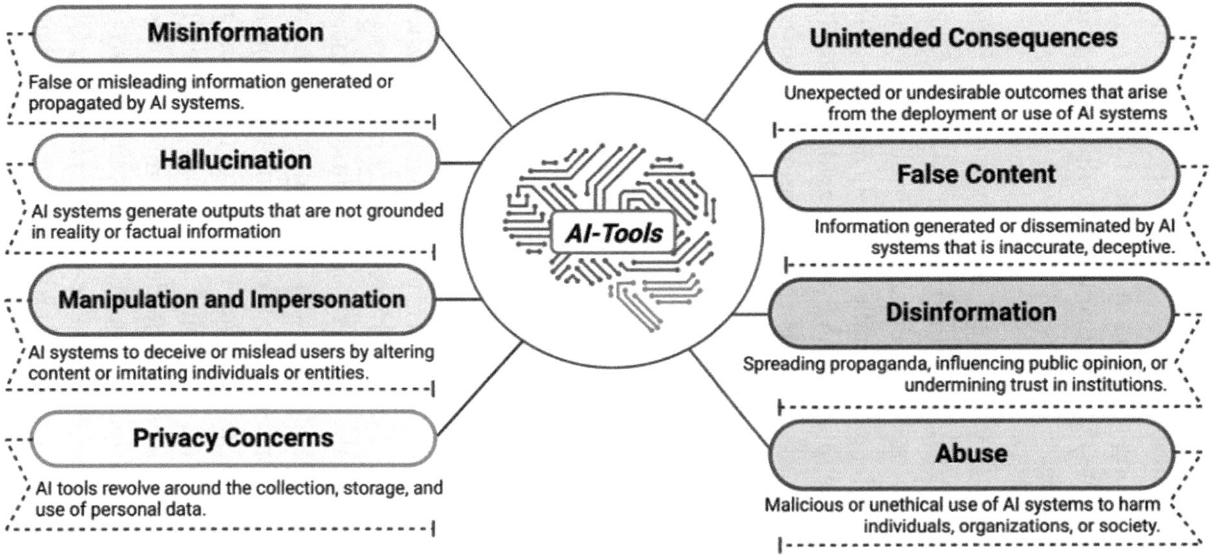

FIGURE 11.2 Content misuse examples of AI generative tools.

is a risk for digital spaces like cybersecurity risks where malicious actors disseminate misinformation for their benefit. There is a need to address these risks that are related to misinformation and disinformation. All stakeholders' involvement is necessary starting from companies, government, media organizations, and the civil society side. So all stakeholders promote literacy among the users and create a culture where the users analyze and investigate the information critically before using or believing in any information.

11.4.2 Identifying and Addressing False or Misleading Content

Identifying and addressing false and misleading information is one of the important factors that need to mitigate the misleading content. Various methods have been adopted to handle this challenge such as fact-checking used to scrutinize the information by using evidence-based strategies to verify its accuracy. Another effort is a collaboration between media organizations and independent fact-checkers and companies. These stakeholders will enhance the scale and efficiency of this process. In addition, promoting literacy is another significant factor where the individuals critically analyze the information they encounter. Educational institutes play their role to liberate the people on how they will identify unreliable sources and assess the information credibility by using evidence and expertise. There is a need to promote critical thinking skills to judge false and misleading content. Technology also plays a very crucial role such as smart algorithms are used in AI to detect and check the false content automatically. Social media platforms should use search engines where they will prioritize authoritative sources to decrease the visibility of misleading information from the search results and the user's side. Additionally, there is a need to provide the construction of contextual information and fact-related information to the users. There is a need for accountability and transparency among content creators and distributor platforms to disclose their content and adopt independent accountability and security policies [12].

Collaborative approaches are needed to combat the misinformation in today's digital landscape. They need to work together to address these issues and develop strategies to handle this complex issue. Different companies have their unique capabilities where they can contribute to addressing this misinformation and misleading issues. For example, media organizations can provide expertise to check the facts whereas technical companies offer some advanced technological tools for data analysis. Collaborative efforts improve the data resource sharing, and coordinated campaigns and strategies will improve the existing strategies to combat the misinformation.

11.5 MANIPULATION AND ABUSE

Manipulation and abuse are leading challenges in the context of misinformation and disinformation. Malicious actors exploit the data by using the digital platform to disseminate false information for different purposes, such as to divert the user's attention to manipulate their opinion and achieve their political gains. One example of common data manipulation is fake news dissemination. Fabricated content is created and disseminated to change the truth and reality through social media platforms [11]. Additionally, these abusive practices are also used to create fake accounts and utilized for different practices. This manipulation also created grassroots movements and public perceptions and their behaviors. Furthermore, misinformation is weaponized for political and social purposes to disseminate propaganda and improve their democratic processes. Social media platforms are used for these activities. The individual is also involved in this manipulation at the personal level, such as being involved in fraud schemes, online harassment, and falling victim to phishing scams. To address this manipulation, proactive measures are required. Proactive measures should be adopted by all stakeholders including government companies where they use strong and transparent algorithms. Collaborative efforts are needed among technology companies, civil society, academia, organizations, and the government side. These stakeholders develop and implement comprehensive methods to counter online abuses. This issue is addressable, if all stakeholders work together and build more defense strategies against the spread of this misinformation and ensure the data integrity.

Acknowledging and addressing the malicious usage of ChatGPT is important. Strong systems and strategies are required for user's privacy, security, and trust on these platforms. Monitoring strategies are required to monitor the ChatGPT interactions for suspicious patterns and detect the potential instances and misuse of these models. Strategies are used to detect misinformation and counter malicious activities, especially for users violating these platform's guidelines and policies [13]. Language models like ChatGPT should be equipped with more capable context to detect abusive behavior such as identifying hate speech, abusive language, and manipulation tactics. Authentication and verification strategies help these models to verify the users and their identity before accessing these models. Content filtering and moderation can help to prevent the misinformation of any sort of inappropriate content and data generated from ChatGPT. These filtering strategies will easily detect hate speech and other forms of abusive content. Educating the users about the potential risks of this misuse and abusive language model interaction also helps to promote and empower the individual to recognize and report any kind of malicious behavior.

Real-time monitoring systems and intervention mechanisms will help to promote the detection responses of these instances. These systems will temporarily suspend and restrict the excess of these models for the users who are involved in harmful behaviors. Reporting and enforcement is another strategy that will help to detect the misuse or abusive language to make these models transparent and enforce the policies for the fear responses. For that reporting mechanism, proactive methods are required to detect and recognize the malicious uses of the ChatGPT platform secure the environment for the users, and improve these models.

11.6 USER CONSENT AND AUTONOMY

Informed consent is one of the important factors and plays an important role in ChatGPT interaction. It has a profound impact on the user's transparency, ethical consideration, legal compliance, and autonomy. Informed consent is one of the factors where the users make their decisions to engage with the AI tools. In this matter, clear information is provided to the users that are related to the nature of interactions and possible implications. Informed consent establishes trust between the users and AI tools, and this transparent method can control and improve the user's confidence by encouraging them to participate with AI generative tools [14]. In addition, informed consent is also aligned with other ethical considerations where the users have rights for their privacy and are aware of any risks that are linked with the data sharing and usage. Informed consent also improves ethical practices, especially in designing AI tools. Informed consent is one of the legal requirements and is directly linked with privacy regulations such as GDPR. Policy. These compliance and regulations do not only protect the user's rights and legal repercussions. It is important to empower the users for their informed content and control all the data interaction between users and ChatGPT. This option will provide the freedom to the users where they will review and manage their data preferences and control the scope and the level of their interaction with AI. The users will also have the option to pause the conversations, specify the preferences, adjust the settings, and specify the preferences for content or topics. By using this strategy, the AI systems are working more respectfully using the user-centric approach to technology. User-friendly interfaces are designed to take the consent of the users which is an integral part of these AI models. The users can easily manage and understand their preferences. It will also provide a clear and accessible information interface where all the things will be mentioned such as the main purpose of the data collection, its usage, and risks involved. The plain language and visual support also enhance the user's comprehension of content-related concepts. It will make the process more transparent and user-friendly. Prioritizing the informed consent also provides the users with control and design initiative. User interfaces are important steps in promoting AI tools and are responsible for all interactions with these tools.

11.7 EMPLOYMENT DISPLACEMENT

Employment displacement is another factor in language models. AI technologies like the ChatGPT model perform different cognitive tasks such as natural language processing customer service in interaction and content generation. These technologies provide awareness benefits in terms of productivity and efficiency. These language models have the potential to disrupt traditional job methods and markets. The impact of these AI generative tools and language models on the workforce is multifaceted. These models streamline and automate traditional processes and enhance productivity in various fields. These models have increased the efficiency and the cost savings for the businesses. However, the automation has potential to eliminate the jobs that were previously performed by humans for well-known and popular jobs like customer services, representative data analysts, and content creators [15]. The AI generative tool replaced these jobs by using their smart AI tools. To address the job displacement issue and its economic implications, there is a need for proactive techniques to prioritize social inclusion and economic growth. There is a need to invest in skills training programs where the workers have enough skills to adopt these new AI tools. The training programs will improve this area, especially for data science for digital marketing or software development. In addition, the companies and the government can implement the policies and take the initiative to support the workers and the communities. The stakeholders may provide job placement services financial assistance and give access to affordable health care and housing. There is a pressing need to create a culture of intrapreneurship and innovation to create new job opportunities and offset the negative impact of employment displacement collaboration between businesses governments stakeholders educational institutes and companies are important to develop comprehensive methods to address employment displacement. These stakeholders will ensure that the AI tools and the advanced technologies are shared equitably across society and mitigate the negative effects on employment displacements.

11.8 DEPENDENCE AND RELIABILITY

Dependence and reliability are important considerations, especially to utilize the AI language models. These models can generate human-like text and perform and automate various tasks. It is important to analyze their reliability and ensure accountability and transparency. An example is ChatGPT output which contains several factors such as this model generates contents and contextually relevant responses and leads to different types of errors. To understand these model's limitations, it is necessary for the developers to test and validate the processes and to assess the model performance with different fields and scenarios. Moreover, feedback and monitoring methods are important for identifying and addressing reliability and accountability issues. So, fully citing the feedback from the users is important to analyze the data and detect the patterns errors, or biases to improve the model based on these insights. Transparency is significant, especially in designing and developing these processes such as training the data to make the methodologies to evaluate the metrics. These strategies will help all the stakeholders to understand the limitations and capabilities of these models. There is a need to prevent the over-reliance on language models to mitigate the risks and improve their decision-making. The users must critically evaluate and verify the ChatGPT output, especially when they are dealing with a highly sensitive context. This strategy will help them to prevent blind reliance on the model recommendation and promote the diverse inclusion of AI development. It will help to mitigate the biases and improve the overall performance of ChatGPT output to ensure the transparency and accountability of these AI models and improve the user's trust and confidence This will provide clear communication strategies to analyze the limitations of the model and provide a mechanism for the users to report these issues and establish the processes and resolve the disputes. Ethical frameworks will provide complete guidelines to the developers to prevent overreliance on the models.

11.9 ENVIRONMENTAL IMPACT

There are various environmental impacts during the training of large language models like ChatGPT. Environmental impact is one of the significant concerns because of different computational resources and energy consumption issues. To design these models, a large amount of data is required. Powerful computing infrastructure and time are also required to design and train the data, which will create significant carbon footprints and environmental strains. During the large language model training, one of the primary environmental factors is energy consumption [16]. The energy-intensive nature of training these AI models will lead to carbon emissions and environmental footprints because these models need high processing resources in the data centers to address these environmental factors and reduce the carbon footprints. There is a need to invest in energy-efficient computing technologies. Energy-efficient computing technologies help to mitigate the environmental impact of AI training such as designing low power processes and renewable energy. For the data centers, optimizing the algorithms and frameworks needed to improve these models. The energy-efficient mechanism will contribute to reducing energy consumption during the model training. There's a need to train the data by reducing the amount of data and minimizing the environmental impact of large-scale data processing. Data compression, data pruning, and data augmentation strategies will help streamline the data training processes and lower energy consumption. The models should be optimized in terms of their performance and efficiency. To reduce computational resources and energy, there are different technologies have been adopted such as the quantization model. Distillation will help to optimize these models without sacrificing and losing their performance. These strategies address the environmental impacts. Lifecycle assessment is another strategy for AI technologies to analyze their different environmental factors starting from manufacturing and deployment of these models. There are several important technologies available to promote renewable energy sources such as solar wind and hydroelectrical power to reduce the reliance and fossil fuel and carbon emissions for sustainable AI research and development. These are all the factors that need to be considered to design and develop the models and adoption of energy-efficient computing frameworks to optimize the data. These strategies will help to improve the model efficiency and minimize the environmental impacts for a more sustainable future for digital innovation.

11.10 EXISTING FRAMEWORKS TO ADDRESS ETHICAL IMPLICATION

Authors in [17] presented a central thematic framework to address the AI ethical concerns. The authors discussed that unfair biased and inequitable data usage has serious ethical and sustainability issues. There are several AI frameworks that have been adopted to tackle these issues. The proposed framework is based on thematic analysis to take AI ethics and sustainability into account by using the DMAXQD Analytics Pro software and identifying some common principles. This study extracted 28 ethical concerns which are disseminated by different groups at agencies. Whereas six ethical toolkits and products are also used to translate the common AI ethics sustainability framework. This framework is validated and presents the findings including privacy autonomy explain ability and biases that need serious attention. There is a pressing need to design and develop a trust-based AI ethics framework to deal with AI solutions that will be helpful for the beneficiaries or the stakeholders.

Authors in [18] reported the opportunities and the risks related to AI and presented the five ethical principles. The principles are the adoption and the development of AI frameworks to deal with ethical concerns and presented the 20 concrete recommendations to develop and assess the AI framework for the stakeholders. They explored the core opportunities and AI tools for society. The findings in this research concluded the need to create the centers, curriculum, and infrastructure to adopt AI tools. There is a need to make equitable programs and policies to adopt these technological innovations and mitigate the risks of AI for society. They highlighted some action points such as assessment development in incentive support to design the AI ethical framework.

Authors in [19] presented the ethical implications of using AI tools. Authors suggested the AI advisory functions for the accounting field and adopted AI tools to increase the processes such as level of accuracy, in-depth insight analysis, business processes, and improve their client services. AI is one of the emerging technology which is based on cognitive skills and the judgment of humans. However, with many uses there are some ethical implications that exist which need serious attention. The authors combined the two pretty strict ethical frameworks to forecast ethical implications in the auditing field and presented a conceptual analysis to explore the social and ethical issues related to AI by using past data. They mentioned that there is a need to formalize the collaboration between the stakeholders to develop the practical policy and the guidelines for an effective governance framework.

Authors in [19] discussed AI development and its applications and its ethical, social, legal, and economic implications. They discussed that the current global implications of AI are based on qualitative and quantitative research methods. The authors conducted the study by using the existing research and reported after the content analysis some insights into AI implications. They mentioned that the existing applications of AI have many limitations and observed that there are major disparities in existing studies. In addition, they also mentioned that existing AI implications must be explored by using multidisciplinary approaches. The stakeholders should develop a trust model for the end users. They also highlighted that the government has a major role in encouraging the development of AI implications by regulating the policies and working on some specific legislation to improve the existing framework and the legislation architectures.

Authors in [20] discussed the AI ethics and the principles for the data science field and proposed the AI ethical principles as a lifecycle of AI-based digital services. In this proposed lifecycle, authors dealt with high-level AI principles and ethical guidelines for trustworthy AI tools. The authors focused on trustworthy AI requirements, including human agency and oversight, privacy and data governance, technical robustness and safety, diversity, transparency, discrimination and fairness, and environmental and societal factors and accountability. The findings of this study are on political implications in data science or software engineering fields. Authors suggested that there is a need to design the real world and the multi-stakeholder projects to consider this principle and the guidelines and fill the gap.

Authors in [21] explored how the principles related to justice, compassion, and ethical responsibility need to be developed to adopt AI technologies. This study is specifically focused on the one religion where they mentioned their significant ethical concerns to deal with carefully and suggested a moral framework for AI technologies. These AI tools have different societal and ethical concerns. The users belong to different background, culture, and religions, and their ethical values are different from each other. This study suggested that there is a need to develop AI frameworks to examine user privacy and bias. Table 11.1 shows the comparison of the discussed literature.

The discussed studies highlighted the different aspects of the ethical implications of AI tools. They suggested different frameworks, lifecycles, and findings to address these ethical concerns in AI adoption. These studies highlighted the complexities and multifaceted nature of AI tools. The ethical models should be designed for multidisciplinary fields to engage all the stakeholders from academia, companies, organizations, and government side.

TABLE 11.1 Comparison of Discussed Studies

Study	Approach	Key Findings	Recommendations
[6]	Thematic analysis using DMAXQD Analytics Pro software.	Identified 28 ethical concerns and common principles. Emphasized the need for trust-based AI ethics frameworks.	Design and develop trust-based AI ethics frameworks.
[7]	Exploration of opportunities and risks related to AI.	Identified the need for creating centers, curriculum, and infrastructure for AI adoption. Emphasized equitable programs and policies to mitigate AI risks.	Create infrastructure for AI adoption, and develop equitable programs and policies.
[8]	Analysis of ethical implications of AI tools, particularly in the accounting field.	Emphasized the need for collaboration between stakeholders to develop effective governance frameworks.	Formalize collaboration between stakeholders, and develop governance frameworks.
[8]	Discussion on AI development and its implications.	Identified limitations and disparities in existing AI applications and studies. Advocated for multidisciplinary approaches and government regulation.	Encourage multidisciplinary approaches, and regulate AI development.
[9]	Discussion on AI ethics and principles for data science.	Focused on trustworthy AI requirements, and political implications, and suggested real-world multi-stakeholder projects.	Design real-world multi-stakeholder projects, and consider ethical principles.
[10]	Exploration of principles related to justice, compassion, and ethical responsibility in adopting AI technologies.	Emphasized the need for developing AI frameworks to examine user privacy and bias.	Develop AI frameworks for user privacy and bias examination.

11.11 STRATEGIES FOR MITIGATING BIAS IN CHATGPT

There are different approaches that have been adopted to mitigate the bias in AI models and categorize them into technical and ethical factors. The commonly adopted strategies are as follows:

- **Diverse Training Data:** The data used to train the ChatGPT models should be diverse and represent different factors such as cultural, demographic, and perspective. This strategy helps to decrease the bias and provides a more inclusive and balanced dataset.
- **Bias Monitoring and Detection:** Well-effective tools are needed to monitor and detect any sort of biases in tool output. These tools are used to analyze and evaluate the responses of biased language. If there is any type of dissimilatory patterns or data is detected, then these tools will take corrective action.
- **Mitigating Strategy:** After detecting the biased data, this strategy ensures to mitigate the data by using adversarial training, and debiasing algorithms. This strategy is also helpful in reducing the impact of biases, especially for ChatGPT model responses. The migration methods should be balanced to minimize the biases without compromising the mode performance.
- **Accountability and Transparency:** This strategy ensures transparency in operating AI generative tools like ChatGPT. Accountability and transparency should be taken into account during data training and in the decision-making process.
- **Human Review and Oversight:** Human review and oversight should be considered to evaluate the AI tool's output. The biased data should be identifiable and corrected through human moderators to remove biased responses.
- **Feedback Mechanism:** The user feedback mechanism should be adopted to collect users' feedback on ChatGPT responses, especially for instances of bias or any offensive content. So this strategy will help to improve the ChatGPT and AI generative tool and address the emerging issues.
- **Diverse Development Team:** There should be diversity and inclusion, especially when developing a team and maintaining the ChatGPT models. So there should be diverse experiences and perspectives among the developers, which can help to identify and address the biases more effectively.

- **Ethical Governess and Guidelines:** So there's a need to establish the ethical guidelines and the governance, architecture, and framework to develop the AI models. These ethical guidelines and governance will ensure transparency and accountability and prioritize fairness, especially for human rights and dignity.

By adopting these discussed strategies, the stakeholders and the developers work together to design and develop AI-generated tools, such as ChatGPT. These tools will be more effective, fair, and respectful of society, especially for diverse perspectives. The collaboration and the collective research and engagement of all stakeholders are important for continuously improving the bias mitigation efforts in AI technologies.

11.12 CONCLUSION AND FUTURE DIRECTIONS

This chapter discussed the ethical concerns that existed in language models and AI generative tools such as transparency, accountability, bias mitigation, empowerment, and regularity oversight. This chapter suggested some improvements by using AI for data protection to improve the overall system's security and the user's privacy. This chapter also explored the various ethical concerns including privacy, bias, security, societal impact, misuse of the content, and manipulating the content. There is a pressing need to address all the ethical concerns with collaborative efforts from policymakers, government organizations, companies, and society at large. The AI generative tools can handle any type of misinformation, abusive language, and biased content, which are harmful to users and society as well.

REFERENCES

1. Y. Chang *et al.*, "A survey on evaluation of large language models," *arXiv preprint arXiv:2307.03109,* 2023.
2. M. S. Aliero, Y. A. Dodo, and K. N. Qureshi, "Novel Machine, and Deep Learning, and Training Techniques for AIoT," in *Artificial Intelligence of Things (AIoT)*: CRC Press, Edited by K. N. Qureshi, 2024, pp. 109–121.
3. F. Fui-Hoon Nah, R. Zheng, J. Cai, K. Siau, and L. Chen, "Generative AI and ChatGPT: Applications, challenges, and AI-human collaboration," *Journal of Information Technology Case and Application Research* vol. 25, pp. 277–304, 2023.
4. M. Gupta, C. Akiri, K. Aryal, E. Parker, and L. Praharaj, "From chatgpt to threatgpt: Impact of generative ai in cybersecurity and privacy," *IEEE Access,* vol. 11, pp. 80218–80245, 2023.

5. C. Novelli, F. Casolari, P. Hacker, G. Spedicato, and L. Floridi, "Generative AI in EU Law: Liability, Privacy, Intellectual Property, and Cybersecurity," *arXiv preprint arXiv:2401.07348,* 2024.
6. U. Ahmad, H. Zaib, and K. N. Qureshi, "Cybersecurity Standards for AIoT Networks," in *Artificial Intelligence of Things (IoT)*: CRC Press, Edited by K. N. Qureshi, 2024, pp. 179–197.
7. K. N. Qureshi, H. J. Hadi, F. Haroon, A. Iftikhar, F. Bashir, and M. N. U. Islam, "5 A Novel Framework for Cyber Secure Smart City," *Security and Organization within IoT and Smart Cities,* p. 75, 2020. https://doi.org/10.1201/9781003018636-5.
8. R. W. Anwar and K. N. Qureshi, "Attack Detection Mechanisms for Internet of Everything (IoE) Networks," in *Cybersecurity Vigilance and Security Engineering of Internet of Everything*: Springer, Edited by K. N. Qureshi, 2023, pp. 41–55.
9. J. M. Blanke, "Protection for 'Inferences drawn': A comparison between the general data protection regulation and the California consumer privacy act," *Global Privacy Law Review,* vol. 1, no. 2, 2020. https://doi.org/10.54648/gplr2020080.
10. C. B. Head, P. Jasper, M. McConnachie, L. Raftree, and G. Higdon, "Large language model applications for evaluation: Opportunities and ethical implications," *New Directions for Evaluation,* vol. 2023, no. 178–179, pp. 33–46, 2023.
11. P. N. Petratos, "Misinformation, disinformation, and fake news: Cyber risks to business," *Business Horizons,* vol. 64, no. 6, pp. 763–774, 2021.
12. L. Yan *et al.*, "Practical and ethical challenges of large language models in education: A systematic scoping review," *British Journal of Educational Technology,* vol. 55, no. 1, pp. 90–112, 2024.
13. E. M. Bosman, C. E. Lyon, M. Burshteyn, and B. S. Kagel, "Deepfake Litigation Risks: The Collision of AI's Machine Learning and Manipulation," *The Journal of Robotics, Artificial Intelligence & Law,* vol. 4, 2021. https://doi.org/10.1201/9781003005629-3.
14. M. L. Jones, E. Kaufman, and E. Edenberg, "AI and the ethics of automating consent," *IEEE Security & Privacy,* vol. 16, no. 3, pp. 64–72, 2018.
15. W. Li *et al.*, "Embracing artificial intelligence (AI) with job crafting: Exploring trickle-down effect and employees' outcomes," *Tourism Management,* vol. 104, p. 104935, 2024.
16. A. S. George, A. H. George, and A. G. Martin, "The environmental impact of ai: A case study of water consumption by chat gpt," *Partners Universal International Innovation Journal,* vol. 1, no. 2, pp. 97–104, 2023.
17. D. Cumming, K. Saurabh, N. Rani, and P. Upadhyay, "Towards AI ethics-led sustainability frameworks and toolkits: Review and research agenda," *Journal of Sustainable Finance and Accounting,* vol. 1, p. 100003, 2024.
18. L. Floridi *et al.*, "AI4People—an ethical framework for a good AI society: opportunities, risks, principles, and recommendations," *Minds and machines,* vol. 28, pp. 689–707, 2018.

19. I. Munoko, H. L. Brown-Liburd, and M. Vasarhelyi, "The ethical implications of using artificial intelligence in auditing," *Journal of Business Ethics,* vol. 167, no. 2, pp. 209–234, 2020.
20. I. Georgieva, C. Lazo, T. Timan, and A. F. van Veenstra, "From AI ethics principles to data science practice: A reflection and a gap analysis based on recent frameworks and practical experience," *AI and Ethics,* vol. 2, no. 4, pp. 697–711, 2022.
21. I. Rahayu, H. Ardiyanti, L. Judijanto, A. Hamid, and E. S. Bani-Domi, "Ethical dilemmas and moral frameworks: Navigating the integration of artificial intelligence in islamic societies," *International Journal of Teaching and Learning,* vol. 1, no. 3, pp. 171–183, 2023.

CHAPTER 12

Security and Privacy Concerns in AI Models

Muhammad Muneer, Faisal Rehman, Muhammad Hamza Sajjad, Muhammad Anwar, and Kashif Naseer Qureshi

12.1 INTRODUCTION

Artificial Intelligence (AI) approaches are now widely used in a variety of applications due to recent technological developments and greater computer power [1]. Many industries, including healthcare, gambling, and finance, have discovered substantial uses for Machine Learning (ML) models. They encourage innovation by making previously difficult jobs and processes automatable. For instance, self-driving vehicle pipelines are developed by manufacturers using deep learning algorithms. This technology has made it possible to build new features and capacities. AlphaStar, a deep reinforcement learning-based AI system from DeepMind, is one exceptional example, which outperformed numerous top-tier human players in the computer game StarCraft II to reach Grandmaster level [2]. Even while ML has numerous advantages and is highly successful, many of the models now in use are weak against hostile data. These carefully constructed inputs are employed by attackers to compromise the availability, confidentiality, or integrity of ML models, which causes them to provide inaccurate predictions [3]. Unfortunately, a large portion of AI

DOI: 10.1201/9781032667911-12

systems lack security concerns in their design, leaving them extremely open to such assaults. ML's training and testing phases are also vulnerable to attacks from the opposite side [4]. To influence input characteristics or data labels during the training phase, attackers might introduce harmful data into the training dataset. These assaults are referred to as "poisoning attacks." Applications that use data from unreliable sources are particularly vulnerable to these attacks. Additionally, the most frequent testing phase assault is an evasion attack, when a model's flaws are exploited to produce adversarial samples, which are then utilized to trick the model during testing [4]. Numerous socioeconomic and environmental issues can be solved thanks to the growing usage of AI techniques in various applications. However, it is essential to do targeted research on protecting these technologies to fulfill their promise fully. The study of adversarial ML has become important, and continuous research in this area will be essential to ensure that transformational AI technologies are widely used. The creation of strong algorithms is essential for assuring a secure and safe future of innovation as AI grows more pervasive in many facets of human activities and lifestyles [5].

Adversarial situations expose weaknesses in the ML models that are the foundation of many AI systems, leaving them open to such assaults. By exposing an ML model's structure or training/testing data, such attacks may aim to compromise data privacy [6]. By leveraging inconsistencies created by the learning process during model training, these instances can also take advantage of the input integrity of ML models. Attacks that jeopardize availability entail antagonistic actions that prevent authorized users from using a model's features or outputs. While the academic community has recently paid special attention to AI systems' privacy and security from cyberattacks, few efforts have been made to analyze the research and experimental findings to give a thorough comparative study of AI privacy and security.

The main objectives of this chapter are as follows:

- A comprehensive analysis of earlier relevant surveys is provided. The identification of shortcomings in previous studies regarding the development of secure AI applications stands as a significant contribution to the field.
- A concise summary of various ML tasks, including deep learning and federated learning, is presented.

- Introduction of an innovative adversarial attack framework that demonstrates sophisticated assaults exploiting AI applications and analyses their associated risks.
- Presentation of an innovative defense architecture illustrating cyber defense techniques for safeguarding AI systems from hostile assaults.
- Discussion of the challenges within this domain and the presentation of recommendations for potential future study avenues.

12.2 RECENT RESEARCH SYNTHESIS

We looked at recent surveys and evaluations to have a better understanding of the scholarly work being done on the security and privacy of AI technology. Using a data-driven approach, the authors in [4] carried out a thorough study that systematically examined the security threats and related protection mechanisms widespread in both the testing and training stages of ML. Due to the dearth of thorough literature studies that include security vulnerabilities and countermeasures for both ML phases, their study is particularly noteworthy. Their pioneering work provides a thorough review of well-known adversarial attack techniques that target ML models as well as efficient defenses. They made a significant contribution by introducing the idea of adversarial ML and thoroughly describing the fundamentals of ML. Although their investigation had a big impact on later research in adversarial ML, it's important to note that the authors didn't examine security risks in the context of reinforcement learning.

A thorough assessment focusing on the security and privacy aspects of ML is discussed in [1], which highlighted the rising understanding of the increasing susceptibility of ML models to a variety of adversarial methods. The adversarial assaults on ML systems and the accompanying defense mechanisms are the main areas of interest. The authors used a thorough strategy, examining the ML threat model across the data pipeline. The authors included efforts on differential privacy in their study, which also covered new assaults on ML during both the testing and training stages. Their evaluation of modern adversarial ML research in [7] takes a unique stance while offering insights into patterns and results. It's crucial to keep in mind that this work didn't focus on specific ML attacks and lacked the thorough coverage frequently found in other review articles in this field.

The analysis of previous research in [8] added to the conversation on security and privacy in deep learning systems across many applications.

Their analysis includes a thorough knowledge of the theoretical underpinnings of deep learning and privacy-enhancing techniques used in the literature. In this study, the concept of secure AI is introduced, and several effective assaults on deep learning models are carefully examined, along with the accompanying defense tactics. Deep learning privacy techniques are classified, and possible research topics are noted to fill up any gaps that are found. A survey of several adversarial assaults and their responses are provided in [9], which discusses various threat models and attack scenarios. The authors provided insights into attack types. Their focus goes beyond particular deep learning applications, and they promote the need for strong deep learning structures as a deterrent to hostile assaults. The chapter falls short, though; it could have included specific instructions for establishing such resilience as a part of potential future research topics.

An overview of the principles of adversarial ML is provided in [10], with an emphasis on current attack and protective tactics in the context of DNN. The study presents metrics that are often used in literature to evaluate input sample perturbations. It is important to keep in mind that the study only concentrates on a tiny percentage of the attack and defensive techniques for DNN that have been reported in the literature, even though adversarial assaults in reinforcement learning are addressed. A summary of popular adversarial attacks and countermeasures directed at deep neural networks. This study emphasizes the outcomes and solution models from the Google Brain-hosted NIPS 2017 Adversarial Learning competition. Despite the survey's insightful findings, it's important to recognize that it doesn't include all adversarial assaults used against deep neural networks.

A thorough summary of adversarial ML research during the past 10 years, focusing especially on cybersecurity and computational vision applications, is provided in [11]. This chapter covers the fundamentals of DL threat models, the evolution and classification of attack techniques, and misunderstandings regarding the security calculation of ML algorithms. This report also encourages the investigation of adversarial attack-resistant ML model architectures. Similar to this, in [12], the authors provided a thorough analysis focusing on current developments in adversarial assaults against deep learning and related protection techniques. This chapter broadens its scope to include attack tactics used in the development and evaluation phases of systems in a variety of fields, including cybersecurity, the cyber-physical world, natural language processing, computer vision, and picture categorization. Additionally, it contains a variety of suggested literary protective plans. This chapter covers several real-world assaults on AI systems and their defenses adequately, but it skips over the privacy issues raised by traditional ML techniques.

In [13], authors added to current evaluations in the field by examining the fundamental elements of adversarial ML and evaluating the security possessions of ML algorithms in hostile settings. Examining recognized adversarial assaults and the related defensive tactics is part of this. The authors focus on adversarial assault strategies against ML models in the visual area and emphasize their use for classification tasks. This chapter covers a wide range of assaults discovered in the literature and carefully assesses the benefits and drawbacks of each of the presented adversarial assaults and countermeasures. By examining assaults in unsupervised and reinforcement learning, fields not covered, our research greatly broadens their reach in comparison. This addition broadens the scope of the research environment we provide.

In [14], the authors discussed a thorough taxonomy and evaluation of assaults on ML systems was completed. This classification, which is based on important assault features, offers insights into the prevalent attack environment and aids in the recommendation of effective defense tactics. This study successfully covers a variety of adversarial assaults in the realms of technology such as spam filtering, image recognition, and intrusion detection [15]. Its exclusion of a study of responses against these adversarial attacks, however, constitutes an inherent weakness. An overview of potential defenses and adversarial attacks in sensor-based, non-camera cyber-physical systems. They concentrate on sensor data from surveillance systems, audio data, text input, and related defensive tactics. To include new adversarial assaults in this area, the essay presents a thorough process that describes adversarial attacks on cyber-physical systems. Notably, only regression and classification problems are included in this chapter.

In the region of DL, the authors [16] contributed to a study that covers the fundamental ideas of adversarial assaults before giving a brief overview of the techniques used in adversarial attacks and the accompanying response tactics. Notably, while the adversarial techniques in this work are examined, the emphasis is mostly on assaults applicable to the field of computer vision and draws from recent works. The field of NLP explores adversarial strategies that target deep learning models. The core elements of DL and adversarial assault techniques for NLP are both thoroughly covered in this overview [17]. Additionally, explanations of both black-box and white-box attack techniques applied to deep-learning NLP models are provided. Additionally, the paper examines two important defense strategies, adversarial training and knowledge distillation, that have been put forth in the literature to improve the resilience of textual deep neural networks. The majority of assaults proven against textual deep neural networks are covered by this study, although it falls short of addressing the whole range of defenses put out in this field.

To fill in the gaps in the literature, new developments in the vulnerability of reinforcement learning to adversarial assaults are needed. Also, need to explore novel techniques that strengthen ML models, particularly in computer vision, cybersecurity, natural language processing, and cyber-physical systems, guarantees a thorough understanding of modern adversarial techniques and strong defenses.

12.3 OVERVIEW OF ML

The idea of ML was initially established with the first collection of algorithms in the 1970s. In ML, patterns are extracted from data to perform a variety of predictive tasks, including predicting, finding anomalies, removing spam, and evaluating credit risk [18]. The main aim is to predict consequences using the input data that is provided. Any ML system's fundamental building block is data. For instance, a computer must be trained on a range of spam emails before it can determine if an email is spam or not; the variety of training data directly affects the prediction's accuracy. Data are often divided into training and test sets when ML is involved. The ML model is created using the training data, and its performance is evaluated using the test data after its prediction accuracy reaches the target level.

Tasks, models, and features are the three main parts of ML [19]. Tasks include the specific issues that ML approaches can solve. ML models are frequently developed to tackle a few jobs, mostly. Models, which are trained on sample data before processing fresh data to make predictions, describe the outcomes of ML. By describing the qualities of the input data and making it easier to spot patterns between input and output data, features are a crucial component of ML. To complete learning-related activities or difficulties, algorithms are used as tools. In [19], ML is the process of consuming acceptable characteristics to create models that are effective at solving a particular issue. Figure 12.1 shows the categories of ML tasks.

The four primary categories of ML tasks are supervised learning, unsupervised learning, semi-supervised learning, and reinforcement learning [20].

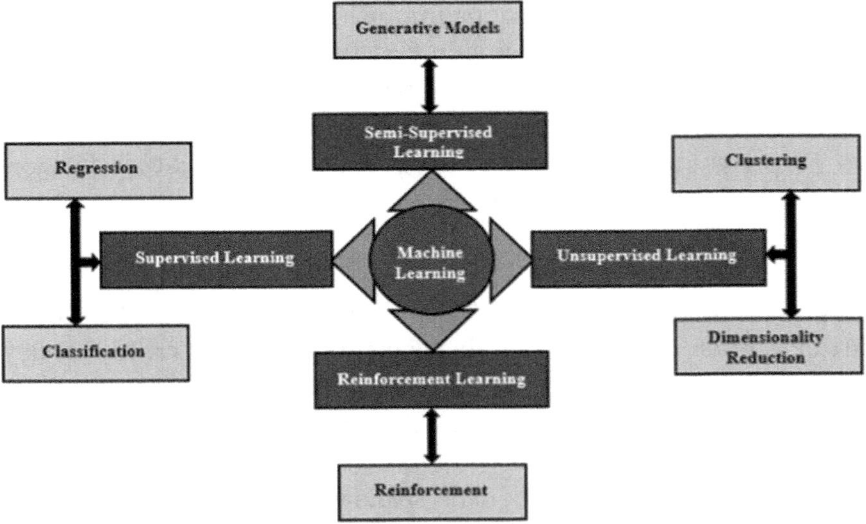

FIGURE 12.1 Categories of ML tasks.

- **Supervised Learning:** A kind of ML called supervised learning employs patterns that an algorithm discovers from a labeled dataset used for training to make predictions or classify data. The exercise dataset for supervised learning is moreover pre-categorized or composed of numbers. Classification and regression are the two primary methods used to categorize tasks that fall within the supervised learning umbrella.

- **Unsupervised Learning:** Unsupervised learning is used when the input data does not have labels and the learning algorithms focus on finding commonalities among the input data's constituent parts to infer important characteristics [20]. Then, potential output labels are generated using these attributes.

- **Semi-supervised Learning:** As a hybrid method to ML, semi-supervised learning combines elements of both supervised and unsupervised learning. By including additional data, as is customary for unsupervised learning, and conversely, it enhances supervised learning.

- **Reinforcement Learning:** In ML, reinforcement learning is a technique that allows a representative to learn in a collaborative setting through repeated efforts and comments obtained from its behaviors and relationships. A situation in which an agent communicates with its atmosphere by engaging in certain behaviors and then getting rewards for those behaviors is the core component of the reward-based learning problem. This learning paradigm's main goal is to equip students with the knowledge and abilities to choose decisions that maximize total return.

12.3.1 Deep Learning Approaches

A specific area of ML known as "deep learning" is based on DANN. The ML algorithms conferred up to this point are sometimes called "Shallow Learning" since they rely on feature developers to extract pertinent qualities from incoming data [21]. Their extremely simple design, which typically involves one layer to translate input data into a context appropriate for the task, is a defining feature of many shallow learning methodologies. On the other hand, modern deep learning algorithms perform difficult learning tasks and feature selection using multiple layers of representations and abstractions of the information being provided in [20]. Numerous shallow learning algorithms have been successful in solving certain issues, but they fall short when faced with more complex challenges, such as those in computer vision and NLP. DL, on the other hand, successfully overcomes this obstacle by building numerous layers of fundamental characteristics to represent complicated ideas. The development of deep learning architectures has been significantly accelerated by the phenomenal increase of data that is now accessible and the improved processing power of computer chips.

12.3.2 Federated Learning

Google made the first mention of federated learning in 2016 [22]. With the use of data that is federated and dispersed throughout a network of nodes, this novel learning strategy makes it easier to train centralized models. The need to incorporate data from users' mobile devices to train models is the main driving force behind federated learning. Privacy issues prevent this data from being stored in centralized data centers [8]. Notably, federated learning offers different privacy advantages over conventional ML strategies. This is because for federated learning only the bare minimum updates needed to improve a particular model are provided, and the training goal affects this data. The more the data available, the more effective the federated learning; the better model performance is the result of more nodes contributing to model training.

12.4 COMPREHENSIVE REVIEW OF AI SYSTEM ATTACKS

Significant concerns are raised by the widespread use of AI technology and its vulnerability to hostile assaults. To purposefully cause inaccurate classifications, adversarial assaults frequently include manipulating the input data of ML or Deep Learning (DL) models. Depending on the job category of the algorithm, these assaults take different shapes. Threats to AI systems essentially all have the same core, although different algorithms can be exploited in different ways. In this part, we start by providing the findings of a thorough literature search that included assaults on AI systems during the previous ten years. We next dive into a detailed study of the assembled literature. A Google Scholar search with certain keywords turned up a collection of research on AI system assaults like attacks and defenses on DL and ML, Membership Inference Attacks, Extraction Attacks, Inversion Attacks, Poisoning Attacks, and Evasion Attacks, which are all examples of adversarial attacks. Understanding the system's attack surface and potential attack vectors that malicious actors may use to infiltrate the system is a vital first step in conducting a thorough evaluation of the privacy and security of AI systems [23].

12.4.1 Attack Surface for AI

The weaknesses that the AI model is vulnerable to throughout the testing and training stages are all included in the attack surface of an AI system. It might alternatively be thought of as a list of inputs that a malicious actor could use to assault the system. A generic data processing pipeline may be used to understand the attack surface of an AI system. The learning algorithm or model, the input data or objects used for training and testing, and the final output data make up this pipeline. The ML model processes the input characteristics during the testing phase to provide class probabilities. An external system is then given actionable instructions to communicate these possibilities. By altering the exercise data, messing with the learning model, or faking the class possibilities, adversaries might try to hack this system.

12.4.2 Attacker's Objectives

The objectives of adversarial assaults can be generally described in terms of the safety features of an IS, such as protecting privacy and keeping data accurate, private, and readily accessible when needed. Attackers who want to compromise the privacy of an AI organization strive to learn more about the internal workings of the dataset or learning model so they can launch more advanced assaults. When it comes to confidentiality violations, attackers may try to find sensitive information by focusing on either the model and its training data or the parameters [1]. An integrity attack, on the other hand, seeks to change the AI's logic, modifying its behavior during the learning or inference phases and affecting the model's outputs. Certainly, take a spam filter as an example. A malicious addition to the training data by an attacker might change the categorization border. As a result of the modified model behavior, this can cause valid emails to be mistakenly categorized as spam. By doing this, the attacker may avoid discovery and keep the system functioning normally. Reduced confidence, induced misclassification, targeted misclassification, and misclassification involving several sources and targets are four aims that, according to [24], threaten the integrity of a deep learning system. Figure 12.2. shows the framework for analyzing adversarial assaults on AI models.

To interfere with the operation of the system, it is possible to target the availability of an AI solution. To make a model unreliable or inconsistent in the intended context, for instance, opponents may flood an AI system with incorrectly categorized objects. Attacks against an ML system's availability might lead to several misclassifications, thereby rendering the organization useless. Attacks on privacy and secrecy sometimes have similar objectives. In such assaults, attackers try to violate an ML system's privacy to force it to provide information about the training data and model. Certainly, there are many other ways that privacy issues may develop. For instance, a hacker may attempt to use a model to expose concealed information about the training set that was not meant to be made public. Additionally, by using the model's predictions, attackers can try to obtain data from the training dataset. These two situations emphasize the possible privacy dangers that come with AI models.

12.4.3 Attacker's Awareness and Abilities

An attacker's understanding of an ML system can be described based on their understanding of the individual modules that make up the system architecture. Understanding the attacker can include different levels of awareness about the system, including training data (B), features (R), learning algorithm (h), the objective function (O), and limitations (**m**). Therefore, adversarial knowledge can be represented by a vector $\theta \in \Theta$, where $\theta = (B, R, h, \mathbf{m})$. In contrast, adversary capabilities indicate the level

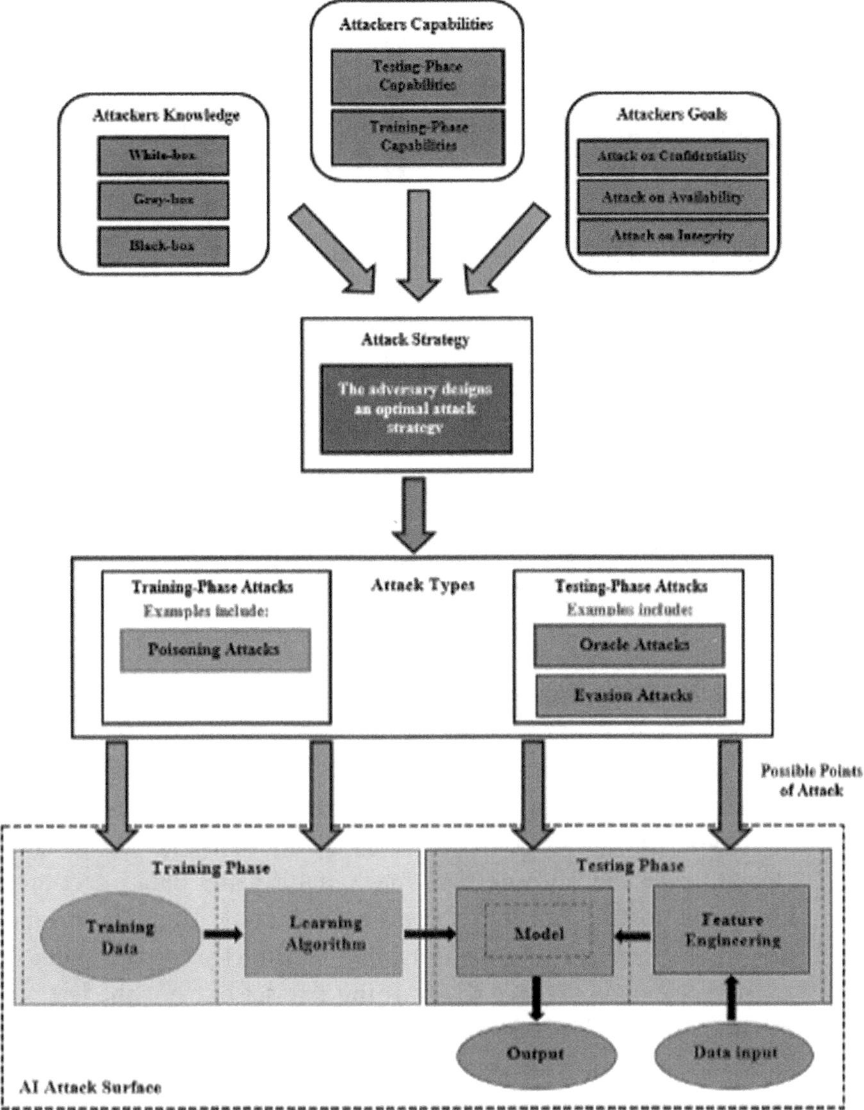

FIGURE 12.2 Framework for analyzing adversarial assaults on AI models.

of knowledge or info about the ML system accessible to the challenger[9]. For example, in the context of the spam detection process mentioned earlier, an adversary with access to or awareness of the spam detection model is used for potentially adversarial grouping. More present to the system than an opponent with limited access to model "tokenized," incoming text or email data.

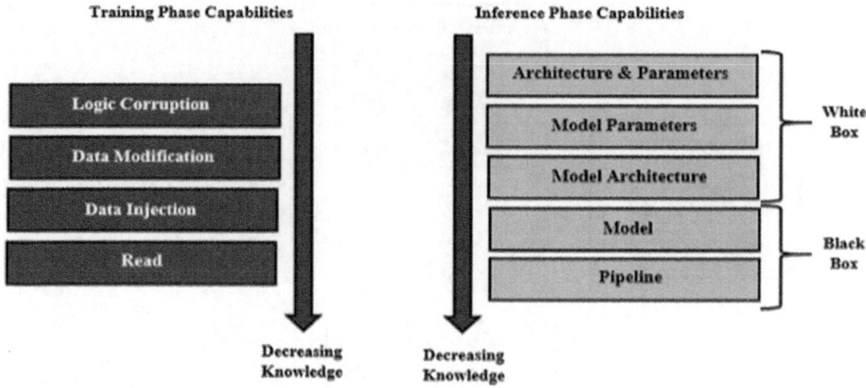

FIGURE 12.3 Attacking techniques.

A classification of adversarial capabilities in ML systems that are aligned with the training and testing phases is shown below, as suggested in [1]. Refer to Figure 12.3 for an illustration of how an adversary's skills and awareness of an ML system are related.

- **Preparing Stage Capabilities:** The goals of attacks made when an ML system is being trained include knowledge gathering, influence, or performance modification. An adversary trying to obtain all or part of the training data is a basic assault during the training phase. Alternatively, an adversary without access to the learning algorithm or the training data may assault using data injection to introduce hostile data into the training dataset. The data or labels (for supervised learning datasets) can be directly contaminated by an adversary who knows the training dataset but not the learning method before it is used to train the target model. The phrase "data/label modification" refers to this type of attack capability. Finally, an adversary familiar with the inner workings of the algorithm can stage an attack that corrupts the logic, by manipulating the learning logic. Attacks on ML systems that target their training phase include very complex threats called logic corruption attacks. The security of the educational process is seriously compromised by these attacks, which are incredibly difficult to defend against.
- **Possibility of Testing Phase:** Attacks during the testing stage are known as exploratory attacks. These attacks are intended to learn more about the learner's condition rather than to change the training process or have an impact on learning. On the other hand, inference

attacks rely on understanding the model and how it is used in the target environment. Adversarial tools used during testing may be broadly characterized as black-box or white-box assaults, as shown in Figure 12.3. The adversary in a white-box assault scenario is fully aware of the model's training data, architecture, parameters, intermediate computations at hidden levels, and any hyper-parameters used for predictions. Therefore, a white-box adversary's information is indicated by the notation θ_{Wb} = (B, R, h, **m**). Contrarily, black-box attacks presuppose the adversary has no information of the model itself but can nevertheless identify weaknesses in the model by using knowledge of prior input/output combinations [1]. When using platforms like ML as a Service (MLaaS) [25], the adversary is restricted from accessing model architecture and parameters. As a result, a black-box adversary's knowledge is defined as θ_{Bb} = (B, R, h, **m**).

12.4.4 Attacking Method

An attack plan is a process used by adversaries to modify training and test datasets to maximize their attack objectives [13]. In this context, considering the opponent's knowledge as ∈ and a set of conflicting examples $B'_c \in \Phi(B_{1D450})$, the opponent's goal can be determined using an objective function $A(B'_c, \theta) \in \mathbb{Q}$, quantifying the impact of an attack using B'_c samples. Therefore, the opponent's optimal attack strategy can be formulated as follows:

$$B_c^* \in argmax\ A\left(B_c^{'}, \theta\right) \quad \ldots\ldots\ldots\ldots\ldots\ldots\ldots\ldots\ldots\ldots\ldots \text{(12.1)}$$

$$B_c^{'} \in \Phi\left(B_c\right)$$

This equation shows how the adversary searches through the set of possible modified samples B'_c to find the best set of perturbed samples B^*_c to maximize the given objective function A to accomplish their attack goals.

12.4.5 Types of Adversarial Attacks

This section elaborates on the methods used by attackers and explores the most common assaults against AI systems.

12.4.5.1 Poisoning Attacks

There are occurrences of poisoning attacks, often known as causative attacks, in the preparation stage. The goal of this kind of attack is to change the decision boundary of the targeted model by manipulating, adding, or deleting samples from the training dataset [1]. These assaults may aim to compromise the accuracy of adaptive, online, or classifiers that have been retrained using test-time data. The issue of poisoning attacks and its connection to the dependability of the enormous volume of data collected by these systems is highlighted in [26] observation that a sizeable portion of data used for training ML systems originates from sources that are not entirely trusted. Although an adversary might not have direct access to an existing training dataset, they can nevertheless add fresh training data through online honeypots and repositories, providing a way to contaminate the training data. According to specific research [26], there are two main possibilities for poisoning attacks that target systems of multiple-class classification:

- **Attacks from Error-Generic Poisoning:** These common poisoning attacks goal to start a denial-of-service attack on the system. Their goal is to produce several incorrect classifications for certain data points, regardless of the classes to which they may belong. The error-generic poisoning assault may be expressed as follows by using the attack strategy formulation in Equation 21.2:

$$B_p^* \in argmax\ A(B_p, \theta) = O(B_{target}, \boldsymbol{m}(B_p)) \qquad \text{.......... (12.2)}$$

$$B_p \in \Phi(B_p)$$

In Equation 12.2, the objective function A, which is defined by the loss function, O calculated on the set of targeted points, B_{target}, that the attacker intends to influence, is what the adversary seeks to maximize. The inserted poisoned points, B_p, have an impact on the parameters ***m***, which in turn have an impact on the loss function, indicated as O.

- **Attacks from Error-Specific Poisoning:** The adversary seeks to cause specific misclassifications using error-specific poisoning attempts [27]. In this case, the following formulation of the offensive approach is appropriate:

$$B_p^* \in argmax\ A(B_p, \theta) = argmax - O(B'_{target}, \boldsymbol{m}(B_p)) \qquad \text{......... (12.3)}$$

$$B_p \in \Phi(B_p) \qquad\qquad B_p \in \Phi(B_p)$$

The same samples as the B'_{target} are used to create B_{target} in this equation, but the labels are chosen by the adversary to produce the intended misclassifications. Since the opponent wants to reduce the loss on the targeted labels, the loss function O has a negative sign. As a result, Equation 12.4 may be expressed similarly as:

$$B_p^* \in argmax\, O\left(B'_{target}, \boldsymbol{m}\left(B_p\right)\right) \quad \text{......................} \quad (12.4)$$

$$B_p \in \Phi\left(B_p\right)$$

12.4.5.2 Oracle Attacks

The Oracle attacks are empirical attacks in which the attackers use samples to collect information on model or training data, and then make a judgment as to its accuracy. The classifiers are often the sole property of an organization to which they belong. This is evidenced by platforms like Microsoft MLaaS, Amazon MLaaS, and Google Expectation API that use ML as a service. Users can input data, train models, and submit queries using these simplified APIs [28]. An adversary can execute an Oracle attack utilizing the learning model's inputs and outputs by using an API. If the adversary is unaware of the classifier, they can launch an oracle assault by feeding it data, watching what happens, and using the input-output pairings as training data for a surrogate model [29]. The surrogate model imitates the actions of the target model. Figure 12.4 illustrates how an oracle attack is carried out. The adversary can give test data to the classifier and record the labels it produces in situations when they are ignorant of the class labels and the classifier's inner workings. Subsequently, this label shall be used for training a substitute classifier that performs similarly to the original one. An adversarial example of an evasion attack on a target model can be produced by this surrogate model. Oracle attacks are composed of several types including member's inference, extraction, and inversion attacks.

The companies that created many ML models view them as their intellectual property. Examples of platforms that fall under the category of ML as a Service (MLaaS) platforms include Microsoft ML, Amazon ML, and Google Prediction API. These platforms are maintained and run by Google, Amazon, and Microsoft, respectively. Users may submit data, train models, and query those using simple APIs in these configurations

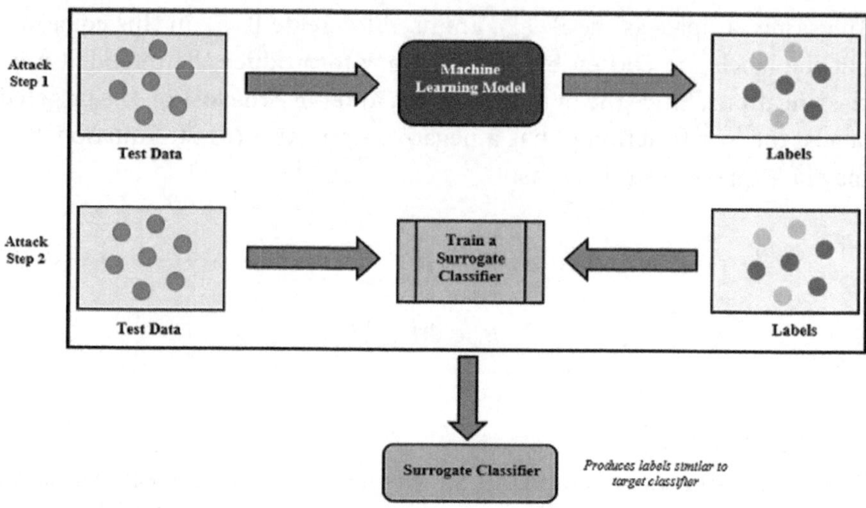

FIGURE 12.4 An Oracle attack's steps.

[28]. An API is used to feed data into an ML model, and then watching the results, an adversary may execute an oracle attack in this situation. In situations when the adversary is unaware of the target model, they can train a surrogate model that mimics the target model's functionality by using the input-output pairs obtained through oracle assaults [29]. The target model may then be the target of adversarial instances produced by this surrogate model. Oracle attacks come in a variety of forms, such as membership inference, inversion, and extraction attacks.

- **Membership Inference Attacks:** Attackers using membership inference attempt to determine if a given data piece was part of the training dataset that was used to determine the model's parameters [1]. Consider an opponent that wants to invade the privacy of a clinical patient as an example. To ascertain if the patient's medical history was included in the dataset used to train a model in charge of making a specific illness diagnosis, they may perform a membership inference attack. Shokri et al.'s studies [28] demonstrate the use of membership inference attacks on black-box models. This is accomplished by taking advantage of the difference in the behavior of the target model while handling data it "knows" through training and data it sees for the first time.
- **Inversion Attacks:** An adversary that can use a model's predictions can carry out model inversion attacks, also known as attacks that extract training data. These extracted inputs pertain to aggregate

representations of data rather than specific individual data points associated with a given class. Model inversion attacks, as illustrated in [30], entail scenarios in which an adversary can determine a patient's genetic marker based on forecasts provided by a linear model intended to predict an appropriate quantity of medicine like Warfarin. The privacy risks connected with allowing ML models that have been trained on sensitive data API access are highlighted by this example [1].

- **Extraction Attacks:** Attacks known as "model extraction" take place when an adversary uses input-output pairs (x_i, y_i) derived from the estimates of an objective model (h) to extract a group of limitations and tries to build an approximation model $\check{h}$, designated as h, that closely mimics the behavior of s. An analysis in [31], where authors demonstrated the efficacy of extraction attacks against two MLaaS models using their openly available APIs. While extraction attacks can make models less profitable, they also raise questions about the privacy of training data and open the door for evasion attempts.

12.4.5.3 Evasion Attacks

The most typical kind of assault against AI systems is evasion attacks [9]. These attacks include an adversary directly altering input samples while they are being tested to avoid detection or deceive the system's categorization. Deep neural networks are especially vulnerable to these dangers, even while a lot of the research on evasion attacks has focused on ML classification tasks [27]. Deep networks, as opposed to conventional linear models, can, nonetheless, express functions that can tolerate adversarial perturbations. Evasion attacks use adversarial instances to exploit weaknesses rather than change a system's fundamental behavior to cause certain misclassifications. The premise behind evasion assaults is that an adversary may build a surrogate classifier using surrogate data without knowing the decision function of the targeted classifier, thereby enabling them to avoid it. Even while adversarial instances are frequently undetectable from typical data sets, they nonetheless provide serious security vulnerabilities for programs utilizing ML [32].

Based on multi-class classifiers, the authors in [33] presented two main types of evasion assaults: error-specific and error-specific evasion attacks. These classifications act as the fundamental ideas from which other evasive assault scenarios may be constructed.

- **Error-Generic Evasion Attacks:** Regardless of the precise output class predicted by the classifier, the adversary's primary goal in these assaults is to introduce misclassification during the testing phase. The predicted class "n" in the context of multi-class classifiers is the class for which the discriminating function "$h_k(x)$" maximizes for a certain input sample "z".

$$n^* = \underset{k=1,\ldots,c}{argmax}\, h_k(z) \quad \text{……………………………… (12.5)}$$

Mathematically speaking, the error-generic evasion attack problem is as follows:

$$z_e^* \in \underset{z_e \in \Phi(z_e)}{argmin}\, h_k(z_e) - \max_{j \neq k} h_j(z_e),\ s.t.\ d(z, z_e) \leq d_{max} \quad \text{……… (12.6)}$$

In the equation, "$h_k(z_e)$" is the discriminant function related to the true class "k" of the source sample "z", and "$\max_{j;\neq k} h_j(z_e)$" represents the maximum discriminant of the false class is not the true value of the class "k" function. The optimization problem is subject to the constraint "$d(z, z_e) \leqslant d_{max}$", which limits the space between the original sample "z" and the control sample "z_e", the perturbation "d_{max}" between them in the input space [27].

12.5 ATTACK DEFENSES FROM AI SYSTEMS

This section categorizes the many types of defenses against attacks on AI systems depending on whether they offer comprehensive protection against hostile instances or just focus on identifying and rejecting them. This classification is shown in Figure 12.5 and provides the basis for an evaluation of the available defense strategies. Whether they handle threats throughout the system's operation's training or testing stages further distinguishes complete defenses [34].

12.5.1 Protections against Training Stage Attacks

Similar to other types of attacks on AI systems, poisoning attacks provide unique difficulties for the learning process. The idea that poisoned

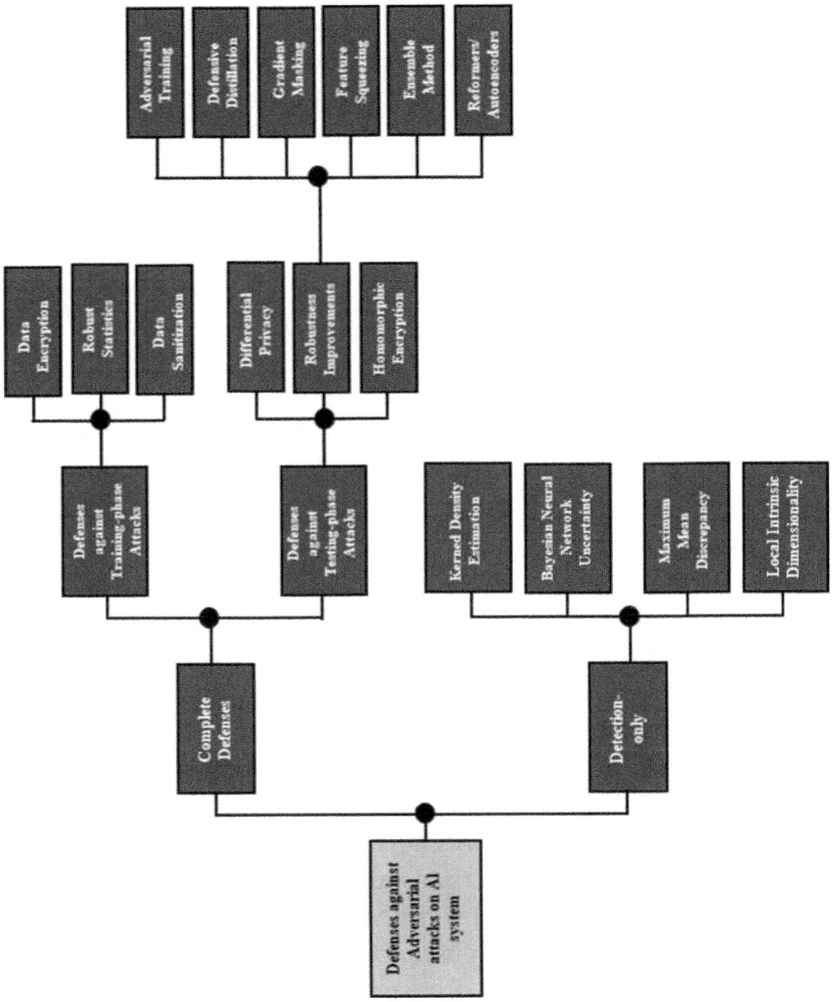

FIGURE 12.5 Organization of AI system assaults defenses.

samples often stray from the expected input distribution is the basis of a common defense against training-time assaults [35]. A handful of the ways that have been suggested as potential responses to training-time assaults are given below.

12.5.1.1 Data Sanitization

To protect against poisoning assaults, data sanitization involves eliminating tainted samples from a preparation dataset before training a model. To prevent poisoning assaults on commercially available anomaly detection classifiers. The Reject on Negative Impact (RONI) strategy is another data sanitization method for thwarting poisoning attempts, according to Nelson et al. [36]. The SpamBayes spam filter was successfully used to screen out dictionary attack messages using this method. While avoiding false flags on legitimate emails, RONI successfully recognized all attack emails.

12.5.1.2 Robust Statistics

This defensive strategy, in comparison to data sanitization, focuses on building robust models that can withstand poisoning attempts rather than trying to identify poisoned samples [37]. Increasing the resilience of pattern recognition models in hostile situations can be achieved by using multiple classifier systems (MCSs). The essential idea is that more than one classifier must be avoided for the organization to become useless. The results of their practical studies suggested that MCS building methods based on randomization might increase the robustness of linear classifiers in combative situations. Furthermore, in [38], the authors showed that "bagging," which is short for bootstrap combining, is a powerful protection mechanism against poisoning attempts, irrespective of the fundamental classification system. Authors in [39], described the process of creating numerous classifier versions and merging their predictions to create an aggregate classifier. This method has been demonstrated to be quite successful, especially when used with classifiers that show large prediction variance with little changes in preparation data.

12.5.2 Protection against Attacks in the Testing Stage

A variety of tactics, including improving model robustness, using distinction privacy, and utilizing homomorphic encryption techniques, are used as defenses against testing-phase assaults.

12.5.2.1 Robustness Improvements

An ML model may have the capacity to recognize and ignore hostile samples to improve its resistance to them. Notably, the training phase, which comes before testing and acts as a protection against assaults during testing, is when various strategies for increasing model resilience are used. Adversarial Training, Ensemble Methods, Reformers/Autoencoders, Gradient Masking, Feature Squeezing, and Defensive Distillation are a few strategies for robustness augmentation that have undertaken extensive investigation [34].

- **Defensive Distillation:** Distillation is a method created expressly to transfer knowledge from a collection of models or a sizable regularized model to a more compact, distilled model while maintaining predicted accuracy. Building on the basic idea put out by Ba and Caruana [40] to improve prediction smoothness by training shallow neural networks to replicate the predictions of deeper ones, the notion was explicitly established in [41], making deep learning practical for computationally limited devices. Defensive distillation is a technique for training models with enhanced resilience to input perturbations. Their research showed that models built on deep neural network distillations had improved resistance to hostile samples. Despite defensive concentration's many benefits and its resilience to adversarial instances [42] empirically showed how their assaults can get beyond defensive distillation by using high-confidence adversarial examples.
- **Feature Squeezing:** A cutting-edge technique called feature squeezing limits the input feature space that may be used by potential adversaries to find hostile cases. To do this, many samples, each of which represents a changed feature vector in the initial input space, are combined into a single sample. Two categories of feature squeezing techniques for image processing: (1) lowering picture color complexity, and (2) median smoothing, which is smoothest to reduce pixel fluctuations. They showed that these techniques reliably and accurately identify hostile inputs. They developed this work in [43], showing that median smoothening is the best squeezing method for fending against the Carlini and Wagner assault. Input characteristics that adversaries rely on for their assaults are disrupted by feature

squeezing, but empirical analyses [44] show that it may also result in a drop in accuracy for categorizing benign inputs.

- **Gradient Masking:** Gradient masking is a method used to make Deep Neural Networks (DNNs) less sensitive to subtle input changes [34]. This defensive strategy computes the initial derivatives of the model's input and then minimizes these derivatives during the learning process [1]. This leads to a technique that hides a model's gradient information, making it more difficult for attackers to exploit the model using gradient-based data. Although it makes sense to reduce a model's sensitivity to minor input modifications as a defense against adversarial attacks, the gradient masking technique has its shortcomings. They showed how a black-box assault may go around it to show off its evasion.
- **Ensemble Method:** Using learning approaches called ensemble methods, supervised learning models' categorization judgments are improved. Even though several ensemble approaches have been described in the literature over the years, their use to strengthen the defenses of ML models against adversarial assaults has only lately drawn attention. A unique method for improving convolutional neural networks' resistance to adversarial assaults was developed in [45]. According to their confidence levels, they advised using an ensemble of many specialized classifiers to recognize and discard hostile cases while accepting benign data. The assurance of predictions for adversarial inputs was demonstrated to be reduced by this method, but partial confidence was maintained for clean samples.
- **Reformers/Autoencoders:** An input x is sent to a reformer, which reconstructs it as x′ before feeding it to a classifier. The prediction accuracy for benign instances is intended to be maintained by an ideal reformer while adversarial examples are changed to resemble benign examples. An encoder plus a decoder makes up an artificial neural network called an autoencoder. Using the concealed data, the decoder reconstructed the output representations that the encoder has learned for the input data. Magnet, a strong protection against adversarial assaults on deep neural networks, was introduced in [46]. A reformer and a detector are both parts of MagNet. The detector finds cases with large reconstruction mistakes by using an autoencoder to calculate reconstruction errors.

12.5.2.2 Distinctive Privacy

DNN can be made more resistant to adversarial assaults by using distinction privacy. In the context of ML, differential privacy entails the addition of randomization to either preparation data or model outputs to prevent the disclosure of private information from specific records within the training data. Using the ideas of differential privacy as a foundation, the authors in [47] proposed PixelDP, a validated protection mechanism against hostile cases. Differential privacy can help protect training data from inversion attacks, which try to recreate training data from model parameters. Comparing PixelDP to other certified defenses, empirical results show that it provides predictions that are more accurate when subjected to the Euclidean Norm attack. It does, however, add some computational burden to the training and testing operations.

12.5.2.3 Homomorphic Encryption

The ability to perform actions on encrypted data without the requirement for data decryption has been made possible by recent developments in homomorphic encryption methods, notably fully homomorphic encryption. Recent literature has suggested this encryption technique as a way to satisfy the privacy-preserving needs for complex data while utilizing analytical models handled by third parties, including in platforms that offer ML as a service. Neural networks known as "crypto-nets" provide encrypted predictions over encrypted data. However, their theoretical and practical investigation showed that in some situations, the computational difficulty of using neural networks to decrypt data makes crypto-nets impractical. CryptoDL is a method for applying deep neural network methods on encrypted data that was developed in [48]. Deep convolutional neural networks trained with homomorphic encryption and low-degree polynomials are the foundational elements of CryptoDL [49]. With regard to effective homomorphic encryption systems, the decision to train CryptoDL using low-degree polynomials overcomes the practical drawbacks of alternative alternatives, such as crypto-nets. Results from experiments support CryptoDL's scalability and efficacy by showing how well it predicts privacy-preserving events in an accurate and timely manner.

12.5.3 Only Detectable Protections

The inadequacies of conventional all-encompassing defenses have led to a lot of research interest in detection-only tactics. Numerous full defenses contain flaws that have been shown by research by Carlini and Wagner

[50], illustrating how easily these defenses may be bypassed or defeated. Defenses based only on detection seek to distinguish between legitimate and malicious instances to successfully weed out malicious inputs and enable reliable categorization. We examine some of the suggested recognition-only defensive strategies below.

12.5.3.1 Detector Based on Kernel Density Estimation (KDE)

To find adversarial instances in DNN, the kernel density estimation approach. To locate points that considerably depart from the data manifold, this method computes kernel density estimates within the feature space of a neural network's last hidden layer. The density estimate is as follows for a particular point z with a projected class t:

$$\hat{K}(z, R_t) = \sum_{z_i \in R_t} k_\sigma(\varnothing(z), \varnothing(z_i)) \quad \text{.............................. (12.7)}$$

where R_t stands for the training group of class, t stands for the tuned bandwidth, and (z) represents the activation vector of the final hidden layer for point x. By conducting trials on the MNIST, CIFAR-10, and SVHN datasets, the effectiveness of this protection against a variety of attacks, including FGSM, BIM, JSMA, and C&W, can be evaluated. . But subsequently, Carlini and Wagner used the C&W method to produce adversarial instances with higher distortion expressly for MNIST, exposing the weakness of the kernel density estimation methods [50].

12.5.3.2. Uncertainty-Based Bayesian Neural Network Detector

Feynman et al. A Bayesian Neural Network Uncertainty Approach to Discrepant Example Detection, focusing on identifying points located in low-confidence regions of the input space. This method can gain additional insight into the reliability of the model, which is not usually possible with distance metric methods such as KDE. To combat overfitting while deep neural networks are being trained, [51] first proposed the dropout technique, which is how the Bayesian uncertainty strategy they suggest introduces randomness into neural networks. The insecurity approximation of the neural network for a given example z* and the random forecast is $\{\hat{v}_1^* \ldots . \hat{v}_t^*\}$ calculated as follows:

$$U(z^*) = \frac{1}{T}\sum_{i=1}^{T} \hat{v}_i^{*T}\hat{v}_i^* - \left(\frac{1}{T}\sum_{i=1}^{T}\hat{v}_i^*\right)^T\left(\frac{1}{T}\sum_{i=1}^{T}\hat{v}_i^*\right) \quad \text{............. (12.8)}$$

Feynman et al. concluded that the Bayesian uncertainty approaches were efficient in recognizing adversarial samples produced by various attack strategies by studying trials carried out using the LeNet CNN trained at a dropout level of 0.5. However, authors in [50] later found that, although initially promising, this approach is also vulnerable to adversary attacks.

12.5.3.3 Detector of Maximum Mean Discrepancy

A statistical hypothesis test was developed in [52] and may be used to locate adversarial cases within an input dataset. To ascertain if samples X_1 and X_2 are taken from the same distribution, the MMD test performs a two-sample statistical hypothesis test. Given a sample X_1 from a distribution p and a sample X_2 from a distribution q, the null hypothesis H_0 $p = q$, while the alternative hypothesis H_A states that $p \neq q$. The arithmetical test takes these two models as input and discriminates between H_0 and H_A.

The MMD test is a kernel-based technique made specifically for high-dimensional data applications. Grosse et al. focused on unbiased MMD's asymptotic distribution while applying it as a detection method. Subsamples are taken from the available data with replacement in this approach, which entails using a subsampling technique. According to the null hypothesis, this method reliably calculates the MMD distribution.

Even though Grosse et al. first noted the MMD's capacity to statistically distinguish hostile cases from clean ones, authors in [50] later established the MMD's shortcomings. By employing focused adversarial instances created with the C&W attack algorithm, they showed that the MMD is ineffective at identifying assaults.

12.5.3.4 Detector Based on Local Intrinsic Dimensionality

Local Intrinsic Dimensionality (LID) was first used in [53] to identify adversarial samples inside a given input dataset. Unlike current detection-only defenses that depend on density distribution, the LID-based detector evaluates the inherent dimensionality of hostile areas inside DNN. The local distance dispersion between an initial point and its surrounding points is examined to achieve this [50]. Ma et al.'s work showed the value of employing LID estimations to locate adversarial cases. To approximate the real distance distribution, they used the Maximum Likelihood Estimator (MLE) of LID.

In formal terms, given a reference data sample $z \in p$, where p denotes the distribution of the data, instead of the maximum likelihood estimate of the local intrinsic dimension (LID), z is mathematically defined as:

$$\widehat{LID}(z) = -\left(\frac{1}{k}\sum_{i=1}^{k}\log\frac{r_i(z)}{r_k(z)}\right)^{-1} \qquad (12.9)$$

Ma et al. provided an example of how Local Intrinsic Dimensionality (LID) properties may be used as a useful tool for identifying adversarial samples produced by different attack techniques. They also emphasized how characteristics might be used to detect hostile instances by characterizing areas that are susceptible to adversarial perturbations.

12.6 CHALLENGES AND RECOMMENDATIONS

We completed a thorough analysis of privacy and security issues in AI systems in the sections above, covering a wide range of ML models. In this section, we examine the difficulties that this field is now facing and make suggestions for prospective directions for future study.

12.6.1 Adversarial Instances' Transferability

Deep neural networks are among the many ML models that are vulnerable to the transferability of hostile samples [50]. This fact suggests that adversarial samples intended to trick one model may easily be used to trick another model. Numerous DNNs have a substantial security risk as a result of the fact that adversarial examples designed to trick one model can successfully target another model without needing to know the target model's parameters. The fundamental causes of this phenomenon are still not fully understood, even though the evaluated studies in this chapter offer verifiable proof of the transferability of adversarial cases. Despite the existence of adversarial cases, experimental results are presented in [54]. The universal transferability hypothesis is not always true, indicating that the transferability trait may not be inherent in non-robust models. i Therefore, to create reliable ML models, research efforts must be focused on understanding the transferability phenomena.

12.6.2 Calculating Robustness of Defense Techniques

The vulnerability of many current defense measures, which have been penetrated or bypassed, is a noteworthy worry in the field of adversarial assaults. This condition makes many worry about how thoroughly the effectiveness of present defensive strategies is assessed. For instance, Carlini and Wagner [50] analyzed 11 ways to find adversarial cases and showed how simply by creating original cost functions, opponents may simply get around these strategies. Their research highlights the inadequacy of present defenses in the face of adaptive attackers and their inadequate security assessments. Differential Privacy has become a tool to strengthen the robustness of DL models against inversion attacks targeting the reconstruction of training data from model parameters. To lessen the exposure of private information available in the training dataset, Differential Privacy provides randomization. Although the usefulness of differential privacy has been demonstrated through several experiments, there are still few real-world evaluations or criteria to determine whether differential privacy boundaries are enough. Similar to this approach, which aims to protect the privacy of complex data used in analytical models, despite achieving high prediction rates in current applications, such as CryptoDL. Despite current applications' high prediction rates, such as those of CryptoDL [48], their accuracy still falls short of many cutting-edge deep models and shows incompatibility with more intricate models. Future research should focus on identifying the particular qualities defensive strategies need to have in instruction to be robust against adaptive attackers.

12.6.3 Effort in Monitoring the Amount of Adversarial Perturbations

The many techniques for creating adversarial instances incorporate minute, scarcely perceptible changes to the input data to change a neural network's predictions. Establishing the precise level of distress required to trick a neural network, however, is a challenging undertaking. This difficulty results from the fact that excessively tiny perturbations might not be able to provide useful adversarial instances, while excessively big perturbations stop being undetectable [55]. Therefore, maintaining control over the size of input perturbations remains a challenge.

12.6.4 Deficiency of Research Attention to Attacks Beyond Categorization Tasks

Applications involving computer vision have shown tremendous success with convolutional neural networks. As a result, a substantial number of adversarial example generation approaches are specifically designed for computer vision tasks, such as object identification and picture recognition, which are crucial parts of ML classification projects [56] [32]. Even while some work has been done to understand adversarial approaches in other ML tasks, classification challenges continue to receive more attention. Research has shown that deep neural networks and other ML models are vulnerable to adversarial assaults. Given this, it is crucial to focus more study on understanding the adversarial risks that cover a variety of ML tasks, including reinforcement learning.

12.6.5 The Conclusion Boundary of the Classifier Is Randomized

The idea has been put out as a defensive strategy, as described in [57], to increase a classifier's security against evasion attempts by inserting randomization into the location of the decision border. Surprisingly, there isn't much research done in the literature currently available on this type of randomization as a workable defensive tactic. While adding randomization might make it harder for attackers to push the decision boundary over a certain point, it's vital to recognize that this strategy has disadvantages as well. Randomization may temporarily raise the classifier's error rate, which would have an impact on how well it performs on clean data.

12.6.6 Evolving Threat of Unknown

For ML systems working in hostile contexts, the existence of "unknown unknowns" poses a significant danger. According to [58], this hazard is representative of its importance in cybersecurity fields like malware and intrusion detection. "Unknown unknowns" frequently show unpredictable behaviors in contrast to "known unknowns," which are used to simulate assaults in adversarial ML. These events may cause situations in which ML models confidently misclassify as a result of inputs materially deviating from the known training data. It is crucial to pursue new research directions in this area to equip ML systems with the capacity to detect "unknown unknowns" using reliable methods like irregularity detection.

12.7 CONCLUSION

The field of adversarial ML has attracted a great deal of attention in recent years. Notably, various research has examined adversarial assaults on ML models in a variety of fields, including cybersecurity, natural language processing, computer vision, and image recognition. Along with this, a variety of defense strategies have been proposed, with focused research efforts aimed at evaluating the effectiveness of these defenses against the always-changing hostile assault scenario. In this chapter, we started our investigation by carefully analyzing recent survey articles that cover many aspects of security and privacy in the field of AI. This study showed that many previous studies have not covered all types of ML task assaults and countermeasures. These studies have mostly concentrated on DNN in the fields of cybersecurity, NLP, and computer vision. We started by giving a theoretical basis embracing several ML task categories to provide the foundations for clarifying adversarial assaults on ML models. We made sure that the more modern deep learning methodology could be distinguished from shallow learning approaches. Then, we unveiled a novel framework made for a thorough analysis of adversarial assaults directed at AI systems. It's critical to recognize that this strategy has limitations even while adding randomization might make it harder for attackers to push the decision boundary over a certain point. The performance of the classifier on clean data may be impacted by randomization. Randomization may potentially temporarily raise the classifier's error rate.

REFERENCES

1. N. Papernot, P. McDaniel, A. Sinha, and M. P. Wellman, "Sok: Security and privacy in machine learning," in *2018 IEEE European Symposium on Security and Privacy (EuroS&P)*, 2018: IEEE, pp. 399–414.
2. S. Risi and M. Preuss, "Behind DeepMind's AlphaStar AI that reached grandmaster level in StarCraft II: Interview with Tom Schaul, Google DeepMind," *KI-Künstliche Intelligenz*, vol. 34, pp. 85–86, 2020.
3. K. N. Qureshi, A. Alhudhaif, S. W. Haider, S. Majeed, and G. Jeon, "Secure data communication for wireless mobile nodes in intelligent transportation systems," *Microprocessors and Microsystems*, p. 104501, 2022. https://doi.org/10.1016/j.micpro.2022.104501.
4. Q. Liu, P. Li, W. Zhao, W. Cai, S. Yu, and V. C. Leung, "A survey on security threats and defensive techniques of machine learning: A data driven view," *IEEE Access*, vol. 6, pp. 12103–12117, 2018.
5. A. Ashfaq, M. Kamran, F. Rehman, N. Sarfaraz, H. U. Ilyas, and H. H. Riaz, "Role of artificial intelligence in renewable energy and its scope in future," in *2022 5th International Conference on Energy Conservation and Efficiency (ICECE)*, 2022: IEEE, pp. 1–6.
6. R. Qazi, K. N. Qureshi, F. Bashir, N. U. Islam, S. Iqbal, and A. Arshad, "Security protocol using elliptic curve cryptography algorithm for wireless sensor networks," *Journal of Ambient Intelligence and Humanized Computing*, vol. 12, pp. 547–566, 2020.
7. S. Thomas and N. Tabrizi, "Adversarial machine learning: A literature review," in *Machine Learning and Data Mining in Pattern Recognition: 14th International Conference, MLDM 2018, New York, NY, USA, July 15–19, 2018, Proceedings, Part I 14*, 2018: Springer, pp. 324–334.
8. H. Bae *et al.*, "Security and privacy issues in deep learning," *arXiv preprint arXiv:1807.11655*, 2018.
9. A. Chakraborty, M. Alam, V. Dey, A. Chattopadhyay, and D. Mukhopadhyay, "Adversarial attacks and defences: A survey," *arXiv preprint arXiv:1810.00069*, 2018.

10. G. Li, P. Zhu, J. Li, Z. Yang, N. Cao, and Z. Chen, "Security matters: A survey on adversarial machine learning," *arXiv preprint arXiv:1810.07339*, 2018.
11. I. Manan, F. Rehman, H. Sharif, C. N. Ali, R. R. Ali, and A. Liaqat, "Cyber security intrusion detection using deep learning approaches, datasets, bot-IOT dataset," in *2023 4th International Conference on Advancements in Computational Sciences (ICACS)*, 2023: IEEE, pp. 1–5.
12. S. Qiu, Q. Liu, S. Zhou, and C. Wu, "Review of artificial intelligence adversarial attack and defense technologies," *Applied Sciences*, vol. 9, no. 5, p. 909, 2019.
13. X. Wang, J. Li, X. Kuang, Y.-a. Tan, and J. Li, "The security of machine learning in an adversarial setting: A survey," *Journal of Parallel and Distributed Computing*, vol. 130, pp. 12–23, 2019.
14. N. Pitropakis, E. Panaousis, T. Giannetsos, E. Anastasiadis, and G. Loukas, "A taxonomy and survey of attacks against machine learning," *Computer Science Review*, vol. 34, p. 100199, 2019.
15. H. Sharif *et al.*, "Application of artificial neural networks in satellite imaging–A systematic review," in *2023 4th International Conference on Computing, Mathematics and Engineering Technologies (iCoMET)*, 2023: IEEE, pp. 1–5.
16. K. Ren, T. Zheng, Z. Qin, and X. Liu, "Adversarial attacks and defenses in deep learning," *Engineering*, vol. 6, no. 3, pp. 346–360, 2020.
17. S. F. Batool, F. Rehman, H. Sharif, M. Jaffer, A. Gul, and S. Butt, "Intrusion detection using deep learning techniques," in *2022 3rd International Conference on Innovations in Computer Science & Software Engineering (ICONICS)*, 2022: IEEE, pp. 1–6.
18. S. Iqbal, H. Maryam, K. N. Qureshi, I. T. Javed, and N. Crespi, "Automised flow rule formation by using machine learning in software defined networks based edge computing," *Egyptian Informatics Journal*, vol. 23, no. 1, pp. 149–157, 2022.
19. P. Flach, *Machine learning: the art and science of algorithms that make sense of data*. Cambridge university press, 2012.
20. I. Lee and Y. J. Shin, "Machine learning for enterprises: Applications, algorithm selection, and challenges," *Business Horizons*, vol. 63, no. 2, pp. 157–170, 2020.
21. G. Apruzzese, M. Colajanni, L. Ferretti, A. Guido, and M. Marchetti, "On the effectiveness of machine and deep learning for cyber security," in *2018 10th international conference on cyber Conflict (CyCon)*, 2018: IEEE, pp. 371–390.
22. B. McMahan, E. Moore, D. Ramage, S. Hampson, and B. A. y Arcas, "Communication-efficient learning of deep networks from decentralized data," in *Artificial intelligence and statistics*, 2017: PMLR, pp. 1273–1282.
23. M. H. Zaib, F. Bashir, K. N. Qureshi, S. Kausar, M. Rizwan, and G. Jeon, "Deep learning based cyber bullying early detection using distributed denial of service flow," *Multimedia Systems*, 2021. https://doi.org/10.1007/s00530-021-00771-z.

24. N. Papernot, P. McDaniel, S. Jha, M. Fredrikson, Z. B. Celik, and A. Swami, "The limitations of deep learning in adversarial settings," in *2016 IEEE European symposium on security and privacy (EuroS&P)*, 2016: IEEE, pp. 372–387.
25. M. Nasr, R. Shokri, and A. Houmansadr, "Comprehensive privacy analysis of deep learning: Passive and active white-box inference attacks against centralized and federated learning," in *2019 IEEE symposium on security and privacy (SP)*, 2019: IEEE, pp. 739–753.
26. L. Muñoz-González and E. C. Lupu, "The security of machine learning systems," *AI in Cybersecurity*, pp. 47–79, 2019.
27. L. Muñoz-González and T. Kenneth, "Co, and Emil C Lupu. Byzantine-robust federated machine learning through adaptive model averaging," *arXiv preprint arXiv:1909.05125*, vol. 90, 2019.
28. R. Shokri, M. Stronati, C. Song, and V. Shmatikov, "Membership inference attacks against machine learning models," in *2017 IEEE symposium on security and privacy (SP)*, 2017: IEEE, pp. 3–18.
29. Y. Shi and Y. E. Sagduyu, "Evasion and causative attacks with adversarial deep learning," in *MILCOM 2017-2017 IEEE Military Communications Conference (MILCOM)*, 2017: IEEE, pp. 243–248.
30. M. Fredrikson, E. Lantz, S. Jha, S. Lin, D. Page, and T. Ristenpart, "Privacy in pharmacogenetics: An {End-to-End} case study of personalized warfarin dosing," in *23rd USENIX Security Symposium (USENIX Security 14)*, 2014, pp. 17–32.
31. F. Tramèr, F. Zhang, A. Juels, M. K. Reiter, and T. Ristenpart, "Stealing machine learning models via prediction {APIs}," in *25th USENIX Security Symposium (USENIX Security 16)*, 2016, pp. 601–618.
32. A. Kurakin, I. Goodfellow, and S. Bengio, "Adversarial machine learning at scale," *arXiv preprint arXiv:1611.01236*, 2016.
33. M. Melis, A. Demontis, B. Biggio, G. Brown, G. Fumera, and F. Roli, "Is deep learning safe for robot vision? adversarial examples against the icub humanoid," in *Proceedings of the IEEE International Conference on Computer Vision Workshops*, 2017, pp. 751–759.
34. E. Tabassi, K. J. Burns, M. Hadjimichael, A. D. Molina-Markham, and J. T. Sexton, "A taxonomy and terminology of adversarial machine learning," *NIST IR*, vol. 2019, pp. 1–29, 2019.
35. A. Iftikhar, K. N. Qureshi, M. Shiraz, and S. Albahli, "Security, trust and privacy risks, responses, and solutions for high-speed smart cities networks: A systematic literature review," *Journal of King Saud University - Computer and Information Sciences*, p. 101788, 2023. https://doi.org/10.1016/j.jksuci.2023.101788.
36. B. Nelson *et al.*, "Misleading learners: Co-opting your spam filter," *Machine Learning in Cyber Trust: Security, Privacy, and Reliability*, pp. 17–51, 2009.
37. F. Rehman, S. I. A. Shah, N. Riaz, and S. O. Gilani, "A robust scheme of vertebrae segmentation for medical diagnosis," *IEEE Access*, vol. 7, pp. 120387–120398, 2019.

38. B. Biggio, I. Corona, G. Fumera, G. Giacinto, and F. Roli, "Bagging classifiers for fighting poisoning attacks in adversarial classification tasks," in *Multiple Classifier Systems: 10th International Workshop, MCS 2011, Naples, Italy, June 15–17, 2011. Proceedings 10*, 2011: Springer, pp. 350–359.
39. L. Breiman, "Bagging predictors," *Machine Learning*, vol. 24, pp. 123–140, 1996.
40. J. Ba and R. Caruana, "Do deep nets really need to be deep?," *Advances in Neural Information Processing Systems*, vol. 27, pp. 1–9, 2014.
41. G. Hinton, O. Vinyals, and J. Dean, "Distilling the knowledge in a neural network," *arXiv preprint arXiv:1503.02531*, 2015.
42. N. Carlini and D. Wagner, "Towards evaluating the robustness of neural networks," in *2017 ieee symposium on security and privacy (sp)*, 2017: IEEE, pp. 39–57.
43. W. Xu, D. Evans, and Y. Qi, "Feature squeezing mitigates and detects carlini/wagner adversarial examples," *arXiv preprint arXiv:1705.10686*, 2017.
44. G. Tao, S. Ma, Y. Liu, and X. Zhang, "Attacks meet interpretability: Attribute-steered detection of adversarial samples," *Advances in Neural Information Processing Systems*, vol. 31, pp. 1–12, 2018.
45. M. Abbasi and C. Gagné, "Robustness to adversarial examples through an ensemble of specialists," *arXiv preprint arXiv:1702.06856*, 2017.
46. D. Meng and H. Chen, "Magnet: a two-pronged defense against adversarial examples," in *Proceedings of the 2017 ACM SIGSAC Conference on Computer and Communications Security*, 2017, pp. 135–147.
47. M. Lecuyer, V. Atlidakis, R. Geambasu, D. Hsu, and S. Jana, "Certified robustness to adversarial examples with differential privacy," in *2019 IEEE symposium on security and privacy (SP)*, 2019: IEEE, pp. 656–672.
48. E. Hesamifard, H. Takabi, and M. Ghasemi, "Cryptodl: Deep neural networks over encrypted data," *arXiv preprint arXiv:1711.05189*, 2017.
49. H. Sharif, F. Rehman, and A. Rida, "Deep learning: Convolutional neural networks for medical image analysis-A quick review," in *2022 2nd International Conference on Digital Futures and Transformative Technologies (ICoDT2)*, 2022: IEEE, pp. 1–4.
50. N. Carlini and D. Wagner, "Adversarial examples are not easily detected: Bypassing ten detection methods," in *Proceedings of the 10th ACM Workshop on Artificial Intelligence and Security*, 2017, pp. 3–14.
51. N. Srivastava, G. Hinton, A. Krizhevsky, I. Sutskever, and R. Salakhutdinov, "Dropout: a simple way to prevent neural networks from overfitting," *The Journal of Machine Learning Research*, vol. 15, no. 1, pp. 1929–1958, 2014.
52. K. Grosse, P. Manoharan, N. Papernot, M. Backes, and P. McDaniel, "On the (statistical) detection of adversarial examples," *arXiv preprint arXiv:1702.06280*, 2017.
53. X. Ma *et al.*, "Characterizing adversarial subspaces using local intrinsic dimensionality," *arXiv preprint arXiv:1801.02613*, 2018.
54. F. Tramèr, N. Papernot, I. Goodfellow, D. Boneh, and P. McDaniel, "The space of transferable adversarial examples," *arXiv preprint arXiv:1704.03453*, 2017.

55. J. Zhang and C. Li, "Adversarial examples: Opportunities and challenges," *IEEE transactions on neural networks and learning systems*, vol. 31, no. 7, pp. 2578–2593, 2019.
56. M. A. Qureshi, K. N. Qureshi, G. Jeon, and F. Piccialli, "Deep learning-based ambient assisted living for self-management of cardiovascular conditions," *Neural Computing and Applications*, 2021/01/07 2021. 10.1007/s00521-020-05678-w.
57. M. Barreno, B. Nelson, R. Sears, A. D. Joseph, and J. D. Tygar, "Can machine learning be secure?," in *Proceedings of the 2006 ACM Symposium on Information, computer and communications security*, 2006, pp. 16–25.
58. B. Biggio and F. Roli, "Wild patterns: Ten years after the rise of adversarial machine learning," in *Proceedings of the 2018 ACM SIGSAC Conference on Computer and Communications Security*, 2018, pp. 2154–2156.

Index

A

Access controls, 72
Accomplished, 47, 57, 60, 99, 130, 140, 143, 221, 308
Accountability, 76, 78, 79, 176, 180, 254, 255, 289
Accuracy, 1, 5, 6, 9, 15, 20, 26, 36, 40, 46, 47, 49, 51, 52, 54, 56, 63, 70, 92, 94, 95, 100–102, 105, 106, 108, 114, 124, 125, 134, 136, 137, 160, 161, 163, 164, 166, 173, 178–180, 183, 194, 198, 199, 208, 219, 226, 229, 231, 233, 234, 237, 238, 249, 253, 256, 266, 280, 286, 298, 306, 307, 313, 314, 319
Acknowledge, 101, 103, 105, 127, 131, 221
Adversarial, 18, 180, 191, 252, 260, 263, 289, 294–298, 301, 302, 304, 307–309, 313–320
Aggregating, 199
AI models, 35, 36, 39, 40, 45–55, 70–80, 89, 91, 101, 110, 121, 170–176, 179–186, 192, 212, 213, 226, 228, 246, 247, 252, 261, 275, 278, 283–285, 288, 290, 302
AI-powered, 28, 30–32, 35, 38, 48, 49, 54, 65, 66, 92, 95, 219, 229
AI tools, 15, 26, 27, 29, 30, 34, 35, 39, 109, 181, 229, 276, 277, 282, 283, 286–288
Algorithms, 19, 26, 28–30, 32, 33, 36–39, 45, 47, 48, 51–56, 58, 60–62, 64–67, 71, 74–76, 78, 87, 90, 93, 100, 101, 103–105, 107–110, 115, 132, 137, 139, 142, 144, 145, 147, 149–151, 157, 159, 161–165, 179–183, 191, 196, 197, 201, 212, 213, 218–222, 224–228, 230, 232, 234–238, 248, 262, 264, 267, 275, 280, 281, 285, 289, 293, 296–301
Analytical, 31, 69, 103, 129, 132, 133, 153, 155, 315, 319
  analytics, 41, 45–55, 57, 62, 68–80, 89, 129–133, 153, 157
Anonymization, 40, 71–73, 177
Architecture, 14, 123, 203, 255
Artificial Intelligence (AI), 8, 25, 86, 114, 129, 170, 191, 218, 293
Attributes, 148, 158, 160, 173, 184, 299
Autogpt, 9
Automated, 9, 11, 29, 30, 33, 35, 49, 57, 66, 87, 111, 129, 132, 140, 143, 181, 210, 224, 227, 258, 259
  data, 49
Automatically, 2, 22, 51–54, 60, 62, 120, 140, 210, 280

B

Biased Results, 78
Biases, 9, 17, 19, 26, 40, 71, 75–77, 171, 179, 180, 183–185, 194, 208, 226, 237–239, 247–250, 253, 257, 260, 264, 267, 268, 274, 284, 286, 289
Bidirectional Encoder Representation (BERT), 1
Big Data, 26, 86, 96, 107, 157, 200, 204, 229, 231

Bigram, 6
Biological Neural Networks (Bnns), 192
Biomedical, 49
Biotechnology, 2
Bootstrap, 199

C

Chatbot, 131
ChatGpt, 2, 10, 11, 19, 206–213, 273–275, 281–285, 288–290
Circumstances, 55, 58, 63, 64, 79, 90, 179, 212, 230, 234
Cognitive, 93, 109, 129, 136, 145, 146, 258, 283, 286
Collaboration, 31, 37, 67, 79, 206, 228, 235, 242, 257, 267
Communication, 37, 211
Comparable, 53, 67, 87, 93, 95, 101, 156, 261
Competitive, 9, 55, 142, 143
Compliance, 19, 72–74, 77, 78, 80, 278, 282
Comprehensive, 1, 11, 15, 23, 38, 39, 55, 58, 60, 67, 76, 80, 86, 89, 91, 102, 116, 119, 121, 123–125, 127, 131, 141, 144, 146, 147, 151, 153, 155, 168, 170, 173–176, 178, 180, 182, 185, 186, 209, 210, 215, 216, 219, 220, 228, 236, 248, 253, 257, 265, 266, 281, 284, 294, 310
  comprehend, 3, 10, 48, 60, 61, 68, 70, 71, 76–79, 126, 133, 135, 136, 140, 146, 172, 195, 206, 208, 209, 220, 221, 226, 227, 236, 238, 256, 264, 273
Computer vision, 61, 62, 192
Conversation, 145, 243, 245, 295
Convolutional Neural Networks (CNNs), 56, 134, 202
COVID-19, 90–92
Crafting, 258
Cross-model, 22
Custom datasets, 14
Customized, 39, 66, 67, 92, 162, 165, 209, 232, 234
Cutting-edge, 62, 63, 67, 95, 115, 121–123, 126, 127, 132, 154, 213, 246, 313, 319
Cybersecurity, 52, 54, 72, 276, 280, 296, 298, 320
Cyber threats, 72

D

Data analytics, 45, 46, 77
Data Dependency, 245
Data Monitoring, 89
Datasets, 1, 4–6, 8, 11, 13, 17, 21, 26, 27, 29, 31, 34, 35, 39, 46, 48–53, 65, 68, 70, 75, 79, 92, 101, 102, 108, 109, 119, 143, 146, 152, 153, 155, 156, 159, 161, 175, 178, 179, 181–184, 186, 196, 197, 202, 205, 208, 212, 227, 228, 230, 233, 257, 274, 275, 304, 305, 316
Deciphering, 226
Decision-making, 11, 17, 25, 30, 31, 33, 39, 41, 46, 57, 62, 63, 68, 69, 71, 75, 76, 78, 86, 91, 93, 105, 109, 110, 115, 125, 127, 130, 133, 137, 147, 150, 153, 154, 162, 173, 176, 179, 183, 184, 194, 200, 212, 218, 225, 227, 233, 236, 237, 253, 254, 256–258, 263, 276, 277, 284, 289
Decision support systems (DSS), 55
Decision tree (DT), 139
Deepfakes, 17
Deep Learning (DL), 27, 45, 100, 222, 296, 301
Democratizing, 101
Dependable, 107, 153, 210
Detection, 32, 35, 52, 54, 58, 61, 62, 66, 87, 99, 104, 106, 139, 159, 162, 165, 166, 182, 192, 197, 202, 204, 214, 223, 233, 264, 267, 276, 277, 282, 297, 303, 309, 312, 315, 317, 320
Diagnoses, 66, 101, 105, 106, 109, 162, 210
Diagnostics/es, 13, 27, 29, 34, 64, 107, 156, 172, 180, 219, 221, 231

Disabilities, 13, 172
Discovery/ies, 4, 25, 29–31, 33–35, 41, 47, 49, 64, 65, 70, 87, 108, 154–162, 208, 210, 213, 219, 227, 229–231, 233, 238, 302
Discriminatory, 17, 75, 77, 260, 274
Distinctive, 70, 213, 227, 259
Distributions, 154
Dynamics, 53, 65, 68, 70, 124, 152, 257

E

E-commerce, 68, 145, 197, 209
Electronic Data Capture (EDC), 90
Electronic Health Record (EHR), 163
Embedded, 129, 151, 250
Encryption, 71, 72, 122–124, 275, 312, 315
Energy, 18–20, 115, 195, 246, 261, 262, 264, 265, 267, 271, 285, 321
Ethical, 17, 21, 26, 36, 39, 41, 46–48, 71–80, 110, 115, 131, 161, 170–186, 206, 213, 220–222, 225, 228, 229, 236, 237, 239, 253, 255, 258–260, 265–267, 273, 274, 278, 282, 286–288, 290
Evolutionary Algorithms (Eas), 147
Evolutionary Computing (EC), 147
Examining, 48, 53, 54, 61, 62, 94, 101, 153, 208, 263, 264, 295, 297
Experiment/s, 13, 25, 29–31, 33, 41, 179, 209, 223, 224, 312, 315, 319
  experimental analysis, 30

F

Fairness, 19, 36, 40, 71–76, 80, 109, 161, 171, 176, 184, 186, 247–249, 255, 264, 267, 274, 287, 290
  and bias, 19
False positive, 98
Filtering, 15, 163, 249, 282, 297
Financial models, 69
Forecast/ing, 31, 32, 46, 47, 55, 65, 68, 78, 91, 96, 99, 121, 149, 151, 153, 157–160, 162, 166, 215, 234, 276, 286, 316
Foreseeing, 55

G

General Data Protection Regulation (GDPR), 73, 277
Generalization, 134, 149, 156, 180, 182
Generative Pre-Trained Transformer (GPT), 4, 8
Genetic data, 27, 39, 46–48
Geneticists, 230
Genetic Programming (GP), 148
GPT-3, 1, 13, 14, 17, 21–23, 210, 244, 245, 252, 254, 255, 260, 261
GPT-4, 2, 11, 18, 182
Gradient Boosting Machine (GBM), 33
Gully Erosion Susceptibility (GES), 158

H

Healthcare, 2, 3, 17, 20, 27–29, 34–36, 49, 61–71, 86, 87, 89–93, 95–98, 108–110, 131, 133, 151, 153, 162–164, 171, 178–181, 192, 197, 202, 209, 219, 225, 229, 230, 232, 233, 293
Hidden, 5, 28, 52, 53, 68, 96, 181, 193, 194, 200–202, 227, 305, 316
Hidden Markov Models (Hmms), 4
Histogram of Oriented Gradients (HOG), 134
Ho Chi Minh City (HCMC), 117
Holistic, 75, 76, 278
HTTP, 121
Human-in-the-Loop (HITL), 173
Humanity, 40, 70, 186
Human knowledge, 218, 220, 223, 225, 226, 239
Hyperplane, 197
Hypothesis, 197, 206, 211, 223, 317, 318

I

Information Retrieval (IR), 9
Information Technology (IT), 165
Innovative, 22, 38, 55, 66, 90, 96, 114, 119, 121, 125, 126, 131, 158–160, 230–232, 238, 295

Interdisciplinary, 31, 67, 69, 71–74, 79, 125, 267
Internet of Things (Iot), 95, 114

K

K-nearest Neighbors (KNN), 33
Knowledge, 3, 8, 10, 11, 17, 25, 26, 30, 31, 37–39, 47, 53, 58, 59, 69–73, 75, 80, 86, 91, 105, 115, 139, 141, 143, 144, 146, 154, 161, 163, 164, 182, 192, 210, 213, 219, 220, 223, 225–231, 233, 235–239, 245, 248, 253, 257–259, 265, 296, 298, 300, 302, 304, 305, 313
Knowledge graphs (Kgs), 9

L

Languages Models (LM), 1
Large amounts, 32, 36, 41, 137, 238
Large-scale models, 246
Latent Semantic Analysis (LSA), 4
Layers, 135, 193, 194, 200, 201, 203, 204, 256, 263, 300
Likelihood, 3, 31, 65, 91, 92, 99, 102, 108, 153, 163, 172, 317

M

Machine Learning (ML), 2, 48, 87, 94, 129, 171, 195, 293
Magnetic Resonance Imaging (MRI), 94, 101, 102
Management systems, 34, 115, 116
Medical records, 36, 94, 153, 164
Memetic Algorithms (MA), 149
Methodologies, 34, 49, 51–53, 70–75, 77–80, 89, 107, 125, 139, 140, 142, 145, 152, 153, 157–161, 179, 181, 185, 186, 225, 227, 238, 284, 300
Misleading, 18, 40, 182, 280
MODIS, 119, 120, 124, 127
Molecular, 36, 65
Multidimensional, 54, 125, 204
Multidisciplinary, 38, 71, 235, 287, 288
Multilayer Perceptron (MLP), 139

N

National Aeronautics and Space Administration (NASA), 119
Natural language generation (NLG), 8
Natural language understanding (NLU), 140
N-gram, 4–6

O

Observations, 4, 125, 258
Online Platform, 37
Openai, 2, 11, 14, 39, 209
Open Source, 13, 14
Ophthalmology, 93, 104, 105
Optical Character Recognition (OCR), 50

P

Parts-of-Speech (POS), 143
Patterns, 2, 3, 5, 8, 11, 15, 27, 28, 31, 32, 34–36, 41, 45, 47, 48, 50, 52–58, 62, 63, 66, 68, 91, 93, 95–97, 106, 115, 125, 136, 140, 146, 151, 156, 159, 183, 191, 192, 197, 203, 211–213, 219–221, 226, 227, 229, 231, 233, 234, 238, 248, 255, 256, 263, 273, 274, 276, 281, 284, 289, 295, 298, 299
Phenomenon, 96, 108, 143, 147, 149, 150, 155, 156, 158, 162, 165, 175, 179, 180, 318
Picture recognition, 56, 134, 145, 146, 319
Plagiarism, 211, 246–251
Political, 48, 67, 69, 281, 287, 288
Predict, 1, 3, 15, 30, 31, 33–35, 46, 49, 51, 55, 94, 98, 133, 154, 155, 157, 160, 225, 227, 229, 230, 276, 298, 309
Predictive analytics, 31, 55
Prejudice, 73, 74, 78, 79, 213, 220
Pre-process, 15, 132, 142
Preprocessing, 47, 49, 58, 75, 76, 80

Privacy, 15, 18, 20, 36, 40, 46, 47, 71–78, 80, 110, 115, 123, 171, 173, 175, 179, 182, 239, 251, 253, 254, 259, 260, 263–265, 267, 273–275, 277, 281, 282, 286–288, 290, 294, 295, 297, 300–302, 308, 309, 312, 315, 318–320
Proactive, 40, 75, 80, 115, 126, 127, 174, 231, 277, 281–283
Problem-solving, 20, 69, 70, 109, 149, 154, 227, 230, 233, 238, 258
Procedure., 55, 109, 135, 153, 232

R

Random Forest (RF), 33, 157, 198
Randomized Controlled Trials (Rcts), 89
Recommendations, 28, 31, 35, 37, 57, 72, 94, 174, 207, 208, 210, 231, 233, 274, 286, 295
Recurrent Neural Network (Rnns)., 3
Regulation, 109, 156, 277, 288
Represent Learning (RL), 145
Research and development, 3, 8, 17, 26, 28, 213, 222, 247, 285
Resembles, 149, 274
Resource-constrained, 101, 246
Resource utilization, 32, 115
Results, 3, 18, 25, 29, 31, 34–36, 39, 40, 46, 49, 51, 52, 59, 67, 75, 78, 89, 94, 96–98, 100, 102, 105, 120, 134, 135, 151–153, 155, 161, 162, 164, 174, 175, 179, 183, 184, 193, 199, 202, 209, 212, 213, 219, 225, 226, 228, 229, 231, 234–239, 262, 280, 295, 308, 312, 315, 318, 319
Revolutionize, 26, 48, 53, 65, 79, 106, 115, 123, 154, 165, 206, 211
Robotics, 92, 94, 145, 147, 166, 293
Rules, 3–5, 40, 71, 73–75, 80, 171, 184, 256, 267

S

Selection process, 40
Semantic, 2–4, 9, 15, 22, 140–144
Shortage, 87
Simulations, 11, 31, 32, 63–65, 133, 161
Smartphone, 98, 107
Social sciences, 2, 27, 28, 67, 170, 224, 225
Sophisticated, 4, 5, 17, 28, 47, 64, 65, 122, 153, 170, 171, 234, 253, 266, 295
Stakeholders, 17, 72, 76–80, 90, 115, 125, 152, 153, 171, 178, 180, 275, 280, 281, 283, 284, 286–288, 290
Standards, 39, 74, 75, 77–80, 90, 121, 124, 171, 174, 178, 179, 185, 211, 242, 267, 277
Structural Causal Models (SCM), 158
Summarization, 3, 5, 8, 9, 21
Supervised, 1, 22, 54, 56, 135, 137, 140, 141, 143, 144, 191, 195, 197, 205, 206, 298, 299, 304, 314
Support Vector Machine (SVM), 139
Sustainability, 18, 20–21, 38, 70, 127, 246, 262, 286

T

Temperature, 18, 27, 118, 119, 122–125, 127
Test hypotheses, 96
Textual, 26, 46, 51, 58–60, 103, 136, 140, 141, 263, 298
Traditional, 3, 8, 9, 13, 27, 29, 33, 34, 98, 102, 108, 124, 131, 137, 141, 152, 170, 172, 176, 182, 183, 203, 213, 227, 228, 276, 283, 297
Transformation, 22, 28, 34, 41, 51, 52, 71, 76, 77, 86, 133, 156, 165, 172, 178, 192, 274
  transformative, 46, 48, 52, 53, 61, 115, 116, 125, 232
Transparency, 17–20, 39, 40, 72, 75–80, 164, 171, 172, 176, 179, 180, 184–186, 254, 256, 260, 263, 265, 275, 278, 280, 282, 284, 287, 289, 290
Trigram, 6
Trusted Artificial Intelligence (TAI), 173

U

Unauthorized, 18, 20, 72, 122, 124, 171, 275
Unbiased, 77, 186, 247, 248, 250, 252, 264, 317

Uncertainty, 19, 94, 101, 105, 159, 163, 174, 235, 316, 317
UNESCO, 119
Unforeseen, 71, 77, 78, 136, 181
Unparalleled, 122, 124, 158, 227

V

Validation, 51, 55, 56, 76, 77, 121, 155, 162, 178, 181, 182, 185, 200, 208, 235, 249
Virtual assistance, 11, 274

W

Wearables, 100
Wifi, 122
Wordnet, 142
Word Sense Disambiguation (WSD), 141

X

X-Ray, 103

For Product Safety Concerns and Information please contact our EU representative GPSR@taylorandfrancis.com Taylor & Francis Verlag GmbH, Kaufingerstraße 24, 80331 München, Germany

Batch number: 10397790

Printed by Printforce, the Netherlands